The Geometry of Lagrange Spaces: Theory and Applications

Fundamental Theories of Physics

An International Book Series on The Fundamental Theories of Physics: Their Clarification, Development and Application

Volume 59

The Geometry of Lagrange Spaces: Theory and Applications

by

Radu Miron
and
Mihai Anastasiei

Faculty of Mathematics,
University "Al. I. Cuza",
Iaşi, Romania

KLUWER ACADEMIC PUBLISHERS
DORDRECHT / BOSTON / LONDON

A C.I.P. Catalogue record for this book is available from the Library of Congress.

ISBN 0-7923-2591-5

Published by Kluwer Academic Publishers,
P.O. Box 17, 3300 AA Dordrecht, The Netherlands.

Kluwer Academic Publishers incorporates
the publishing programmes of
D. Reidel, Martinus Nijhoff, Dr W. Junk and MTP Press.

Sold and distributed in the U.S.A. and Canada
by Kluwer Academic Publishers,
101 Philip Drive, Norwell, MA 02061, U.S.A.

In all other countries, sold and distributed
by Kluwer Academic Publishers Group,
P.O. Box 322, 3300 AH Dordrecht, The Netherlands.

Printed on acid-free paper

Contents

Preface

The methods coming fr m differential geometry are of considerable scientific interest due to the realization that even those simple ones are very useful in understanding a vast range of fundamental natural phenomena. Today more and more elaborate geometrical methods are used in Physics, Chemistry, Biology, Ecology etc.

Very often a starting point in studying natural phenomena is a variational problem formulated for a convenient Lagrangian.

From a formal viewpoint a Lagrangian is a smooth real function defined on the total space TM of the tangent bundle to a manifold M satisfying some regularity conditions. The geometrization of such a Lagrangian, which will be extensively discussed in this book, is an old idea which comes from Theoretical Mechanics. It involves in a high degree the geometry of the manifold TM fibred over M. For instance, a full and interesting geometrical meaning for all concepts related to a Lagrangian is obtained by interpreting the Lagrange geometry as a geometry of an almost Kählerian structure on TM.

Since the definition of a Lagrangian may be enlarged such that more general bundles than the tangent bundle are involved, it is natural to study first the geometry of the total space E of a vector bundle over the manifold M. The geometry of the manifold E has some special features due to the fact it is fibred over M, each fibre being a linear space. A direct study of it raises a lot of difficulties especially when the analysis is made in local coordinates as is needed for the main applications.

The experience of the authors shows that the concept of non-linear connection is essential for overcoming such difficulties.

A non-linear connection is a distribution on E supplementary to the vertical distribution on E. It generalizes the usual notion of linear connection. The use of a non-linear connection is of help in keeping the geometrical meaning of various concepts on E and controlling the geometrical character of the operations with them.

The main purpose of the book is three-fold:

a) An extensive presentation of the geometry of the total space of a vector bundle.

b) A detailed exposition of the Lagrange geometry.

c) A description of the most important applications.

The book contains new methods for the study of the total space of a vector bundle and for constructing geometrical models for applications in various fields. We mention that all applications are at the level of the geometrical models i.e. they are not concrete ones in a sense of including numerical computations. However, the models which have been proposed are enough convenient as some computations made by G.S. Asanov regarding the precession of a gyroscope and the perihelion shifts of planets have shown.

The material is arranged so as to proceed from general to special or particular. There are no exercises, but some examples worked out in all details are included. Coordinate-free expressions are almost always translated into coordinates in order to help readers to relate to classical treatises and to facilitate the applications to Theoretical Physics.

Concerning prerequisites, the reader should have a good foundation in general manifold theory, as well as some general background on geometrical models in Physics.

Our intention was to provide an unitary text in order to be useful for scientists as well as partially for graduate students.

In chapter I is defined a very general notion of fibre bundle. In particular, principal fibre bundles and vector bundles are discussed. The emphasis is here on vector bundles since these bundles will frequently be used in the book. Topics such as morphisms of vector bundles, vector subbundles, operations with vector bundles and sections in a vector bundle are presented. The main examples are the tangent bundle, cotangent bundle and frame bundle of a manifold. There are very few proofs. The reader is referred to some well-known books where these topics are contained.

Chapter II deals mainly with the non-linear connections in vector bundles. A non-linear connection in the vector bundle (E,p,M) is defined as a splitting of a short exact sequence of vector bundles over E. Next it is shown it can be thought of as a distribution, called horizontal, which is supplementary to the vertical distribution on E, i.e. to the distribution defined by the kernels of the differential the projection p. Characterisations of a non-linear connection using its local representation, a pair of projectors or an almost product structure on E are provided. The horizontal lift defined by a non-linear connection is studied. Its curvature is introduced, too. In particular, the linear connections in the vector bundle (E,p,M) are considered. Finally the morphisms preserving a non-linear connection are discussed. Here all results are carefully proved since these are basic in understanding the next chapters.

Chapter III is concerned with the geometry of the total space E of a vector bundle (E,p,M). First, assuming E endowed with a non-linear connection, the linear connections on E preserving the horizontal and vertical distributions on E are introduced, under the name of d-connections (d - means "distinguished"). Some characterisations of them are given. Then their local representation is found. Next the torsion and curvature of a d-connection is expressed in a frame adapted to the given non-linear connection and to the vertical distribution. The structure equations of d-connections are derived. Finally, introducing a metrical structure on E, metrical d-connections are considered, their existence is proved and a canonical one is established.

Chapter IV is devoted to geometrical theory of the embeddings of vector bundles. In this theory some ideas used by the first author in studying the subspaces in Lagrange spaces are extended.

The moving frame method, as well as tensorial and exterior calculus are used. First, the local form of an embedding of a vector bundle into a vector bundle is given. Then a convenient moving frame is considered and an induced non-linear connection, as well as a relative covariant derivative are introduced and studied. The corresponding Gauss-Weingarten formulae are established and the Gauss-Codazzi equations are derived. Some interesting particular cases are pointed out.

In chapter V the Einstein equations on the total space of a vector bundle are written and studied. These equations established by the first author for the Lagrange spaces and by the second author for a vector bundle improve those empirically proposed by physicists Y. Takano and H. Ishikawa. The idea is to attach the usual Einstein equation to the canonical metrical d-connection on E and then to express it in a special frame on E. One obtains a set of four equations, two of them being very similar to the classical Einstein equations. A special attention is paid to the case when the fibres of E are 1-dimensional, since that four equations may be used in formulating a kind of projective unitary theory. The conservation laws are also studied explicitly for some particular metrics on E.

In chapter VI a description of the ways of reaching a higher degree of generality in the gauge fields theory, pointed out by G.S. Asanov, is given by taking as foundation our theory of vector bundles. First, a very general gauge transformation is discussed. A gauge covariant derivative is defined and gauge metrical d-connections are studied. A way to derive the generalized Einstein Yang-Mills equations is sketched.

The manifold TM, the total space of the tangent bundle to M, carries a lot of interesting geometrical structures. Our chapter VII deals with the study of the geometry of TM. Noticing that many facts in this geometry may be easily derived by particularising ones from the geometry of E we pay great attention to these which are specific to TM.

Thus by considering the non-linear connections on TM we study in detail their relations to semisprays and sprays on TM. A class of d-connections which are compatible with the almost complex structure deriving from a given non-linear connection is discussed. Also, metrical d-connections on TM are analyzed.

Chapter VIII is concerned with the Finsler spaces. The geometry of these spaces is now well studied as can be seen from a book by M. Matsumoto [97]. We have considered it is useful to provide an introduction in this subject from our standpoint as an illustration of the theory developed till now as well as a motivation for the geometry of the Lagrange spaces. The definition of the Finsler spaces is followed by a description of the Cartan non-linear connection and a study of geodesics of a Finsler space. Then the metrical Cartan connection is introduced and its structure equations as well as the Bianchi identities are established. Remarkable d-connections due to L. Berwald, H. Rund, M. Hashiguchi are discussed. Next, it is shown that when M is a Finsler manifold, TM has an almost Kählerian structure by the help of which Finsler geometry is recast. This is called the almost Kählerian model of a Finsler space. Finally, some basic results on subspaces in Finsler spaces are proved.

In chapter IX the main subject of the book, that is, the theory of Lagrange spaces is presented. These spaces were introduced in 1975 by J. Kern [82], but their geometry was developed ten years later using the general theory of vector bundles. The initiative to study these spaces was due to the first author who organized since 1980 a National Seminar on Finsler and Lagrange spaces which is held at the University of Braşov, Romania, once every two years. Many Romanian geometers have contributed to this Seminar. Also, it benefits by the presence of many geometers from abroad.

Now the geometry of the Lagrange spaces is well stated and many people work out its applications in various directions. First, we introduce the notion of the regular Lagrangian and Lagrange space. Using the Euler-Lagrange equations, derived from the usual variational problem associated to a regular Lagrangian, a canonical non-linear connection for a Lagrange space is introduced. Then a canonical metrical d-connection is constructed. Using these connections a geometrical model for the gravitational and electromagnetic fields is provided. The Einstein equations and the generalized Maxwell equations are introduced and discussed. The conservation laws are studied, too.

The whole theory is then applied to the well-known Lagrangian from electrodynamics:

$$L(x,y) = mc\gamma_{ij}(x)y^iy^j + \frac{2e}{m}A_i(x) + U(x).$$

In particular, the so-called almost Finslerian-Lagrange spaces are studied in order to provide new geometrical models useful in applications. The legitimacy, from geometrical point of view, of the whole theory is gained by providing an almost Kählerian structure on TM in which the geometry of a Lagrange space is recast.

The most important generalization of a Lagrange space is studied in Chapter X, under the name generalized Lagrange space. This generalization is due to the first author, [107]. More precisely, a generalized Lagrange space retains from the notion of Lagrange space only the metric structure, neglecting the concept of Lagrangian itself. Such a space is a pair $(M,g_{ij}(x,y))$, where the d-tensor field g_{ij} is not provided by a Lagrangian, that is does not exists a Lagrangian L such that

$$g_{ij}(x,y) = \frac{1}{2}\frac{\partial^2 L(x,y)}{\partial y^i \partial y^j}.$$

These spaces are more and more important for applications in theoretical physics as some recent researches have pointed out. In order to study them we fix a non-linear connection and construct a metrical d-connection. The structure equations of this d-connection are established. Special attention is paid to the h-covariant constant d-tensor fields as they enter in many places in the next chapters.

Models for gravitational and electromagnetic fields in this more general framework are provided. The geometrical support is given here by an almost Hermitian structure on TM, which is not reducible to a Kählerian one except if the generalized Lagrange spaces reduce to a Lagrange one. It means the methods of the symplectic geometry can not be used in order to study geometrical models based on the generalized Lagrange spaces.

We think the power of our theory is mainly due to this fact since, as we shall point out in the next chapters, there exist some important geometrical models which use generalized Lagrange spaces.

In chapter IX, the generalized Lagrange spaces with the fundamental metric tensor field

$$g_{ij}(x,y) = e^{\sigma(x,y)}\gamma_{ij}(x,y)$$

where $\gamma_{ij}(x,y)$ is the metric tensor of a Finsler space, are studied. First, it is shown these spaces are not reducible to Lagrange spaces. Then some important classes of these spaces are taken into consideration. The first of them, obtained when γ_{ij} is a pseudo-Riemannian metric, gives a new model for a space-time in which the EPS conditions from general relativity are verified.

A theory of the gravitational and electromagnetic fields related to this is developed. The second one is characterised by $g_{ij}(x,y)=g_{ij}(y)$, that is, provides locally Minkowski generalized Lagrange spaces. This class contains the Antonelli spaces which have been proposed to study a geometrical model from ecology.

Chapter XII is concerned with a new geometrical model for the relativistic optics of dispersive media. This is obtained by introducing a generalized Lagrange metric of a special form with the property that its restriction to the image of a cross-section in the tangent bundle defined by a vector field (thought of as the field of speeds) is just the Synge metric from relativistic optics. Then this generalized Lagrange metric is studied by using the theory from chapter X. A post-Newtonian estimation based on it is described. Gravitational and electromagnetic fields, as well as the almost Hermitian model are considered again. Special attention is paid to the locally Minkowski case.

Until now we have considered the Lagrangians which do not depend explicitly on time. It means the systems to which we applied the geometry were autonomous. But there exist nonautonomous systems, too. These are generally described by Lagrangians which depend explicitly on time. Thus the problem to geometrize time dependent Lagrangians is quite natural. Formally, a time dependent Lagrangian is a real function on $\mathbb{R}\times TM$. In chapter XIII we introduce and study a geometrization of such Lagrangians by using the vector bundle theory expounded in chapter III. First, we treat, in general, the non-linear connection on $\mathbb{R}\times TM$ and then we derive a canonical one from a time dependent Lagrangian by using the Euler-Lagrange equations. Next we consider metrical d-connections on $\mathbb{R}\times TM$ and we find there exists a family of such connections. A canonical one is pointed out. The homogeneous case leads to the rheonomic Finsler spaces. Basic results on these spaces are established. A geometrical legitimacy for the whole theory is obtained here by constructing a metrical almost contact structure on $\mathbb{R}\times TM$ which appears as a model for a rheonomic Lagrange space.

Finally, we remark that the bibliography is not an exhaustless one. It contains papers and books quoted in the text or closely related to each topic. A subject index is included.

The origin of this book may be found in a monograph, quoted as [109], published in Romanian by the Romanian Academy in 1987. Many colleagues from abroad have asked us to translate it in English. This was our starting point but in fact we have realized a new book with respect to the theory and applications.

Acknowledgements. The authors are indebted to all the members of the Romanian Seminar on Finsler and Lagrange spaces, as well as to the attendants at the Romanian-Japanese Colloquium in 1984.

It is a pleasure for us to give many thanks especially to Professors M. Matsumoto, M. Hashiguchi, Y. Ichijio, G.S. Asanov, P.L. Antonelli and R. Ingarden for valuable discussions, collaborations and their comments on the contents of this book.

Our warmest thanks are extended to Prof.Dr. R.G. Beil who suggested and encouraged us to realise this book.

The editing and word-processing of the text were made by Dr. Mihai Postolache, "Politehnica" University of Bucharest, Romania. The authors are very grateful to him.

We should like to express our deep gratitude to the publishers for their unfailing support.

Faculty of Mathematics
University "Al. I. Cuza" Iaşi
Iaşi, Romania

R. Miron
M. Anastasiei

Chapter I

Fibre Bundles. General Theory

The notion of locally trivial fibre bundle naturally generalizes that of the manifold. The theory of these bundles was developed in fifties especially by Ch. Ehresmann but its origins are found in the works of E. Cartan. The fibre bundles were used to get global results in differential geometry as well as to build geometrical models for modern physical theories. As one of the our main purposes is to describe such geometrical models, it is natural to begin with a recalling of the main notions and results from the theory of the fibre bundles. This is the object of the first chapter.

In §1 we introduce the notion of locally trivial fibre bundle. The main examples as principal fibre bundles and vector bundles are described in §2 and §3, respectively. Then we pay a great deal of attention to the vector bundles. Besides the examples as tangent and cotangent bundles we discuss morphisms of vector bundles in §4, vector subbundles in §5, operations with vector bundles in §6, sections in vector bundles in §8. The principal bundle associated to a vector bundle is constructed in §7. The notion of G-structure is introduced, too.

In general the proofs are omitted. The reader is referred to some well-known books containing the theory of fibre bundles.

§1. Fibre Bundles

Let be G a Lie group which acts differentiably and effectively on a manifold F. Then every element $g \in G$ defines a diffeomorphism $L_g: F \to F$.

Unless otherwise stated, all manifolds are assumed to be smooth (i.e. C^∞), finite dimensional, Hausdorff, paracompact and connected; all mappings are smooth.

Definition 1.1. *A (locally trivial) fibre bundle consists of the data (E,p,M,F,G), where M is a manifold, $p: E \to M$ is a surjective mapping and the following conditions are satisfied:*

a) The manifold M can be covered by a set $\mathscr{E}$ of open sets $U, V, W, \ldots$ such that for every such an open set U there exists a bijective mapping $\varphi_U: p^{-1}(U) \to U \times F$ so that $p(\varphi_U^{-1}(x,y)) = x$, $x \in U$, $y \in F$,

b) If $x \in U \cap V \neq \Phi$, then $\varphi_{V,x} \circ \varphi_{U,x}^{-1}: F \to F$ is an L_g with $g \in G$. Here $\varphi_{U,x}$

denotes the restriction of φ_U *to* $p^{-1}(x)$ *and* $U, V \in \mathscr{E}$,

c) The mapping $g_{VU}: U \cap V \to G$ *defined by* $g_{UV}(x) = \varphi_{V,x} \circ \varphi_{U,x}^{-1}$ *is smooth.*

Some remarks on this definition are in order. The set $E_x = p^{-1}(x)$ which is called the fibre in $x \in M$ of the given fibre bundle is bijectively mapped on F by $\varphi_{U,x}$. One takes on E_x the topology and the manifold structure for which $\varphi_{U,x}$ is a diffeomorphism. The uniqueness of these structures (up to diffeomorphisms) is assured by the condition b).

Let $(U_\alpha, \varphi_\alpha)_{\alpha \in I}$ be a smooth atlas on M such that the covering $(U_\alpha)_{\alpha \in I}$ refines the covering U,V,W,... and let $(V_\beta, \psi_\beta)_{\beta \in J}$ be a smooth atlas on F. Let us consider the open sets $U_\alpha \times V_\beta, U_\alpha \subset U$ and let us define as open sets in E the sets $W_{\alpha\beta} = \varphi_U^{-1}(U_\alpha \times V_\beta) \subset p^{-1}(U)$. One obtains a topology on E for which the bijections $\varphi_U, \varphi_V, \ldots$ become homeomorphisms. Let us denote by n and m the dimensions of the manifolds M and F, respectively, and let us define the mappings

$$\phi_{\alpha\beta}: W_{\alpha\beta} \to \mathbf{R}^{n+m},\ \phi_{\alpha\beta} = (\varphi_\alpha \times \psi_\beta) \circ \varphi_U^{\alpha\beta},$$

where $\varphi_U^{\alpha\beta}$ is the restriction of the mapping φ_U to $W_{\alpha\beta}$. Then the set $(W_{\alpha\beta}, \phi_{\alpha\beta})_{(\alpha,\beta) \in I \times J}$ is a smooth atlas on E.

Considering E endowed with the manifold structure defined by this smooth atlas, the mappings $\varphi_U, \varphi_V, \ldots$ become diffeomorphisms and the mapping $p: E \to M$ is a submersion. The manifold structure on E defined in the above is unique (up to diffeomorphisms) and dim E $= n+m$.

As to the locally trivial fibre bundle (E,p,M,F,G) the following labels are used: E is the total space, M is the base space, p is the projection of E on M, F is the type fibre, E_x is the local fibre in the point $x \in M$ and G is the structural group. The pairs (U, φ_U) are bundle charts and the set of bundle charts is a bundle atlas. The functions $g_{VU}: U \cap V \to G$ are called structural functions. These have the following properties:

(1.1) $g_{WV}(x) g_{VU}(x) = g_{WU}(x),\ x \in U \cap V \cap W,$

(1.2) $g_{UU}(x) = e$; (e is the neutral element of G), $x \in U,$

(1.3) $g_{VU}^{-1}(x) = g_{UV}(x),\ x \in U \cap V,$

(1.4) $\varphi_U^{-1}(x, y) = \varphi_V^{-1}(x, g_{VU}(x) y),$

where the notation $L_g(y) = gy,\ g \in G,\ y \in F$ was used.

Putting

$$\varphi(x) = (x^1, \ldots, x^n),\ x \in M,\ \psi_j(y) = (y^1, \ldots, y^m),\ y \in F$$

the coordinates of a point $u \in W_{\alpha\beta} \subset E$ are (x^k, y^a), $k = 1,2,\ldots,n$, $a = 1,2,\ldots,m$. Finally we note that a transformation of local coordinates on E is as follows:

$$(1.5)\quad \phi'_{\alpha\beta} \circ \phi^{-1}_{\alpha\beta} = (\varphi'_\alpha \times \psi'_\beta) \circ \varphi^{\alpha\beta}_V \circ (\varphi^{\alpha\beta}_U)^{-1} \circ (\varphi^{-1}_\alpha \times \psi^{-1}_\beta).$$

By (1.4), this can be written in the form

$$(\phi'_{\alpha\beta} \circ \phi^{-1}_{\alpha\beta})(x^k, y^a) = (\varphi'_\alpha \circ \varphi^{-1}_\alpha(x^k)), \psi'_\beta(g_{VU}(\varphi^{-1}_\alpha(x^k))\psi^{-1}_\beta(y^a)).$$

Example 1.1. The data (MxF, pr_1,M,F,G), where pr_1 is the projection on the first term of the Cartesian product MxF, give a fibre bundle. It is called the product fibre bundle. Its bundle atlas has only a local chart (M, id_{MxF}), where id_{MxF} denotes the identity mapping on MxF.

Other examples, much more significant, such as principal bundles and vector bundles, will be presented in the next section.

§2. Principal Fibre Bundles

We begin with those fibre bundles for which the type fibre coincides with the structural group.

Definition 2.1. *The fibre bundle (E,p,M,F,G) is called a (locally trivial) principal bundle if F = G and the action of G on G is given by the left translations:*$L_g(a) = ga$, $a, g \in G$.

A principal bundle will be denoted by (P, π, M, G).

The definition 2.1 is equivalent to the following one (cf. [89], [124]).

Definition 2.2. *A principal bundle is the data* (P, π, M, G), *where P and M are manifolds and* π *is a smooth surjective mapping such that the following conditions are satisfied:*

a) G acts at the right on the manifold P as a Lie group of transformations: for every $g \in G$ *we have* $R_g : P \to P$, *denoted as* $R_g(u) = ug$, $u \in P$. *The orbits of this action are the local fibres on which the action is simple transitive.*

b) For every $x \in M$ *there exists an open set* $U \subset M$ *containing it and a smooth map* $\varphi_U : \pi^{-1}(U) \to G$ *which satisfies the condition* $\varphi_U(ug) = \varphi_U(u)g$; *the mapping* $\psi_U : \pi^{-1}(U) \to U \times G$, $\psi_U(u) = (\pi(u), \varphi_U(u))$ *is a diffeomorphism.*

The structural functions of the principal bundle (P, π, M, G) are

$$g_{VU}: U \cap V \to G,\ g_{VU}(\pi(u)) = \varphi_V(u)(\varphi_U(u))^{-1},\ u \in \pi^{-1}(U \cap V).$$

Definition 2.3. *A morphism of the principal bundles (P,π,M,G) and (P',π',M',G') is a pair (f,f') of mappings for which the following conditions are fulfilled:*

a) $f: P \to P'$ is a smooth mapping,
b) $f': G \to G'$ is a morphism of Lie groups,
c) $f(ug) = f(u)f'(g),\ u \in P,\ g \in G.$

The morphism (f,f') induces a smooth mapping

$$f'': M \to M',\ f''(\pi(uG)) = \pi'(f(u)G'),\ u \in P.$$

Definition 2.4. *The principal bundles (P,π,M,G) and (P',π',M',G') are isomorphic if there exists a morphism (f,f') with f a diffeomorphism and f' an isomorphism of Lie groups.*

An isomorphism $(f,f'):(P,\pi,M,G) \to (P,\pi,M,G)$ is called an automorphism of (P,M,π,G).

Definition 2.5. *The principal bundle (P,π,M,G) is said to be a subbundle of the principal bundle (P',π',M',G') if $P \subset P'$, $G \subset G'$ and the pair (i,i') of the inclusion mappings $i: P \to P'$, $i': G \to G'$ is a morphism of these principal bundles.*

For every open set U in M, we have that $(\pi^{-1}(U), \pi_{|\pi^{-1}(U)}, U, G)$ is a subbundle of the principal bundle (P,π,M,G). This subbundle will be denoted by $(P,\pi,M,G)_{|U}$.

Definition 2.6. *A subbundle (P,π,M,G) of the principal bundle (P',π',M,G') is called a reduction of the structural group G' to G.*

Definition 2.7. *One says that the principal bundles (P,π,M,G) and (P',π',M,G') are locally equivalent if for every $x \in M$ there exists an open set containing it such that the principal bundles $(P,\pi,M,G)_{|U}$ and $(P',\pi',M,G')_{|U}$ to be isomorphic.*

It is obvious that $(M\text{x}G, pr_1, M, G)$ is a principal bundle. It is called the trivial principal bundle.

If one writes in all detail the Definition 2.1, it comes out that any principal bundle is locally equivalent to the trivial principal bundle. This explains the label of the locally trivial for a principal bundle.

Example 2.1. The principal bundle of frames on a manifold. A frame in a point $x \in M$ is a set of n linearly independent vectors tangent to M in x. Equivalently, a frame in $x \in M$ is a basis of the tangent space $T_x M$ to M in x. The set L(M) of all frames in all points of M can be naturally endowed with a manifold structure. In order to do, let $(U_\alpha, \varphi_\alpha)$ be a local chart on M and $\bar{U}_\alpha$ be the set of frames in the points of U_α. Let us consider the mapping $\bar{\varphi}_\alpha : \bar{U}_\alpha \to \mathbb{R}^{n+n^2}$ defined by

$$\bar{\varphi}_i(x, \underset{1}{X}_x, \ldots, \underset{n}{X}_x) = (x^1, \ldots, x^n, \underset{1}{X}^1_x, \ldots, \underset{1}{X}^n_x, \ldots, \underset{n}{X}^1_x, \ldots, \underset{n}{X}^n_x),$$

where $(x^1, \ldots, x^n) = \varphi_\alpha(x)$ and $\underset{a}{X}^1_x, \ldots, \underset{a}{X}^n_x$, $a = 1, 2, \ldots, n$ are the coordinates of the tangent vector $\underset{a}{X}_x$ in the local chart $(U_\alpha, \varphi_\alpha)$.

If (U_β, φ_β) is another local chart such that $U_\alpha \cap U_\beta \neq \Phi$ and $\varphi_\beta(x) = (x^{1'}, \ldots, x^{n'})$, then the change of coordinates $\varphi_\beta \circ \varphi_\alpha^{-1}$ is of the form

$$(2.1) \quad x^{k'} = x^{k'}(x^1, \ldots, x^n), \quad \mathrm{rank}(\partial x^{k'}/\partial x^k) = n, \quad k, k' = 1, \ldots, n.$$

Let us note that $\bar{\varphi}_\alpha(\bar{U}_\alpha)$ are open sets in $\mathbb{R}^{n+n^2}$. We consider on L(M) the topology for which all mappings $\bar{\varphi}_\alpha$ are homeomorphisms. The pairs $(\bar{U}_\alpha, \bar{\varphi}_\alpha)_{\alpha \in I}$ give rise to a smooth atlas on L(M) since the union of all $\bar{U}_\alpha$ is L(M) and the mappings $\bar{\varphi}_\beta \circ \bar{\varphi}_\alpha^{-1}$ have the local form

$$(2.2) \quad x^{k'} = x^{k'}(x^1, \ldots, x^n), \quad \underset{a}{X}^{k'} = \sum_{k=1}^{n} \frac{\partial x^{k'}}{\partial x^k} \cdot \underset{a}{X}^k, \quad a, k' = 1, 2, \ldots, n$$

which shows us these are smooth. It follows that $\bar{\varphi}_\alpha$ are diffeomorphisms and the structure manifold defined in this way is unique (up to diffeomorphisms). The principal bundle $(L(M), \pi, M, GL(n, \mathbb{R}))$ of frames on M, denoted by L(M) for the sake of brevity, has L(M) as total space, M as base manifold and the projection π assigns to each frame its origin $x \in M$. The structural group of this principal bundle is the general linear group $GL(n, \mathbb{R})$. A bundle chart associated to the local chart $(U_\alpha, \varphi_\alpha)$ on M is $(\pi^{-1}(U_\alpha), \Phi_\alpha)$, where

$$\Phi_\alpha : \pi^{-1}(U_\alpha) \to U_\alpha \times GL(n,\mathbb{R}),\ \Phi_\alpha\Big(\sum_{k=1}^{n} X_a^{k}\Big(\frac{\partial}{\partial x^{k}}\Big)_x\Big) = (x, X_a^{k}),\ x \in U_\alpha .$$

It comes out that

$$(\Phi_\beta \circ \Phi_\alpha^{-1})(x, X_a^{k}) = \Big(x, \sum_{k=1}^{n}\Big(\frac{\partial x^{k'}}{\partial x^{k}}\Big)_x X^{k}\Big).$$

Thus the structural functions $g_{\alpha\beta} : U_\alpha \cap U_\beta \to G$ of the principal bundle L(M) are $g_{\alpha\beta}(x) = \Big(\frac{\partial x^{k'}}{\partial x^{k}}\Big)_x$.

Using L(M) we can give

Definition 2.8. *A G-structure on a manifold M is a subbundle (P,π,M,G) of the principal bundle L(M) i.e. a reduction of the structural group $GL(n,\mathbb{R})$ to a subgroup G of it.*

Almost all remarkable geometrical structures may be thought of as G-structures. Thus a Riemannian structure is an O(n)-structure, where O(n) is the orthogonal group. A symplectic structure is an Sp(n)-structure, where Sp(n) is the symplectic group. The almost product structures, the almost complex structures, the f-structures are also G-structures.

We shall present now the standard construction by which one associates to a principal bundle (P,π,M,G) and a manifold F on which G acts smoothly and effectively, a locally trivial fibre bundle.

Let $L : G \times F \to F$, $L(g,f) = gf$, $g \in G$, $f \in F$ be the action (on the left) of G on F. We consider the product manifold PxF and we define an action (on the right) of G on it by

$$(P \times F) \times G \to P \times F,\ (u,f)g = (ug, g^{-1}f),\ u \in P.$$

We endow with the quotient topology the quotient set E = (PxF)/G. An element of the set E is

$$[u,f] = \{(ug, g^{-1}f), g \in G\}.$$

The mapping $p_E : E \to M$, $p_E([u,f]) = \pi(u)$ is well-defined since $\pi(ug) = \pi(u)$ for every $g \in G$, continuous and surjective. Let (U, ψ_U) be a bundle chart of (P,π,M,G). Setting $\psi_U(u) = (\pi(u), \varphi_U(u))$ for every $u \in p_E^{-1}(U)$, the action of G on PxF may be expressed as

$(u,f)g = (\pi(u), \varphi_U(u)g, g^{-1}f)$. We define a mapping $p_E^{-1}(U) \to U \times F$ by $[u,f] \to (\pi(u), \varphi_U(u)f)$. This is well-defined and bijective. Thus we can transport the manifold structure of UxF on $p_E^{-1}(U)$ and thus this mapping becomes a diffeomorphism. These local manifold structures determine a manifold structure on E. Indeed, if(U, ψ_U) and (V, ψ_V) are two bundle charts of (P,π,M,G), for every $x \in U \cap V \neq \Phi$ and $f \in F$ we have

$$(x,f) \in (U \cap V) x F \to [\psi_U^{-1}(x,e),f] \to (\pi(\psi_U^{-1}(x,e)), \psi_V(\psi_U^{-1}(x,e)f) = (x, g_{VU}(x)f) \in (U \cap V) \times F,$$

where $g_{VU}(x) = (\psi_V \circ \psi_U^{-1})(x,e)$. Recall e is the neutral element of G. Thus the changes of local coordinates on E are smooth. We have now all elements which allow us to assert that (E,p_E,M,F,G) is a locally trivial fibre bundle. The structural functions of it coincide to those of (P,π,M,G). The following two theorems (for the proofs see S. Kobayashi and K. Nomizu [89]) underline the important role the structural functions are playing.

Theorem 2.1. *Let M be a manifold,* $(U_\alpha)_{\alpha \in I}$ *an open covering of it and G a Lie group. Assuming there are given for every* $\alpha, \beta \in I$ *some functions* $g_{\beta\alpha}: U_\alpha \cap U_\beta \to G$ *which verify the equalities:*

$$(2.3)\quad g_{\gamma\beta}(x) g_{\beta\alpha}(x) = g_{\gamma\alpha}(x),\ x \in U_\alpha \cap U_\beta \cap U_\gamma,$$

then there exists a principal bundle such that $(g_{\beta\alpha})$ *are its structural functions associated to the covering* $(U_\alpha)_{\alpha \in I}$.

Theorem 2.2. *Let* (P', π', M, G') *be a principal bundle and G be a Lie subgroup of* G'. *A necessary and sufficient condition to exist a reduction* (P,π,M,G) *of* G' *to G is to exist a system of structural functions* $(g'_{\beta\alpha})$ *of* (P', π', M, G') *associated to an open covering* $(U_\alpha)_{\alpha \in I}$ *of M, whose values belong to G.*

§3. Vector Bundles

Another important class of locally trivial fibre bundles is that of vector bundles. It can be used for a fruitful theory of some geometrical models for gravitation and electromagnetism.

Definition 3.1. *A locally trivial fibre bundle (E,p,M,F,G) is called a vector bundle if F is a linear space and G is the Lie group of all automorphisms of F.*

In what follows we shall consider only the vector bundles whose type of fibre is a real linear space of finite dimension, say m. Concretely, we shall put $F = \mathbb{R}^m$ and then $G = GL(m,\mathbb{R})$. The vector bundle $(E,p,M,\mathbb{R}^m,GL(m,\mathbb{R}))$ will be denoted shortly by (E,p,M).

The structure of the vector bundle (E,p,M) is given by a set of bundle atlases, called also vectorial atlases, which are equivalent each to other. Two vectorial atlases are equivalent if their union is again a vectorial atlas. In conformity with the Definition 1.1 a vectorial atlas is $(U_\alpha, \varphi_\alpha, \mathbb{R}^m)_{\alpha\in I}$ such that the following conditions hold:

a) $(U_\alpha)_{\alpha\in I}$ is an open covering of the manifold M,

b) The mappings $\varphi_\alpha : p^{-1}(U_\alpha) \to U_\alpha \times \mathbb{R}^m$, $\alpha \in I$, are bijective and

$$p(\varphi_\alpha^{-1}(x, f) = x,\ x \in M,\ \ f \in \mathbb{R}^m.$$

c) For every pair $(\alpha,\beta) \in I \times I$ with $U_\alpha \cap U_\beta \neq \Phi$ there exists a smooth mapping $g_{\beta\alpha} : U_\alpha \cap U_\beta \to GL(m,\mathbb{R})$ such that $\varphi_{\alpha,x}^{-1} = \varphi_{\beta,x}^{-1} \circ g_{\beta\alpha}(x)$ for every $x \in U_\alpha \cap U_\beta$, where $\varphi_{\alpha,x}^{-1}$ is the restriction of φ_α^{-1} to $\{x\}$ x $\mathbb{R}^m$.

As in the general case of fibre bundles the manifold structure of E is induced by the manifold structure of M and by the vector bundle structure of (E,p,M). We exhibit here this structure since it will be frequently used in the following. Let $(U_\alpha, \psi_\alpha)_{\alpha\in I}$ be an atlas on M and $(U_\alpha, \varphi_\alpha, \mathbb{R}^m)_{\alpha\in I}$ be a vectorial atlas for (E,p,M). Note that such a choice of a vectorial atlas is always possible. Then $(p^{-1}(U_\alpha), h_\alpha)_{\alpha\in I}$ where

$$h_\alpha : p^{-1}(U_\alpha) \to \mathbb{R}^n \times \mathbb{R}^m,\ h_\alpha(u) = (\psi_\alpha(p(u)), \varphi_{\alpha,p(u)}(u))$$

is a smooth atlas on the total space E. Indeed, if (U_β, ψ_β) is another local chart on M with $U_\alpha \cap U_\beta \neq \Phi$ and $(U_\beta, \varphi_\beta, \mathbb{R}^m)$ is a bundle chart, then

$$(3.1)\ \ (h_\beta \circ h_\alpha^{-1})(x,y) = ((\psi_\beta \circ \psi_\alpha^{-1})(x), g_{\beta\alpha}(\psi_\alpha^{-1}(x)y)),\ \ (x,y) \in \mathbb{R}^n x \mathbb{R}^m.$$

Hence $h_\beta \circ h_\alpha^{-1}$ are smooth mappings.

We shall denote by $M_a^{a'}(x)$ the entries of the matrix associated with the linear operator $g_{\beta\alpha}(\psi_\alpha^{-1}(x))$. Then (3.1) takes the form

$$(3.2)\ \ (h_\beta \circ h_\alpha^{-1})(x,y) = (x^{k'}(x^1,\ldots,x^n), M_a^{a'}(x)y^a),\ \ (x,y) \in \mathbb{R}^n x \mathbb{R}^m.$$

Here we have used Einstein's convention on summation over repeated indices. This will be used throughout in the sequel. Therefore, the transformations of coordinates $(x^k, y^a) \to (x^{k'}, y^{a'})$ on E are of the form

$$\text{(3.3)} \qquad \begin{aligned} &x^{k'} = x^{k'}(x^1, \dots, x^n), \ \operatorname{rank}\left(\frac{\partial x^{k'}}{\partial x^k}\right) = n, \\ &y^{a'} = M_a^{a'}(x) y^a, \ (M_a^{a'}(x)) \in GL(m, \mathbf{R}). \end{aligned}$$

The properties (1.1) and (1.2) of the structural functions of a fibre bundle become in this context:

$$\text{(3.4)} \ M_{a'}^{a''}(x')M_a^{a'}(x) = M_a^{a''}(x), \text{ with } \psi_i^{-1}(x) = \psi_j^{-1}(x') = \psi_k^{-1}(x'') \in U_\alpha \cap U_\beta \cap U_\gamma,$$

$$\text{(3.5)} \ M_a^{a'} = \delta_a^{a'}, \text{ (Kronecker' symbol)}, \ \psi_\alpha^{-1}(x) \in U_\alpha = U_\beta.$$

Some geometrical properties of the base manifold M are carried by a vectorial atlas to the manifold E. Thus if M is paracompact, then E is paracompact. Consequently, if M is paracompact, then E admits partitions of unity. A special partition of unity on E may be derived from a partition of unity on M (cf. M. Anastasiei [7]).

Example 3.1. Tangent bundle to a manifold. Let T_xM be the tangent space to M in a point $x \in M$. Let us consider the set $TM = \bigcup_{x \in M} T_xM$ and the mapping $\tau: TM \to M$, $\tau(X_x) = x$, where $X_x \in T_xM$.

If $(U_\alpha, \varphi_\alpha)_{\alpha \in I}$ is an atlas on M, let us define a bundle chart $(U_\alpha, \varphi'_\alpha, \mathbf{R}^n)$ by

$$\varphi'_\alpha : \tau^{-1}(U_\alpha) \to U_\alpha x \mathbf{R}^n, \ \varphi'_\alpha = (x, X_x^1, \dots, X_x^n)$$

if $X_x = X_x^k \left(\frac{\partial}{\partial x^k}\right)_x$, k=1,2,...n. The mapping $\varphi'_{\alpha,x}$ is clearly bijective. Let us consider another bundle chart $(U_\beta, \varphi'_\beta, \mathbf{R}^n)$ with $U_\alpha \cap U_\beta \neq \Phi$ and let us compute $\varphi'_\beta \circ \varphi'^{-1}_\alpha$. We have

$$(\varphi'_\beta \circ \varphi'^{-1}_\alpha)(x, X_x^1, \dots, X_x^n) = \varphi'_\beta\left(X_x^k \left(\frac{\partial}{\partial x^k}\right)_x\right) = \varphi'_\beta\left(\frac{\partial x^{k'}}{\partial x^k} X_x^k \left(\frac{\partial}{\partial x^{k'}}\right)_x\right) = \left(x, \left(\frac{\partial x^{k'}}{\partial x^k}\right)_{\varphi_\alpha(x)} X_x^k\right).$$

Hence the structural functions are as follows:

$$g_{\beta\alpha}: x \to \left(\frac{\partial x^{k'}}{\partial x^{k}}\right)_{\varphi_\alpha(x)} \in GL(n,\mathbb{R}),\ x \in U_\alpha \cap U_\beta.$$

It is clear that they are smooth. Thus (TM,τ,M) is a vector bundle. It is called the tangent bundle to M or over M.

By (3.3) the transformations of coordinates $(x^k,y^k) \to (x^{k'},y^{k'})$ on the manifold TM are of the form

$$\text{(3.6)}\quad \begin{aligned} &x^{k'} = x^{k'}(x^1,\ldots,x^n),\ \operatorname{rank}\left(\frac{\partial x^{k'}}{\partial x^k}\right) = n,\\ &y^{k'} = \frac{\partial x^{k'}}{\partial x^k} y^k,\ k,\ k' = 1,2,\ldots,n. \end{aligned}$$

If we look on the Example 2.1, we remark the structural functions of the principal bundle of frames over M are defined by the same matrices $\left(\frac{\partial x^{k'}}{\partial x^k}\right)$. This fact suggests the existence of a relation between these two bundles. Indeed, the fibre bundle associated to L(M) by the construction given in the end of §2 with $F = \mathbb{R}^n$ and $G = GL(n,\mathbb{R})$ is just the tangent bundle (TM,τ,M) if we identify the coset $[(\underset{k}{X})_x,(z^i)]$ to the vector $\sum_{k,s=1}^{n} z^k \underset{k}{X}{}^s\left(\frac{\partial}{\partial x^s}\right)_x$, where $(\underset{k}{X})_x$ is a frame in the point $x \in M$ and (z^i) is a vector in $\mathbb{R}^n$.

Example 3.2. Cotangent bundle over a manifold. Let T_x^*M be the dual of the tangent space T_xM, $x \in M$. We set $T^*M = \bigcup_{x\in M} T_x^*M$ and we define $\tau^*: T^*M \to M$ by $\tau^*(\omega_x) = x$, where $\omega_x \in T_x^*M$. Let $(U_\alpha,\psi_\alpha)_{\alpha\in I}$ be an atlas on M. We define a bundle chart $(U_\alpha,\psi'_\alpha,\mathbb{R}^n)$, where

$$\psi'_\alpha: \tau^{*-1}(U_\alpha) \to U_\alpha \times \mathbb{R}^n,\quad \psi'_\alpha(\omega_x) = (x,\omega_{1,x},\ldots,\omega_{n,x})$$

if $\omega_x = \omega_{k,x}(dx^k)_x$, $k=1,2,\ldots,n$. The mappings $\psi'_{\alpha,x}$ are obviously bijective. Let $(U_\beta,\psi'_\beta,\mathbb{R}^n)$ be another bundle chart with $U_\alpha \cap U_\beta \neq \Phi$. We have

$$(\psi'_\beta \circ \psi'^{-1}_\alpha)(x,\omega_{1,x},\ldots,\omega_{n,x}) = \psi'_\beta(\omega_x) = \psi'_\beta(\omega_{k,x}(dx^k)_x) =$$

$$= \psi'_\beta(\omega_{k,x}\frac{\partial x^k}{\partial x^{k'}}(dx^{k'})_x) = (x, \frac{\partial x^k}{\partial x^{k'}}\omega_{k,x}),\ k,k'=1,2,\ldots,n,$$

where $(\frac{\partial x^k}{\partial x^{k'}})$ is the inverse matrix of the matrix $(\frac{\partial x^{k'}}{\partial x^k})$. Thus a structural function associate to $x \in U_\alpha \cap U_\beta$ the matrix $(\frac{\partial x^k}{\partial x^{k'}})_{\varphi_\alpha(x)}$ from $GL(n,\mathbb{R})$. Such a function is clearly smooth. Thus (T^*M,τ^*,M) is a vector bundle. It is called the cotangent bundle over M.

Finally, we note the possibility of considering the product or trivial vector bundle $(M\mathrm{x}\mathbb{R}^n, pr_1, M)$.

§4. Morphisms of Vector Bundles

It is now natural to deal with the notion of morphism of vector bundles. We limit ourselves to vector bundles with the same base.

Definition 4.1. *A morphism of the vector bundles (E,p,M) and (E',p',M) is a mapping $f:E \to E'$, with the property that for every point $x \in M$ there exists a bundle chart $(U,\varphi,\mathbb{R}^m)$ of (E,p,M) and a bundle chart $(U',\varphi',\mathbb{R}^{m'})$ of (E',p',M) around $x \in M$ and a smooth mapping $h_{U'U}: U \cap U' \to L(\mathbb{R}^m,\mathbb{R}^{m'})$ such that*

$$f_x \circ \varphi_x^{-1} = \varphi_x'^{-1} \circ h_{U'U}(x),\ x \in U \cap U',$$

where f_x is the restriction of f to the local fibre E_x.

This definition shows us that $f_x: E_x \to E'_x$ is a linear mapping and $h_{U'U}(x) = \varphi'_x \circ f_x \circ \varphi_x^{-1}$. One checks without difficulties that f is smooth.

A morphism of vector bundles is a monomorphism (epimorphism) if f_x is a monomorphism (epimorphism) of vector spaces for every point x from base manifold M.

If f is a bijective morphism of the vector bundles (E,p,M) and (E',p',M) then the mappings f_x are isomorphisms for every $x \in M$ and $f^{-1}: E' \to E$ is a morphism of vector bundles whose associated functions are $h_{U'U}^{-1}(x)$. In this case f is called an isomorphism of the vector bundles (E,p,M) and (E',p',M). Two vector bundles are called isomorphic if there exists an isomorphism of them.

Let (E,p,M) be a vector bundle and $h: M' \to M$ be a mapping from a manifold M' to M. Let us set

$$h^*E = \{(x',u) \in M'\mathrm{x}E,\ h(x') = p(u)\}$$

and let us define the mapping

$$h^*p: h^*E \to M', \ h^*p(x',u) = x'.$$

Then (h^*E, h^*p, M) is a vector bundle. It is called the *induced vector bundle by h* or the *pull-back* of the vector bundle (E,p,M) by h. Every bundle chart $(U,\varphi,\mathbb{R}^m)$ of the vector bundle (E,p,M) determines a bundle chart $(h^{-1}(U),\varphi',\mathbb{R}^m)$ of the induced vector bundle, where

$$\varphi':(h^*p)^{-1}(h^{-1}(U)) \to h^{-1}(U_\alpha)x\mathbb{R}^m, \ \varphi'(x',u) = (x',\varphi_{h(x)}(u)).$$

The structural functions of (h^*E, h^*p, M') are $g^*_{\beta\alpha}(x') = g_{\beta\alpha}(h(x'))$. Note also that its local fibre in $x' \in M'$ is $(h^*E)_{x'} = \{x'\}xE_{h(x)}$.

If M' is a submanifold of M, the considerations made in the above may be applied to the inclusion mapping $i:M' \to M$. The pull-back of (E,p,M) by i will be called the restriction of it to M' and will be denoted by $(E_{|M'}, p, M')$. Finally, we note that if h is the identity mapping, then the pull-back of the vector bundle (E,p,M) by h coincides with itself and if h is a constant map the corresponding pull-back is a product (trivial) vector bundle.

If f and f' are two morphisms of the vector bundles (E,p,M) and (E',p',M') the sum $f + f'$ as well the product af, where a is a real function on E may be defined fibrewise. Checking step by step the conditions from the Definition 4.1 one obtains

Proposition 4.1. *The sum $f + f'$ and the product af, where f,f' are morphisms of the same vector bundles and a is a real function on their base manifold are again morphisms of those vector bundles.*

Now we shall give a proof of the existence of a morphism of any two vector bundles.

Theorem 4.1. *If (E,p,M) and (E',p',M) are two vector bundles whose base manifold M is paracompact, then there exists a morphism of vector bundles $f:E \to E'$.*

Proof. Let $(U_\alpha,\varphi_\alpha,\mathbb{R}^m)_{\alpha\in I}$ be a bundle atlas of (E,p,M). For every U_α any smooth map $U_\alpha \to L(\mathbb{R}^m,\mathbb{R}^{m'})$ defines, via convenient local charts, a morphism of vector bundles $f_\alpha: E_{|U_\alpha} \to E'_{|U_\alpha}$. Let now $(a_\alpha)_{\alpha\in I}$ be a partition of unity subordinated to the open covering $(U_\alpha)_{\alpha\in I}$ of M. By the Proposition 4.1, $a_\alpha f_\alpha: E_{|U_\alpha} \to E'_{|U_\alpha}$ are morphisms of vector bundles. Every $a_\alpha f_\alpha$ can be prolonged to E taking $(a_\alpha f_\alpha)_x = 0$ if $x \notin U_\alpha$ since $\mathrm{supp}\, a_\alpha \subset U_\alpha$. Applying again the Proposition 4.1 one obtains that the mapping $f: E \to E'$ defined by $f_x = \sum_\alpha (a_\alpha f_\alpha)_x$ (this sum is finite) is a morphism of vector bundles.

§5. Vector Subbundles

Let $\tilde{E}$ be a subset of the total space of a vector bundle $\xi = (E,p,M)$ having the property that every point $x \in M$ there exists a bundle chart $(U,\varphi,\mathbf{R}^m)$ around it such that $\varphi(p^{-1}(U) \cap \tilde{E}) = U \times \mathbf{R}^p, 0 < p < m$. It follows immediately that $\tilde{\xi} = (\tilde{E}, p_{|\tilde{E}}, M)$ is a vector bundle which has the type fibre $\mathbf{R}^p$. It is called a vector subbundle of $\xi = (E,p,M)$. It is also easy to check that $\tilde{E}$ is a submanifold of E and the inclusion map $i: \tilde{E} \to E$ is a morphism of vector bundles.

In order to obtain a first example of vector subbundle, let us consider a monomorphism $f: E' \to E$ of the vector bundles $\xi = (E',p',M)$ and $\xi = (E,p,M)$. Then the image $f(E')$ is the total space of a vector subbundle of ξ which is isomorphic to ξ'. This vector subbundle will be denoted by Im f and will be called the image of the monomorphism f.

Another important example of vector subbundle is provided by an epimorphism of vector bundles.

If $f: E' \to E$ is an epimorphism of the vector bundles $\xi' = (E',p',M)$ and $\xi = (E,p,M)$, then the set $\mathrm{Ker}\, f = \bigcup_{x \in M} \mathrm{Ker}\, f_x$, where $\mathrm{Ker}\, f_x$ is the kernel of the linear map f_x, is the total space of a vector subbundle of ξ'. This vector subbundle will be called the kernel of the bundle epimorphism f and will be denoted by Ker f.

If a vector bundle ξ and a vector subbundle ξ' of it are given, a new vector bundle, called the quotient of ξ by ξ' may be constructed, Namely, if E_x and E'_x are local fibres in the point $x \in M$ of ξ and ξ', respectively, considering the quotient vector space E_x/E'_x one checks that the set $E/E' = \bigcup_{x \in M} E_x/E'_x$ has a vector bundle structure over base M and type fibre $\mathbf{R}^m/F'$, where F' is the type fibre of ξ'. This vector bundle will be denoted by E/E' or ξ/ξ'.

The sequence of vector bundles over a manifold M:

$$(5.1)\quad 0 \longrightarrow \xi' \xrightarrow{\;f\;} \xi \xrightarrow{\;g\;} \xi'' \longrightarrow 0$$

is said to be *exact* if f is a monomorphism, g an epimorphism and Im f = Ker g.

If the sequence (5.1) is exact, then it is clear that the vector bundle ξ'' can be identified with the *quotient vector* bundle ξ/ξ'.

§6. Operations with Vector Bundles

Let $\xi = (E,p,M)$ and $\xi' = (E',p',M')$ be two vector bundles. It is clear that $\xi x \xi' = (ExE',pxp',MxM')$ is a vector bundle. This vector bundle is called the product of the vector bundles ξ and ξ'..

Let us assume that the vector bundles ξ and ξ' have the same base manifold M and let us consider the set $E \oplus E' = \bigcup_{x \in M} E_x \oplus E'_x$, where $\oplus$ denote direct sum of linear spaces, together with the map $p \oplus p'$, which associates to a pair $(u,u') \in E_x \oplus E'_x$ the point $x \in M$. It is not difficult to check that the triad $(E \oplus E', p \oplus p', M)$ is a vector bundle with the type fibre $\mathbb{R}^m$ x $\mathbb{R}^{m'}$. This vector bundle is called the Whitney sum of the vector bundles ξ and ξ' and will be denoted by $\xi \oplus \xi'$. Its structural functions are defined by the matrices of the form $\begin{bmatrix} g(x) & 0 \\ 0 & g'(x) \end{bmatrix}$, where $x \to g(x)$ and $x \to g'(x)$ are the structural functions of ξ and ξ', respectively.

The local fibres E_x and E'_x of ξ and ξ', respectively, being linear spaces, the linear space $L(E_x,E'_x)$ of all linear maps from E_x to E'_x may be considered. Then

$$L(\xi,\xi') = L(E,E') = (\bigcup_{x \in M} L(E_x,E'_x), \rho, M),$$

where the mapping ρ associates to $z \in L(E_x,E'_x)$ the point $x \in M$, is a vector bundle, called the vector bundle of linear maps from ξ to ξ'. Its type fibre is the linear space $L(\mathbb{R}^m,\mathbb{R}^{m'})$ of all linear maps from $\mathbb{R}^m$ to $\mathbb{R}^{m'}$, while the value of its structural function in a point $x \in U_\alpha \cap U_\beta$ is given by a mapping $h \to g'(x) \circ h \circ g(x)$, where $h \in L(\mathbb{R}^m,\mathbb{R}^{m'})$. More generally, if the q vector bundles $\xi_\alpha = (E_\alpha,p_\alpha,M)$, $\alpha = 1,2,\ldots,q$ are given, one may construct the vector bundle of the multi-linear mappings which are ξ' valued, denoted by

$$L(\xi_1,\ldots\xi_q;\xi') = (L(E_1,\ldots,E_q;E'),\rho_q,M),$$

where $L(E_1,\ldots,E_q;E') = \bigcup_{x \in M} L(E_{1,x},\ldots,E_{q,x};E'_x)$ and ρ_q is defined similar to ρ. When

$\xi'=(M\times\mathbb{R},pr_1,M)$ and $E_1 = E_2 = \ldots = E_q = E$, we shall put $L(E_1,\ldots,E_q,M\times\mathbb{R}) = L_q(E,\mathbb{R})$ and $L_1(E,\mathbb{R})$ will be denoted by $\xi^* = (E^*,\rho_1,M)$. This last vector bundle is called the *dual* of the vector bundle $\xi = (E,p,M)$. Its fibre type is $(\mathbb{R}^m)^*$. The value of its structural function in a point $x \in M$ is given by a map $\theta \to \theta \circ g(x), \theta \in (\mathbb{R}^m)^*$.

Also, using q copies of the vector bundle $\xi = (E,p,M)$ one may construct the vector bundle of the *alternate* q-linear mappings which are ξ' valued, denoted by $A_q(E,E')$ and, particularly, $A_q(E,\mathbb{R})$ may be considered.

The *tensor product* of the vector bundles $\xi = (E,p,M)$ and $\xi'=(E',p',M)$ is the vector bundle

$$\xi\otimes\xi' = E\otimes E' = (\bigcup_{x\in M} E_x\otimes E'_x, p\otimes p', M)$$

where $p\otimes p'$ maps $u\otimes u' \in E_x\otimes E'_x$ to $x\in M$. Obviously, the operation of tensor product $\otimes$ may be extended to a finite number of vector bundles.

Let $\xi'=(E',p',M)$ be a vector subbundle of the vector bundle $\xi = (E,p,M)$ and let us consider an exact sequence of vector bundles over M

$$(6.1)\quad 0 \longrightarrow \xi' \xrightarrow{\ i\ } \xi \xrightarrow{\ g\ } \xi'' \longrightarrow 0,$$

where i is the inclusion map. We shall say the exact sequence (6.1) is splitting on the left if there exists a morphism of vector bundles $h:\xi\to\xi'$ such that $h \circ i$ = identity on ξ'. It follows easily that Ker h is isomorphic to ξ'' and $\xi = \xi'\oplus\xi''$. Moreover, it results that the exact sequence (6.1) is splitting on the right i.e. there exists a morphism of vector bundles $f:\xi''\to\xi$ such that $g \circ f$ = identity on ξ''. It is also clear that $i\circ h + f\circ g$ = identity on ξ. Conversely, if the exact sequence (6.1) is splitting on the right it is also splitting on the left.

Using a partition of unity on M one proves

Theorem 6.1. *Every exact sequence (6.1) of vector bundles over a paracompact manifold M is splitting on the left (right).*

§7. Principal Bundle Associated with a Vector Bundle

Let F be the type fibre of the vector bundle $\xi = (E,p,M)$ and $(M\times F,pr_1,M)$ be the trivial vector bundle. Let us consider the vector bundle of the linear mappings $(L(M\times F,E),\rho,M)$ and let be

$$P = \bigcup_{x \in M} \{u/u:F \to E_x \text{ is an isomorphism of linear spaces}, \rho(u) = x\}.$$

If $(U_\alpha)_{\alpha \in I}$ are the domains of the bundle charts of a bundle atlas of the vector bundle $(L(MxF,E),\rho,M)$ then $P = \bigcup_{\alpha \in I} \rho^{-1}(U_\alpha) \cap P$. The set $\rho^{-1}(U_\alpha) \cap P$ is diffeomorphic to $U_\alpha x \, Aut(F)$, where Aut(F) is the group of isomorphisms of F. Hence it is open and consequently, it is a submanifold of P. Denote by π the restriction of the map ρ to P. The group Aut(F) acts (on the right) on P as follows:

$$R_g:P \to P, \; R_g(u)f = u(g(f)), g \in Aut(F), f \in F.$$

The orbits of this action are the local fibres on which Aut(F) acts simple transitive. The diffeomorphisms $\rho^{-1}(U_\alpha) \cap P \to U_\alpha x \, Aut(F)$ help us to check the condition b) from the Definition 2.2. Thus we conclude that $(P,\pi,M,Aut(F))$ is a principal bundle. It is called the *principal bundle of frames* associated with the vector bundle (E,p,M). Every element u of its total space being an isomorphism $u:F \to E_x$, may be identified to a basis of E_x, namely the image by u of a fixed basis of F. If we consider $F = \mathbb{R}^m$ with the canonical basis $e_a = (0,\ldots,1,\ldots0)$, $a=1,\ldots m$, u may be identified with the basis $u(e_a)$ of E_x. This identification justifies the term of frame in $x \in M$ for $u \in E_x$ as well as the term of principal bundle of frames for $(P,\pi,M,Aut(F))$.

The construction made in the above may be particularly applied to the tangent bundle (TM,τ,M). The principal bundle obtained in such a way is isomorphic to the principal bundle of frames on the manifold M described in the Example 2.1. The specified isomorphism is obtained by taking $\varphi:GL(n,\mathbb{R}) \to GL(n,\mathbb{R})$ the identity map and defining $f:P \to L(M)$ as the map which associates to each isomorphism $u:\mathbb{R}^n \to T_xM$ the frame $u(e_i)$ in T_xM. We shall identify these principal bundles via the isomorphism we just described.

§8. Sections in Vector Bundles

Let $\xi = (E,p,M)$ be a vector bundle and U an open subset of M.

Definition 8.1. *A mapping $s:U \to E$ with the property $p \circ s =$ identity on U is called a section of ξ over U.*

We shall denote by $S(U,\xi)$ or $S(U,E)$ the set of sections of ξ over U. We define the sum of two sections of ξ over U and the product of a section of ξ by a real function f on U, pointwise as follows:

(8.1) $(s+s')(x) = s(x) + s'(x)$, $(fs)(x) = f(x)s(x)$.

One checks without difficulties that $S(U,\xi)$ is an $\mathcal{F}(U)$- module.

The existence of sections of ξ on open subsets of M can be established by using the bundle charts of ξ. Thus, if $(U,\varphi,\mathbb{R}^m)$ is a bundle chart of ξ, the mappings: $U \to E$, $s(x) = \varphi^{-1}(x,v)$, where v is a fixed vector in $\mathbb{R}^m$, is a section of ξ over U.

Let us note that if we take v as (e_a), $a=1,\ldots,m$ the elements of a basis of $\mathbb{R}^m$, we obtain the sections $s_a(x) = \varphi^{-1}(x,e_a)$ which are linearly independent on U. Conversely, if m linear independent sections of ξ over U are given, then a local chart $(U,\varphi,\mathbb{R}^m)$ of ξ is obtained by putting

$$\varphi: p^{-1}(U) \to U \times \mathbb{R}^m,\ \varphi(\lambda^a(x)s_a(x)) = (x,\lambda^a(x)e_a),\ \lambda^a \in \mathcal{F}(U).$$

The sections of ξ over M are called global sections of ξ. The set $S(M,\xi)$ or $S(M,E)$ of all global sections of ξ is an $\mathcal{F}(M)$-module. It is nonvoid, when the base manifold M is paracompact.

Every morphism f of vector bundles $\xi = (E,p,M)$ and $\xi' = (E',p',M)$ induces an $\mathcal{F}(M)$-linear mapping $F: S(M,E) \to S(M,E')$, $F(s)(x) = f(s(x))$, $x \in M$. This is true, of course, for all open subsets of M. The sections of the tangent bundle (TM,τ,M) are called vector fields on M and the sections of the cotangent bundle are called 1-forms on M.

Finally, we make some comments on G-structures. By the Theorem 2.2 and the fact that the structural functions of the frame bundle coincide to those of the tangent bundle, a G-structure on M, i.e. a reduction of $GL(n,\mathbb{R})$ to G does exist if and only if the tangent bundle (TM,τ,M) does admit a vector bundle atlas whose structural functions are G-valued. If this happens we shall say shortly that the tangent bundle admits a reduction to G. In terms of local sections this condition is equivalent to the existence of an open covering (U_α) of the manifold M such that for every U_α there exists a system of local vector fields $\underset{\alpha}{X}_i$, $i=1,\ldots n$ which satisfy the following conditions:

1. $(\underset{\alpha}{X}_i(x))$ is a frame in $x \in U_\alpha$,
2. $\underset{\beta}{X}_j(x) = \sum_{i=1}^{n} g_{\beta\alpha}(x)^i_j \underset{\alpha}{X}_i$, $x \in U_\alpha \cap U_\beta \neq \Phi$, $j=1,\ldots,n$,
3. The matrix $(g_{\beta\alpha}(x)^i_j)$ belongs to G.

The frames $(\underset{\alpha}{X}_i(x))$ are said to be adapted to the G-structure which we have considered.

Definition 8.2. *A G-structure on M is called integrable if the natural frames* $(\frac{\partial}{\partial x^i})_x$ *are adapted to it for every* $x \in M$.

Definition 8.3. *Let* $\xi = (E,p,M)$ *be a vector bundle with type fibre* $\mathbb{R}^m$ *and G a Lie subgroup of* $GL(m,\mathbb{R})$*. One says that* ξ *admits a G-structure if there exists a reduction of* $GL(m,\mathbb{R})$ *to G for the principal bundle of frames of* ξ.

The existence of a G-structure for a vector bundle (E,p,M) does not depend on the existence of a G-structure on the base manifold M.

Before ending this chapter we mention that the Definition 4.1 can be formulated with some minor changes for vector bundles over different manifolds M and M′ assuming there exists a mapping $f_0: M \to M'$. In this case a morphism of two vector bundles (E,p,M) and (E′,p′,M′) appears as a pair (f,f_0). The results regarding the vector bundle morphisms from this chapter remains true, with the corresponding modifications, for the pairs (f,f_0).

Later special attention will be paid to those vector bundle morphisms, which will be called embeddings, that is, the pairs (f,f_0) for which f_0 are embeddings and f are injective mappings whose restrictions to fibres are monomorphisms.

Chapter II

Connections in Fibre Bundles

The theory of connections is an important field of differential geometry. It was initially developed in order to solve pure geometrical problems. Later, it was pointed out that its ingredients could serve as models for objects from the physical world. For instance the gauge fields may be thought of as the connections coefficients in a fibre bundle while the gauge tensor is interpreted in terms of the curvature and torsion tensors of that connection. In order to avoid too much generality, we shall not introduce the general notion of connection in a fibre bundle. For our purpose it is sufficient to present the theory of connections in vector bundles.

However, in this framework we shall begin with a very general notion of connection, the so-called non-linear connection, because of its importance in the development of the geometry of Finsler, Lagrange and Hamilton spaces, as well as for its useful physical interpretations. The linear connections will appear as particular cases. A local characterisation of a non-linear connection is given in §2 and other characterisations of it are provided in §3. Horizontal lift is introduced in §4 while the curvature is treated in §5. Morphisms preserving a non-linear connection are discussed in §6.

§1. Non-linear Connections in Vector Bundles

The concept of non-linear connection came from Finsler geometry. As a set of coefficients it appear in a book by E. Cartan [34], and then in a more explicit fashion in some papers of A. Kawaguchi [76], [77], where the importance of this notion is for the first time recognized.

The global formulation of it is due to W. Barthel [25]. Then it was studied by J. Vilms, [160], J. Grifone [47], etc.

We shall present the notion of non-linear connection and its main properties on a way suggested by the afore mentioned papers.

Let $\xi = (E,p,M)$ be a vector bundle whose type fibre is $\mathbb{R}^m$ and $p^T:TE\to TM$ be the differential of the mapping p. The mapping p^T is a fibre-preserving morphism of the tangent bundle (TE,τ_E,E) to E and of tangent bundle (TM,τ,M) to M. The kernel of this vector bundle morphism, which is a vector subbundle of the vector bundle (TE,τ_E,E), will be denoted by (VE,τ_V,E) and will be called the *vertical subbundle* over E. Its total space is $VE = \bigcup_{u\in E} V_u$, where $V_u = \mathrm{Ker}\, p_u^T$, $u\in E$. Let us denote by $i: VE \to TE$ the inclusion mapping. We recall that the local coordinates of a point u of E are (x^i, y^a), $i=1,2,\ldots,n$, $a=1,2,\ldots,m$, where, in fact $x^i = x^i \circ p$ if we think x^i as real functions on $U\subset M$ and the transformation of coordinates $(x^i, y^a)\to(x^{i'}, y^{a'})$ are given by the formula (3.3), Ch.I.

A vector X_u tangent to E in the point $u\in E$ will be locally represented as (x^i, y^a, X^i, A^a), or shortly as (x,y,X,A), where $(X^i)\in\mathbb{R}^n$ and $(A^a)\in\mathbb{R}^m$ are defined by the equality $X_u = X^i\frac{\partial}{\partial x^i} + A^a\frac{\partial}{\partial y^a}$. It results that $p^T(x,y,X,A) = (x,X)$. Thus the submanifold VE contains the elements which are locally represented as (x,y,o,A) and the local fibres of the vertical subbundle are isomorphic to $\mathbb{R}^m$. Since $p^T(\frac{\partial}{\partial y^a}) = 0$ it comes out that $(\frac{\partial}{\partial y^a})$ is a local basis of the vertical distribution $u\to V_uE$ on E. Thus this distribution is integrable.

Let us consider the following exact sequence of vector bundles over E:

$$(1.1)\quad 0 \longrightarrow VE \xrightarrow{\ i\ } TE \xrightarrow{\ p!\ } TE/VE \longrightarrow 0.$$

Definition 1.1. *A non-linear connection in the vector bundle $\xi = (E,p,M)$ is a splitting on the left of the exact sequence (1.1) i.e. a morphism of vector bundles $C: TE\to VE$ such that $C\circ i$ is the identity on VE.*

The kernel of the morphism C is a vector subbundle of (TE,τ_E,E). It will be called the *horizontal subbundle* and will be denoted by (HE,τ_H,E). It is obvious that the vector bundle (TE,τ_E,E) is the Whitney sum of the vertical and horizontal subbundles. Consequently, a vector subbundle (HE,τ_H,E) of (TE,τ_E,E) such that $TE = HE\oplus VE$ defines a splitting on the left of the exact sequence (1.1). Indeed, we can define C fibrewise as the projection on the second term of the direct sum $HE\oplus VE$. Thus we have proved

Theorem 1.1. *There exists a non-linear connection in ξ if and only if there exists a vector subbundle (HE,τ_H,E) of the tangent bundle over E such that*

$$TE = HE\oplus VE.$$

Corollary 1.1. *A non-linear connection in ξ is a distribution $u\to H_uE$ on E such that*

$$(1.2)\quad T_uE = H_u\oplus V_uE,\ u\in E.$$

Note that if a non-linear connection C in ξ is given, then a morphism

$$D:TE/VE \to TE$$

such that $p! \circ D$ is the identity on TE/VE is defined. Moreover, ImD is just the horizontal subbundle and

$$i \circ C + D \circ p! = id_{TE},$$

where id_{TE} means the identity on TE.

The existence of the non-linear connection in ξ is assured under natural geometrical assumptions.

Theorem 1.2. *Let (E,p,M) be a vector bundle over a paracompact manifold M. Then there exist non-linear connections in* ξ.

Proof. As we have noted in §3 of Ch.1, if M is a paracompact manifold then E is so. Thus the Theorem 6.1, Ch.I applies to the exact sequence (1.1) of vector bundles over E and it follows the sequence admits a splitting on the left. Hence a non-linear connection in ξ does exist.

§2. Local Representation of a Non-linear Connection

Let (U,φ) be a local chart on M. The exact sequence (1.1) can be locally written as follows:

(2.1) $0 \to \varphi(U) \times \mathbb{R}^m \times 0 \times \mathbb{R}^m \to \varphi(U) \times \mathbb{R}^m \times \mathbb{R}^n \times \mathbb{R}^m \to \varphi(U) \times \mathbb{R}^m \times \mathbb{R}^n \to 0.$

Here we have identified the coset $[(x,y,X,A)]$ in TE/VE to (x,y,X). Thus the map p! has the local form $p!(x,y,X,A) = (x,X)$. A mapping $C:TE \to VE$ which defines a non-linear connection is locally expressed as

$$C(x,y,X,A) = (x,y,0,C_\varphi(x,y,X,A)),$$

where $C_\varphi:\varphi(U) \times \mathbb{R}^m \times \mathbb{R}^n \times \mathbb{R}^m \to \mathbb{R}^m$ is a linear mapping with respect to the variable X and A. Since C is a splitting of the exact sequence (1.1), we have $C_\varphi(x,y,0,A) = A$ from which we deduce that the mapping $\dot{C}_\varphi(x,y,X,A) = C_\varphi(x,y,X,A) - A$ does not depend on A. It is linear with respect to X so that we can write

(2.2) $C_\varphi(x,y,X^i,A^a) = (A^a + N_i^a(x,y)X^i)$.

The real functions $N_i^a(x,y)$ defined on $p^{-1}(U)$ are called the *local coefficients* of the non-linear connection C.

Let us seek for the law of transformation of the functions (N_i^a) under a change of coordinates on E. A change of local coordinates on the total space of the vector bundle(TE,τ_E,E) is, by virtute of (3.3), Ch.I, of the form:

$$\text{(2.3)}\quad \begin{aligned} &x^{i'} = x^{i'}(x^1,\ldots,x^n),\quad y^{a'} = M_a^{a'}(x)y^a,\\ &X^{i'} = \frac{\partial x^{i'}}{\partial x^i}X^i,\quad A^{a'} = \frac{\partial M_a^{a'}(x)}{\partial x^i}y^a X^i + M_a^{a'}(x)A^a. \end{aligned}$$

We shall write the inverse of mapping (3.3), Ch.I as follows:

(2.4) $x^i = x^i(x^{1'},\ldots,x^{n'}),\ y^b = M_{a'}^b(x')y^{a'}$,

where $(M_{a'}^b(x'))$ is the inverse or the matrix $(M_c^{a'}(x))$.

Then we have $(x,y,X,A)\overset{C_\varphi}{\rightarrow}(A^a + N_i^a(x,y)X^i)$. On the other hand

$$(x,y,X,A) \overset{(2.3)}{\rightarrow} (x^{i'}, M_a^{a'}(x)y^a, \frac{\partial x^{i'}}{\partial x^i}X^i, \frac{\partial M_a^{a'}(x)}{\partial x^i}y^a X^i + M_a^{a'}(x)A^a) \overset{C_\varphi}{\rightarrow}$$

$$(\frac{\partial M_a^{a'}(x)}{\partial x^i}y^a X^i + M_a^{a'}(x)A^a + N_{i'}^{a'}(x',y')\frac{\partial x^{i'}}{\partial x^i}X^i) \overset{(2.4)}{\rightarrow}$$

$$(A^b + M_{a'}^b(x)(\frac{\partial M_a^{a'}(x)}{\partial x^i}y^a + N_{i'}^{a'}\frac{\partial x^{i'}}{\partial x^i})X^i).$$

Thus it results: $M_{a'}^b(x')(\frac{\partial M_a^{a'}(x)}{\partial x^i}y^a + N_{i'}^{a'}\frac{\partial x^{i'}}{\partial x^i}) = N_i^b$. Therefore we proved

Proposition 2.1. *Under (2.3) the local coefficients $N_i^a(x,y)$ of a non-linear connection C transform as follows:*

$$\text{(2.5)}\quad N_{i'}^{a'}\frac{\partial x^{i'}}{\partial x^i} = M_a^{a'}(x)N_i^a - \frac{\partial M_a^{a'}(x)}{\partial x^i}y^a.$$

Now we are in a position to give a new characterisation of a nonlinear connection in a vector bundle ξ.

Theorem 2.1. *To give a non-linear connection in ξ is equivalent to give a set of real functions $(N_i^a(x,y))$ defined on every domain of local chart on E, which under the local transformation (2.3) satisfies (2.5).*

Proof. If a non-linear connection in ξ is given, as we have seen before, a set of functions (N_i^a) satisfying (2.5) is obtained. Conversely, if such a set is given, we define C_φ by (2.2) and the splitting C as the mapping which locally is $(x,y,X,A) \to (x,y,0,C_\varphi(x,y,X,A))$. This definition is correct because of (2.5).

Sometimes we shall denote a non-linear connection by N to remember the functions $(N_i^a(x,y))$.

Definition 2.1. *A non-linear connection in the vector bundle ξ is said to be homogeneous (resp. linear) if the functions* $N_i^a(x,y)$ *are homogeneous (resp. linear) with respect to* $y \equiv (y^j)$.

Proposition 2.2. *Every homogeneous connection in ξ is a linear connection in ξ.*

Proof. This is nothing but a consequence of the fact that every homogeneous real function on $\mathbf{R}^n$ which is differentiable in 0 is a linear function.

Remark 2.1. The proof of the Proposition 2.2 shows us that in order to have also the concept of homogeneous connection, which is essential in the Finsler geometry, we must avoid the null-section in the vector bundle ξ i.e. to consider the non-linear connections in the bundle $(E-0, p_{|E-0}, M)$, where E-0 is the manifold E without the image of the null-section.

§3. Other Characterisations of a Non-linear Connection

Let $\chi(M)$ be the $\mathcal{F}(M)$-module of the vector fields on M and $\chi(E)$ be the $\mathcal{F}(E)$-module of the vector fields on the total space E of the vector bundle $\xi = (E,p,M)$. Assuming ξ is endowed with a non-linear connection $C: TE \to VE$, $C \circ i =$ identity on VE, the vector bundle morphism C induces a morphism $C: \chi(E) \to S(VE)$. We shall denote always by the same letter the morphism between spaces of sections induced by the morphisms between respective vector bundles.

Now we define a morphism $v: \chi(E) \to \chi(E)$ as follows: $v(X) = C(X)$ if X is a vertical vector field and $v(X) = 0$ if X is an horizontal vector field. It easily results the following properties of v:

1. $v(\chi(E)) \subset S(VE)$,
2. $v(X) = X \Leftrightarrow X$ is a vertical vector field.

Conversely, a morphism $v:\chi(E) \to \chi(E)$ with the properties 1. and 2. defines a non-linear connection in the vector bundle ξ since Ker v defines a distribution on E which is supplementary to the vertical distribution. Therefore we have proved

Theorem 3.1. *A non-linear connection in the vector bundle $\xi = (E,p,M)$ is characterised by a morphism $v:\chi(E) \to \chi(E)$ with the properties 1. and 2. from the above.*

By the definition of v it comes out that $v^2 = v$ i.e. v is a projector. It is called the vertical projector of the non-linear connection N in ξ.

Now we shall prove

Theorem 3.2. *A non-linear connection in the vector bundle $\xi = (E,p,M)$ is characterized by a morphism $h:\chi(E) \to \chi(E)$ with the properties:*

a) $h^2 = h$, b) Ker $h = S(VE)$.

Proof. Given a non-linear connection C in the vector bundle ξ, we put $h = I - v$, where v is the vertical projector of C and I is the identity mapping on $\chi(E)$. It results immediately that $h^2 = h$ and

$$\text{Ker } h = \{X / h(X) = 0\} = \{X / X = v(X)\} = S(VE).$$

Conversely, given a morphism h with the properties a) and b) we set $v = I - h$. It follows easily that v is a vertical projector which defines a non-linear connection in the vector bundle ξ.

Corollary 3.1. *The morphism h has the properties*

3. $h(\chi(E)) \subset s(HE)$
4. $h(X) = X \Leftrightarrow X$ is an horizontal vector field.

The morphism h is called the horizontal projector of a non-linear connection C in a vector bundle ξ.

Theorem 3.3. *A non-linear connection in a vector bundle $\xi = (E,p,M)$ is characterised by an almost product structure P on E whose distribution of eigensubspaces which correspond to the eigenvalue -1 coincide to the total vertical distribution on E.*

Proof. Given a non-linear connection in ξ whose vertical projector is v, we put $P = I-2v$. It easily results that $P^2 = I$, hence P is an almost product structure on E. Moreover $\{X/P(X)=-X\}=S(VE)$. Conversely, an almost product structure P with property $\{X/P(X)=-X\}=S(VE)$ be given, by putting $v=\frac{1}{2}(I-P)$ one obtains the vertical projector of a non-linear connection in ξ.

Finally, let us note the following equalities

(3.1) $P=2h-I,\ P=h-v.$

The existence of a non-linear connection in $\xi=(E,p,M)$ induces a decomposition $\chi(E)=S(HE)\oplus S(VE)$. We shall set $X=hX+vX,\ X\in\chi(E)$.

Let (e_a), $a=1,2...,m$ be m local sections of ξ which are linearly independent on an open subset $U\subset M$. Then every section of ξ over U is $s=s^a(x)e_a$, $x\in U$. A non-linear connection in ξ induces a covariant derivative of the section in ξ defined locally as follows

(3.2) $D_X s = X^i(\frac{\partial s^a}{\partial x^i} + N_i^a(x,s(x))X^i)e_a,\ X = X^i(\frac{\partial}{\partial x^i})_x \in T_xM.$

By (3.2) one gets

Proposition 3.1. *The covariant derivative on the sections of ξ has the properties:*

1. $D_{X+Y}s = D_X s + D_Y s$,
2. $D_{fX}s = fD_X s,\ X, Y\in \chi(M)$.

If N is homogeneous we also have

3. $D_X(fs) = (Xf)s + fD_X s, s\in S(E), f\in \mathcal{F}(M)$.

If, moreover, N is linear then 1.,2.,3. and

4. $D_X(s_1+s_2) = D_X s_1 + D_X s_2, s_1, s_2\in S(E)$

hold.

Let $c:[a,b]\to M, t\to c(t)$ be a smooth curve on M and let be $\dot{c}:[a,b]\to TM, \dot{c} = c^T(t,1)$ the tangent field along on the curve c. Putting $\frac{Ds}{dt} = D_{\dot{c}}s$ we shall say that the section s in

ξ is parallel along c if $\frac{Ds}{dt}=0$. By (3.2) we have:

Proposition 3.2. *A local section $s:U\to E$, $U\subset M$ in ξ is parallel along c with $c([a,b])\subset U$ if and only if*

$$(3.3)\quad \frac{ds^a}{dt}+N_i^a(x,s(x))\frac{dx^i}{dt}=0,\quad s=s^a(x)e_a,\ x\in U.$$

Let us consider a linear connection C in the vector bundle ξ. Its local coefficients are of the form:

$$(3.4)\quad N_i^a=K_{bi}^a(x)y^b,$$

where by (2.4) the real functions K_{bi}^a satisfy the following law of transformation under (3.3), Ch.I:

$$(3.5)\quad K_{b'i'}^{a'}\frac{\partial x^{i'}}{\partial x^i}M_b^{b'}(x)=M_a^{a'}(x)K_{bi}^a-\frac{\partial M_b^{a'}}{\partial x^i}.$$

The covariant derivative defined by a linear connection satisfies the properties 1.-4. from the Proposition 3.1.

Conversely, a mapping $D:\chi(M)\times S(E)\to S(E)$, $(X,s)\to D_Xs$ satisfying 1.-4. from the above mentioned Proposition defines a linear connection in the vector bundle ξ. Indeed, if $s=s^a(x)e_a$ is a local section in ξ, putting

$$(3.6)\quad D_{\frac{\partial}{\partial x^i}}s_b=K_{bi}^a(x)s_a,$$

one easily obtains the real functions K_{bi}^a satisfy (3.5). By the Proposition 2.1 these functions define a linear connection in ξ. Hence we may identify a linear connection in $\xi$to D as in the above.

A linear connection in ξinduces a linear connection in its dual ξ^* since, as is easily to check, the mapping

$$D^*:\chi(M)\times S(E^*)\to S(E^*),\ (X,\omega)\to D_X^*\omega,$$

defined by

(3.7) $(D^*_X)\omega)(s) = X\omega(s) - \omega(D_X s), \omega \in S(E^*), s \in S(E)$

satisfies the equalities 1.-4. from the Proposition 3.1.

The curvature of a linear connection D in a vector bundle is defined by

(3.8) $R(X,Y)s = D_X D_Y s - D_Y D_X s - D_{[X,Y]}s,\ X,Y \in \chi(M), s \in S(E).$

The mapping $R(X,Y):S(E) \to S(E)$ is linear in the both variables and skew symmetric. Thus it appears as a 2-form which is $L(\xi;\xi)$-valued.

Recall that a linear connection D in ξ induces an operator of covariant differentiation on q-forms which are ξ-valued by the following formula:

$$(3.9)\quad \begin{aligned} d\Phi(X_0,\ldots,X_q) &= \sum_{i=0}^{q} (-1)^i D_{X_i}\Phi(X_0,\ldots,\hat{X}_i,\ldots,X_q) + \\ &\sum_{0\le i<j\le q} (-1)^{i+j}\Phi([X_i,X_j],X_0,\ldots,\hat{X}_i,\ldots,\hat{X}_j,\ldots,X_q), \end{aligned}$$

where Φ is a q-form whose values are sections in ξ and the superscript "^" means the omission of the argument indicated by it.

§4. Vertical and Horizontal Lifts

We shall deal with the operation of the lifting of the sections in the vector bundles $\xi=(E,p,M)$, $\xi^*=(E^*,p^*,M)$, (TM,τ,M) and (T^*M,τ^*,M) to the geometrical objects on the total space of these bundles. Thus, if $s:M\to E$ is a section in ξ given locally as $s=s^a(x)e_a$, $x\in U\subset M$, it can be lifted to a vector field s^v of E given locally by $s^v = s^a(x)\dfrac{\partial}{\partial y^a}$.

Proposition 4.1. *The mapping* $s \to s^v$ *has the properties:*

1. $(fs)^v = (f\circ p)s^v, f\in \mathcal{F}(M)$,
2. $(s+t)^v = s^v + t^v$,
3. $[s^v,t^v]=0,\ s,t\in S(E)$.

Proof. A direct calculation by using the local coordinates.

The vector field s^v is called the vertical lift of s. Analogously, every 1-form ω given locally as $\omega = \omega_i dx^i$ on M can be lifted to an 1-form ω^h on E by the formula

$\omega^h = \omega_i(x)dx^i$, where (dx^i, dy^a) is the dual frame to $(\frac{\partial}{\partial x^i}, \frac{\partial}{\partial y^a})$ on E.

It is easily to prove

Proposition 4.2. *The mapping* $\omega \to \omega^h$ *has the properties:*

1. $(f\omega)^h = (f \circ p)\omega^h$,
2. $(\omega + \pi)^h = \omega^h + \pi^h$, $f \in \mathcal{F}(M)$, $\omega, \pi \in \Lambda^1(M)$.

Now let us assume that $\xi = (E,p,M)$ is endowed with a non-linear connection N. Then the tangent bundle (TM, τ, M) is isomorphic to the horizontal bundle(HE, τ_{HE}, E), the isomorphism being the restriction of the mapping p^T. It follows that every vector field X on M determine an unique vector field X^h on E such that $p^T(X^h) = X$. The vector field X^h will be called the *horizontal lift* of the vector field X. In order to derive a local representation for X^h let us put $(\frac{\partial}{\partial x^i})^h = \frac{\delta}{\delta x^i}, i = 1, \ldots, n$. Thus a local basis $(\frac{\delta}{\delta x^i})$ of H_uE is determined and it is obvious that, generally, we have $\frac{\delta}{\delta x^i} = X_i^j \frac{\partial}{\partial x^j} + A_i^a \frac{\partial}{\partial y^a}$. But the condition $p^T(\frac{\delta}{\delta x^i}) = \frac{\partial}{\partial x^i}$ gives $X_i^j = \delta_i^j$ (Kronecker symbol) and the fact that $\frac{\delta}{\delta x^i}$ are the kernel of the mapping C implies $A_i^a = -N_i^a$. Thus we have:

$$(4.1) \quad \frac{\delta}{\delta x^i} = \frac{\partial}{\partial x^i} - N_i^a(x,y)\frac{\partial}{\partial y^a}.$$

By the definition of the horizontal lifting we easily deduce:

Proposition 4.3. *The mapping* $X \to X^h$ *has the properties:*

1. $(fX)^h = (f \circ p)X^h$,
2. $(X+Y)^h = X^h + Y^h$, $X, Y \in \chi(M)$.

By virtute of the Proposition 4.3 and the formula (4.1) we have

Proposition 4.4. *If* $X = X^i(x)\frac{\partial}{\partial x^i}$, *then*

$$X^h = X^i\left(\frac{\partial}{\partial x^i} - N_i^a(x,y)\frac{\partial}{\partial y^a}\right).$$

Remark 4.1. Generally, the bracket of two horizontal vector fields is not an horizontal one.

Let us note that $\left(\frac{\delta}{\delta x^i}, \frac{\partial}{\partial y^a}\right)$ is a basis of T_uE. We say it is an adapted basis to the direct decomposition $T_uE = H_uE \oplus V_uE$ (or to the horizontal and vertical distribution N and V) induced by the non-linear connection N.

Proposition 4.5. *The adapted basis* $\left(\frac{\delta}{\delta x^i}, \frac{\partial}{\partial y^a}\right)$ *transforms under a change of coordinates (3.3) Ch.I, as follows:*

$$(4.2) \quad \frac{\delta}{\delta x^i} = \frac{\partial x^{i'}}{\partial x^i}\frac{\delta}{\delta x^{i'}}, \quad \frac{\partial}{\partial y^a} = M_a^{a'}(x)\frac{\partial}{\partial y^{a'}}.$$

Proof. A straightforward calculation by using (2.3), (2.5) and (4.1).

The Proposition 4.5 shows us that the adapted basis $\left(\frac{\delta}{\delta x^i}, \frac{\partial}{\partial y^a}\right)$ leads to simple local expressions of the geometrical objects on the total space E of a vector bundle $\xi = (E,p,M)$.

The decomposition $T_uE = H_uE \oplus V_uE$ determined by a non-linear connection in ξ induces a decomposition

$$(T_uE)^* = (V_uE)^{\perp} \oplus (H_uE)^{\perp}, \ u \in E,$$

where $(V_uE)^{\perp}$ denotes the subspace of 1-forms which vanish on horizontal vectors and $(H_uE)^{\perp}$ denotes the subspace of 1-forms which vanish on vertical vectors.

Let $(dx^i, \delta y^a)$ be the dual basis of the adapted basis $\left(\frac{\delta}{\delta x^i}, \frac{\partial}{\partial y^a}\right)$. Then we have

$$(4.3) \quad \delta y^a = dy^a + N_i^a(x,y)dx^i.$$

For every section $\tilde{\omega}$ in ξ^* given locally as $\tilde{\omega} = \omega_a e^a$, where (e^a) is the dual basis of (e_a), we define the vertical lift of it as the 1-form on E given by

$$(4.4) \quad \tilde{\omega}^V = \omega_a \delta y^a.$$

It is easy to check that the mapping $\tilde{\omega}\to\tilde{\omega}^V$ is $\mathcal{F}(M)$-linear.

Note that $\tilde{\omega}^V$ vanishes on horizontal vectors and ω^h vanishes on vertical vectors.

§5. Curvature of a Non-linear Connection

Let N be a non-linear connection in the vector bundle $\xi=(E,p,M)$, whose local coefficients are $(N_i^a(x,y))$. Of course, these coefficients satisfy (2.5). Deriving it with respect to (y^b) it comes out that the set of real functions $\hat{N}^a_{b\alpha}$, $\alpha=1,\dots,n+m$, where $\hat{N}^a_{bi}=\dfrac{\partial N_i^a}{\partial y^b}$ and $\hat{N}^a_{bc}(x,y)=0$ satisfies the law of transformation of the local coefficients of a linear connection in the vertical vector bundle (VE,τ_V,E). The covariant derivative with respect to this connection is given as follows:

$$(5.1)\quad \hat{D}_{\frac{\partial}{\partial x^i}}\frac{\partial}{\partial y^a}=\hat{N}^b_{ai}\frac{\partial}{\partial y^b},\quad \hat{D}_{\frac{\partial}{\partial y^a}}\frac{\partial}{\partial y^b}=0.$$

Definition 5.1. *The linear connection $\hat{D}$ is called the Berwald connection associated to the non-linear connection N.*

If $Z=X^i\dfrac{\partial}{\partial x^i}+Y^a\dfrac{\partial}{\partial y^a}$ is a vector field on E and $B=B^a(x)\dfrac{\partial}{\partial y^a}$ is a section in the vertical bundle (VE,τ_V,E), by using the properties of $\hat{D}$ one obtains:

$$(5.2)\quad \hat{D}_Z B=[X^i(\frac{\partial B^a}{\partial x^i}+\frac{\partial N_i^a}{\partial y^b}B^b)+Y^b\frac{\partial B^a}{\partial y^b}]\frac{\partial}{\partial y^a}.$$

By the last equality one gets

Proposition 5.1. *The following equalities hold good*

(5.3) $\hat{D}_{X^h}A^v=[X^h,A^v]$, $X\in\chi(M), A\in S(E)$,

(5.4) $\hat{D}_Z U-\hat{D}_U Z=[Z,U]$, *if U,Z are vertical vector fields,*

(5.5) $\hat{D}_Z U=v[Z,U]$, *if U is a vertical vector field and Z is an horizontal one.*

Definition 5.2. *The absolute differential of a q-form Φ whose values are in the vertical bundle is the $q+1$ - form $\nabla\Phi$ given by*

(5.6) $(\nabla\Phi)(Z_0,\ldots,Z_q)=d\Phi(hZ_0,\ldots,hZ_q), Z_0,\ldots,Z_q\in\chi(E),$

where d is the operator of the exterior differentiation associated to the Berwald connection, in conformity with (3.9).

It is obvious that the splitting mapping $C:TE\to VE$ may be viewed as a vertical valued 1-form so that it is natural to set

Definition 5.3. *The curvature of a non-linear connection in a vector bundle $\xi=(E,p,M)$ is the vertical valued 2-form $\Omega=-\nabla C$.*

Let v be the vertical projector associated to the non-linear connection C and N_v be the Nijenhuis tensor field associated to it i.e.

(5.7) $N_v(X,Y)=[vX,vY]+v[X,Y]-v[vX,Y]-v[X,vY],\quad X,Y\in\chi(E)$

Theorem 5.1. *The following structure equation holds good*

(5.8) $\Omega=N_v.$

Proof. If X and Y are vertical fields, the both members of (5.8) vanish because of $v[X,Y]=0$ and $vX=X$, $vY=Y$. When X is an horizontal vector field and Y a vertical one again the both members of (5.8) vanish. Finally, if X and Y are horizontal vector fields then

$$-\Omega(X,Y)=dC(X,Y)=\hat{D}_X CY-\hat{D}_Y CX-C[X,Y]=-C[X,Y],$$

since C vanishes on horizontal vector fields, while $N_v(X,Y)=v[X,Y]$ by (5.7). Thus $\Omega(X,Y)=N_v(X,Y)$ since v coincides to C when C is viewed as morphism of the spaces of sections.

Setting $\Omega=\frac{1}{2}\Omega^a_{ij}dx^i\wedge dx^j\otimes\frac{\partial}{\partial y^a}$, by virtute of (5.7) the structure equation (5.8) takes the local form

(5.9) $\Omega^a_{ij}=\frac{\partial N^a_i}{\partial x^j}-\frac{\partial N^a_j}{\partial x^i}+N^b_i\hat{N}^a_{bj}-N^b_j\hat{N}^a_{bi}.$

Remark 5.1. By (5.7) and $v=I-h$ it comes out that $N_h=N_v$. Thus the structure equations (5.8) takes also the form $\Omega=N_h$.

Remark 5.2. The tensor field N_h was taken by J. Grifone [46] as defining the curvature of a non-linear connection in the tangent bundle over a smooth manifold.

A direct calculation in the local coordinates leads to

Proposition 5.2. *The following equality holds good*

(5.10) $[X^h,Y^h]=[X,Y]^h-\Omega(X^h,Y^h), X,Y\in\chi(M).$

An important result is given by

Theorem 5.2. *The following Bianchi identity holds good*

(5.11) $\nabla\Omega=0.$

Proof. It suffices to prove that $\nabla\Omega(X,Y,Z)=0$ for every horizontal vector fields X,Y,Z on E. We have

$$(\nabla\Omega)(X,Y,Z)=d\Omega(X,Y,Z)=\sum D_X(\Omega(Y,Z))-\sum\Omega([X,Y],Z)=$$

$$=-\sum\hat{D}_X C[Y,Z]+\sum C([X,Y],Z),$$

where $\sum$ means cyclic sum over X,Y,Z. The vector field C[Y,Z] is vertical and X is horizontal hence, by (5.5): $\hat{D}_X C[Y,Z]=C(X,[Y,Z])$ and now it results $\nabla\Omega=0$ in virtute of the Jacobi identity.

The structure equation (5.8) allows us to give a characterisation of the integrability of the horizontal distribution $u\to H_u, u\in E$.

Theorem 5.3. *The horizontal distribution of a non-linear connection C is integrable if and only if the curvature 2-form Ω identically vanishes.*

Proof. From the structure equation (5.8) written in the form $\Omega(X,Y)=v[hX,hY], X,Y\in S(HE)$, it follows that if $\Omega=0$ then the vector field [hX,hY] is an horizontal one and conversely, if the bracket of two horizontal vector fields is an horizontal one then $\Omega = 0$.

Corollary 5.1. *The horizontal distribution of a non-linear connection is integrable if and only if one of the following equivalent conditions hold good:*

1. $N_v = 0$*; 2.* $N_h = 0$*; 3.* $N_P = 0$*.*

Proof. The conditions 1.-3. are equivalent each to others by virtute of the equations $N_v = N_h = 4N_P$. The structure equation (5.8) and the Theorem 5.3 end the proof.

Note that the structure equation (5.8) may have also the form $\Omega = 4N_P$. Now we can give an answer to a question raised by the Remark 4.1.

Proposition 5.3. *The equality* $[X^h, Y^h] = [X, Y]^h$ *holds for every* $X, Y \in \chi(M)$ *if and only if the horizontal distribution of the non-linear connection C is integrable.*

Proof. By (5.10) the equality under discussion holds if and only if $\Omega(X^h, Y^h) = 0$ and by the Theorem 5.3 this happens if and only if the horizontal distribution on E is integrable.

§6. Affine Morphisms of Vector Bundles

The behaviour of the non-linear connections under morphisms of vector bundles is given by

Theorem 6.1. *Let* $f: E \to E'$ *be a morphism of vector bundles* $\xi = (E, p, M)$ *and* $\xi' = (E', p', M)$ *such that f to be isomorphism on fibres. Then:*

1. The vertical subspace in every point $u \in E$ *is isomorphic to the vertical subspace in* $f(u) \in E'$.

2. Every non-linear (resp. homogeneous, linear) connection in ξ' *induces a non-linear (resp. homogeneous, linear) connection in* ξ.

Proof. The assertion 1. follows quickly from the equality $p'^T \circ f^T = p^T$ and from the fact that f is an isomorphism on fibres. In order to prove 2. let C' be the left splitting mapping which is defining a non-linear (resp. homogeneous, linear) connection in ξ'. Putting $C = (f^T_{|VE})^{-1} \circ C' \circ f^T$ one obtains a mapping which is defining a nonlinear (resp. homogeneous, linear) connection in ξ.

Definition 6.1. *The non-linear (resp. homogeneous, linear) connection C induced*

by C' as in the previous theorem is called the pull-back of C' by f.

It will be set $C = f^* C'$.
Taking into account the Theorem 6.1 we may formulate also

Definition 6.2. *A morphism $f: E \to E'$ of the vector bundles $\xi = (E,p,M)$ and $\xi' = (E',p',M)$ endowed with the non-linear connection C and C', respectively, which is an isomorphism on fibres is said to be an affine morphism if $f^* C' = C$.*

Some characterisations of the affine morphisms are given in

Proposition 6.1. *Let $f: E \to E'$ be a morphism of the vector bundles ξ and ξ' endowed with the non-linear connections C and C', respectively. Let v (resp. v'), h (resp. h') and P (resp P') be the vertical projector, horizontal projector and the almost product structure associated to C (resp. C'). Let us assume f is an isomorphism on fibres. Then f is an affine morphism if and only if one of the following equivalent conditions holds good:*

1. $f^T \circ v = v' \circ f^T$,
2. $f^T \circ h = h' \circ f^T$,
3. $f^T \circ P = P' \circ f^T$.

Proof. The mutual equivalence of conditions 1.,2.,3. is immediate if we take into account that $h = I - v$ and $P = I - 2v$. As it results from the proof of the Theorem 6.1, f is an affine morphism if and only if $C = (f^T_{|V})^{-1} \circ C' \circ f^T$. But this equality is clearly equivalent to 1.

We leave to reader the proof of the following

Proposition 6.2. *Let $f: E \to E'$ be an affine morphism of the vector bundles ξ and ξ', endowed with the non-linear connections C and C' and let Ω and Ω' be their curvature 2-forms, respectively. Then $\Omega = 0$ implies $\Omega' = 0$. If, moreover, f^T is a surjective mapping then $\Omega = 0$ if and only if $\Omega' = 0$.*

Chapter III

Geometry of the Total Space of a Vector Bundle

The geometry of the total space of the tangent bundle to a smooth manifold is very rich. It contains a lot of geometrical objects of theoretical interest and of a great importance in constructing of various geometrical models useful in Physics. For the similar reasons the geometry of the total space of a vector bundle was intensively studied in the last fifteen years.

In this chapter we shall present the main results from the geometry of the total space of a vector bundle. In order to avoid long computations and the neglect of geometrical meanings as may happen in a classical approach, the notion of non-linear connection will be systematically used. The idea of using the non-linear connections in this framework comes from Finsler geometry and was proposed by R. Miron, [105,107].

Thus, as a first step, assuming there exists a non-linear connection N in the vector bundle (E,p,M) we shall study in §1 those linear connections, on the total space E, which preserve by parallelism the horizontal and vertical distributions on E. These connections will be called d-connections (or M-connections). Then, we study the h- and v- covariant derivations determined by a d-connection, a local representation of d-connections in §2, torsion and curvature in §3 and structure equations in §4. Next, in §5, we shall assume there exists a metrical structure on E and we shall study the metrical d-connections by using a non-linear connection canonically derived from the given metrical structure.

§1. d-Connections on the Total Space of a Vector Bundle

Let $\xi = (E,p,M)$ be a vector bundle endowed with a non-linear connection N. As we have seen in the previous chapter to give N is equivalent to giving the so-called horizontal distribution on E, $u \to H_uE$, which is supplementary to the vertical distribution $u \to V_uE$, $u \in E$.

The non-linear connection N induces a decomposition of every $X \in \chi(E)$ as a sum of its horizontal and vertical parts, that is:

$$(1.1)\quad X = hX + vX.$$

Definition 1.1. *A d-connection on E is a linear connection on E which preserves by parallelism the horizontal and vertical distributions on E.*

As usual by a linear connection on the manifold E we understand a linear connection in the tangent bundle (TE, τ_E, E).

Thus, an operator of covariant derivation $D: \chi(E) \times \chi(E) \to \chi(E)$, $(X,Y) \to D_X Y$ can be associated to it by adapting the formula (3.2), Ch. II. By the Proposition 3.1, Ch. II, the operator D has the following properties (Koszul's axioms):

1. $D_{X+Y}Z = D_X Z + D_Y Z$,
2. $D_{fX} Y = f D_X Y$,
3. $D_X(fY) = X(f)Y + fD_X Y$,
4. $D_X(Y+Z) = D_X Y + D_X Z$, $X,Y,Z \in \chi(E)$, $f \in \mathcal{F}(E)$.

As we have seen in Ch. II, these properties of D characterise a linear connection in a vector bundle.

We identify a linear connection on E with the operator D of covariant derivation associated to it.

By the general theory of connections on manifolds the horizontal and vertical distributions are preserved by parallelism if for every $X \in \chi(E)$, D_X carries the horizontal vector fields to the horizontal vector fields and the vertical vector fields to the vertical vector fields. Thus $D_X hY$ is always an horizontal vector field and $D_X vY$ is always a vertical one.

Theorem 1.1. *A linear connection D on E is a d-connection if and only if one of the following conditions holds:*

(i) $vD_X hY = 0$, $hD_X vY = 0$,
(ii) $D_X Y = hD_X hY + vD_X vY$,
(iii) $D_X v = 0$, $D_X h = 0$,
(iv) $D_X P = 0$,

for $X, Y \in \chi(E)$, where v, h, P are the vertical projector, horizontal projector and almost product structure associated to N, respectively.

Proof. If D is a d-connection, from the fact that $D_X hY = vD_X hY + hD_X hY$ belongs to S(HE) and $D_X vY = vD_X vY + hD_X vY$ belongs to S(VE) it follows (i) and it is clear that (i) implies that D is a d-connection. Now $D_X Y = hD_X hY + vD_X hY + hD_X vY + vD_X vY$ shows that (i) is equivalent to (ii). Then $D_X v = 0$ is equivalent to $D_X vY = vD_X Y$ and $D_X h = 0$ is equivalent to $D_X hY = hD_X Y$ and by these equations it becomes obvious that (iii) is equivalent to (i). From P = h - v it follows that (iii) implies (iv) and from $h = \frac{1}{2}(I+P)$, $v = \frac{1}{2}(I-P)$ it results that (iv) implies (iii). Thus the theorem is proved.

Remark 1.1. The almost product structure P is a G-structure on E, where G = GL(n,R) x GL(m,R). By (iv) of the Theorem 1.1 a linear connection on E is a d-connection if and only if it is a G-connection with respect to the G-structure P (cf.

Y.Ichijyo, R. Miron [64]).

The operator D induces an operator of covariant derivation on 1-forms, which we shall denote by the same letter D, given by

(1.2) $(D_X\omega)(Y) = D_X\omega(Y) - \omega(D_XY)$, $\omega \in \Lambda^1(E)$, $X, Y \in \chi(E)$.

As we have seen in Ch. II, the non-linear connection N induces a decomposition of every $\omega \in \Lambda^1(E)$ as a sum of its horizontal and vertical components, that is:

(1.3) $\omega = h\omega + v\omega$.

Theorem 1.2. *A linear connection D on E is a d-connection if and only if one of the following conditions holds:*

(i) $vD_Xh\omega = 0$, $hD_Xv\omega = 0$,
(ii) $D_X\omega = hD_Xh\omega + vD_Xv\omega$, $X \in \chi(E)$, $\omega \in \Lambda^1(E)$.

Proof. If D is a d-connection, replacing in (1.2) the 1-form ω by $h\omega$ and Y by vY one obtains the first equation from (i). Then replacing in (1.2) the 1-form ω by $v\omega$ and Y by hY one obtains the second equation from (i). Conversely, by $(D_Xh\omega)(vY) = 0$ we have $h\omega(D_XvY) = 0$, hence D_XvY is a vertical vector field. Analogously, from $hD_Xv\omega = 0$ we deduce that D_XhY is an horizontal vector field. Thus D is a d-connection. The equations (i) are equivalent to the equation (ii) as it follows easily from the identity

$$D_X\omega = hD_Xh\omega + vD_Xh\omega + hD_Xv\omega + vD_Xv\omega.$$

Thus the theorem is proved.

The decompositions (1.1) and (1.3) induce similar decompositions for every tensor field on E. Thus if W is a tensor field of type (p + r, q + s) on E, then

$$W(h\omega_1 + v\omega_1, \ldots, h\omega_p + v\omega_p, hX_1 + vX_1, \ldots, hX_q + vX_q,$$
$$h\omega_{p+1} + v\omega_{p+1}, \ldots, h\omega_{p+r} + v\omega_{p+r}, hX_{q+1} + vX_{q+1}, \ldots, hX_{q+s} + vX_{q+s})$$

is clearly a sum of $2^{p+r+q+s}$ terms, every one being what we shall call a d-tensor field on E . More precisely, we introduce:

Definition 1.2. *A tensor field W of type (p + r, q + s) on E is said to be a d-tensor field or a M-tensor field of type* $\begin{pmatrix} p & r \\ q & s \end{pmatrix}$ *on E if*

$$
\begin{aligned}
&W(\omega_1,\dots,\omega_p,X_1,\dots,X_q,\omega_{p+1},\dots,\omega_{p+r},X_{q+1},\dots X_{q+s}) = \\
(1.4)\quad &= W(h\omega_1,\dots,h\omega_p,hX_1,\dots,hX_q,v\omega_{p+1},\dots,v\omega_{p+r},vX_{q+1},\dots,vX_{q+s}), \\
&\omega_1,\dots,\omega_{p+r}\in\Lambda^1(E),\ X_1,,X_{q+s}\in\chi(E).
\end{aligned}
$$

Thus an horizontal vector field is a d-tensor field of type $\begin{pmatrix} 1 & 0 \\ 0 & 0 \end{pmatrix}$ and a vertical vector field is a d-tensor field of type $\begin{pmatrix} 0 & 1 \\ 0 & 0 \end{pmatrix}$. If $\omega \in \Lambda^1(E)$ then $h\omega$ is a d-tensor field of type $\begin{pmatrix} 0 & 0 \\ 1 & 0 \end{pmatrix}$ and $v\omega$ is a d-tensor field of type $\begin{pmatrix} 0 & 0 \\ 0 & 1 \end{pmatrix}$.

Locally, a d-tensor field of type $\begin{pmatrix} p & r \\ q & s \end{pmatrix}$ may be written in the form:

$$(1.5)\quad W = W^{i_1\dots i_p\, a_1\dots a_r}_{j_1\dots j_q\, b_1\dots b_s}\frac{\delta}{\delta x^{i_1}}\otimes\dots\otimes\frac{\partial}{\partial y^{a_1}}\otimes\dots\otimes dx^{j_1}\otimes\dots\otimes\delta y^{b_1}\otimes\dots\otimes\delta y^{b_s}.$$

By (4.2), Ch. II, the local components $W^{\dots}_{\dots}$ of W transform, under a change of coordinates on E, as follows

$$(1.6)\quad W^{i_1'\dots a_1'\dots}_{j_1'\dots b_1'\dots} = \frac{\partial x^{i_1'}}{\partial x^{i_1}}\dots M^{a_1'}_{a_1}\dots\frac{\partial x^{j_1}}{\partial x^{j_1'}}\dots M^{b_1}_{b_1'}\dots\ W^{i_1\dots a_1\dots}_{j_1\dots b_1\dots}.$$

It is clear that the d-tensor fields on E are nothing but the elements of the tensorial algebra of the vector bundle $HE\oplus VE\to E$. Conversely, all elements of this algebra are d-tensor fields on E.

The simplicity of the transformation law (1.6) shows us it is convenient to use d-tensor fields on E and we shall do this in what follows.

We shall associate to every d-connection on E two new operators of h- and v-covariant derivation in the algebra of d-tensor fields on E. Indeed, it is clear that any d-connection D on E leads to a decomposition

$$(1.7)\quad D_X Y = D_{hX}Y + D_{vX}Y,\ X,Y\in\chi(E).$$

For each $X\in\chi(E)$ we set

$$(1.8)\quad D^h_X Y = D_{hX}Y,\ D^h_X f = (hX)f,\ f\in\mathcal{F}(M),\ Y\in\chi(E).$$

Then, if $\omega \in \Lambda^1(E)$ putting $(D^h_X\omega)(Y) = hX\omega(Y) - \omega(D^h_X Y)$, it appears possible to define the action of the operator D^h_X for any d-tensor field by requiring D^h_X preserves the type of d-tensor fields, is R-linear, satisfies the Leibnitz rule with respect to tensor product and commutes with all contractions. We shall keep the notation D^h_X for this operator on the algebra of the d-tensor fields. We call it the operator of h-covariant derivation.

Analogously, if for each $X \in \chi(E)$ we set

(1.9) $D^v_X Y = D_{vX} Y$, $D^v_X f = (vX)f$, $f \in \mathcal{F}(M)$, $Y \in \chi(E)$

and we extend D^v_X to any d-tensor field on a similar way as D^h_X we obtain an operator on the algebra of the d-tensor fields on E which will be also denoted by D^v_X and will be called the v-covariant derivation.

By the definition of the operators D^h_X and D^v_X we get

Theorem 1.3. *The operator of h-covariant derivation D^h_X, $X \in \chi(E)$ has the following properties*

1. $vD^h_X hY = 0$, $hD^h_X vY = 0$, $D^h_X f = (hX)f$,
2. $D^h_X Y = hD^h_X hY + vD^h_X vY$,
3. $vD^h_X h\omega = 0$, $hD^h_X v\omega = 0$,
4. $D^h_X\omega = hD^h_X\omega + vD^h_X\omega$, $X, Y \in \chi(E)$, $\omega \in \Lambda^1(E)$, $f \in \mathcal{F}(E)$.

Theorem 1.4. *The v-covariant derivation D^v_X has the following properties:*

1. $vD^v_X hY = 0$, $hD^v_X vY = 0$, $D^v_X f = (vX)f$,
2. $D^v_X Y = hD^v_X hY + vD^v_X vY$,
3. $vD^v_X h\omega = 0$, $hD^v_X v\omega = 0$,
4. $D^v_X\omega = hD^v_X h\omega + vD^v_X\omega$, $X, Y \in \chi(E)$, $\omega \in \Lambda^1(E)$, $f \in \mathcal{F}(E)$.

The other properties of the d-connections will be deduced from their local form.

§2. Local Representation of d-Connections

Now we shall see what local form a d-connection on E may take. Let (x^i, y^a), $i=1,...,n$; $a=1,2,...,m$ be the local coordinates on E with respect to a local chart $(p^{-1}(U),h)$, where (U,φ) is a local chart on M.

The local coefficients of a linear connection on E with respect to the local frame $(\frac{\partial}{\partial x^i}, \frac{\partial}{\partial y^a})$ naturally separate into eight groups, each of them having complicated transformation laws under a change of coordinates on E. If we assume that E is endowed with a non-linear connection N, then the adapted local frame $(\frac{\delta}{\delta x^i}, \frac{\partial}{\partial y^a})$ to N and V may be used. The local coefficients of a linear connection on E, with respect to this adapted local frame split also into eight groups, which transform much simpler when a change of coordinates on E is performed.

The local coefficients of a d-connection on E with respect to the local adapted frame $(\frac{\delta}{\delta x^i}, \frac{\partial}{\partial y^a})$, as we shall see, separate only in four groups. The laws of transformation of two of them are similar to the law of transformation of a linear connection in the vector bundle ξ and on M, respectively, and the other two transform as the local components of d-tensor fields. These local coefficients are provided by the operators of h- and v-covariant derivation, as we shall see in the following.

Let D be a d-connection on the total space E of the vector bundle (E, p, M). The operator of h-covariant derivation D^h_X is well determined by its restrictions to real functions and vector fields on E. Thus there exists an unique system of functions $(L^i_{jk}(x,y), \tilde{L}^a_{bk}(x,y))$ such that

$$(2.1)\quad D^h_{\frac{\delta}{\delta x^k}} \frac{\delta}{\delta x^j} = L^i_{jk}(x,y)\frac{\delta}{\delta x^i}; D^h_{\frac{\delta}{\delta x^k}} \frac{\partial}{\partial y^b} = \tilde{L}^a_{bk}(x,y)\frac{\partial}{\partial y^a},$$

$$(2.1)'\quad D^h_{\frac{\delta}{\delta x^k}} f = \frac{\delta f}{\delta x^k} = \frac{\partial f}{\partial x^k} - N^a_k(x,y)\frac{\partial f}{\partial y^a}, f \in \mathcal{F}(E).$$

The real functions $(L^i_{jk}(x,y), \tilde{L}^a_{bk}(x,y))$ are called the local coefficients of D^h_X in the local chart $(p^{-1}(U),h)$ with respect to the local frame $(\frac{\delta}{\delta x^i}, \frac{\partial}{\partial y^a})$. It is clear that if $X = X^k\frac{\delta}{\delta x^k}$ then $D^h_X = X^k D_{\frac{\delta}{\delta x^k}}$.

Now let W be a d-tensor field say of type $\begin{pmatrix} 1 & 1 \\ 1 & 1 \end{pmatrix}$ given by

$$(2.2)\quad W = W^{ia}_{jb}\frac{\delta}{\delta x^i}\otimes\frac{\partial}{\partial y^a}\otimes dx^j\otimes\delta y^b .$$

Its h-covariant derivative is

$$(2.3)\quad D^h_X W = X^k\ W^{ia}_{jb|k}\frac{\delta}{\delta x^i}\otimes\frac{\partial}{\partial y^a}\otimes dx^j\otimes\delta y^b$$

where

$$(2.4)\quad W^{ia}_{jb|k} = \frac{\delta W^{ia}_{jb}}{\delta x^k} + L^i_{hk}W^{ha}_{jb} + \tilde{L}^a_{ck}W^{ic}_{jb} - L^h_{jk}W^{ia}_{hb} - \tilde{L}^c_{bk}W^{ia}_{jc}$$

since

$$D^h_{\frac{\delta}{\delta x^k}}dx^j = -L^j_{ik}dx^i,\ D^h_{\frac{\delta}{\delta x^k}}\delta y^a = -\tilde{L}^a_{bk}\delta y^b.$$

It is obvious that similar formulae to (2.3) and (2.4) can be written for d-tensor fields of any type.

Analogously, since the v-covariant derivation D^v_X is also determined by its restrictions to real functions and vector fields on E, there exists an unique system of real functions $(\tilde{C}^i_{jc}(x,y),\ C^a_{bc}(x,y))$ such that

$$(2.5)\quad D^v_{\frac{\partial}{\partial y^c}}\frac{\delta}{\delta x^j} = \tilde{C}^i_{jc}(x,y)\frac{\delta}{\delta x^i};\ D^v_{\frac{\partial}{\partial y^c}}\frac{\partial}{\partial y^b} = C^a_{bc}\frac{\partial}{\partial y^a},$$

$$(2.5)'\quad D^v_{\frac{\partial}{\partial y^c}}f = \frac{\partial f}{\partial y^c},\ f\in\mathcal{F}(E).$$

These functions are called the local coefficients of the v-covariant derivation D^v_X in the local frame $(\frac{\delta}{\delta x^i},\frac{\partial}{\partial y^a})$. Also is clear that $D^v_{X^a\frac{\partial}{\partial y^a}} = X^a D^v_{\frac{\partial}{\partial y^a}}$.

If we take $X = Y^c\frac{\partial}{\partial y^c}$ the v-covariant derivative of the d-tensor field W given by (2.2) is as follows:

$$(2.6)\quad D^{v}_{X}W = Y^{c}W^{ia}_{jb}|_{c}\frac{\delta}{\delta x^{i}}\otimes\frac{\partial}{\partial y^{a}}\otimes dx^{j}\otimes\delta y^{b},$$

where

$$(2.7)\quad W^{ia}_{jb}|_{c} = \frac{\partial W^{ia}_{jb}}{\partial y^{c}} + \tilde{C}^{i}_{hc}W^{ha}_{jb} + C^{a}_{dc}W^{id}_{jb} - \tilde{C}^{h}_{jc}W^{ia}_{hb} - C^{d}_{bc}W^{ia}_{jd}$$

since

$$D^{v}_{\frac{\partial}{\partial y^{b}}}dx^{j} = -\tilde{C}^{j}_{ib}dx^{i},\quad D_{\frac{\partial}{\partial y^{b}}}\delta y^{c} = -C^{c}_{db}\delta y^{d}.$$

It is clear that similar formulae to (2.6) and (2.7) can be written for any d-tensor field.

By (2.3) and (2.6), taking $X = X^{i}\frac{\delta}{\delta x^{i}} + Y^{a}\frac{\partial}{\partial y^{a}}$ we have

$$(2.8)\quad D_{X}W = (X^{k}W^{ia}_{jb|k} + Y^{c}W^{ia}_{jb}|_{c})\frac{\delta}{\delta x^{i}}\otimes\frac{\partial}{\partial y^{a}}\otimes dx^{j}\otimes\delta y^{b}.$$

Here and throughout from now on the indices i,j,k,... run from 1 to n and the indices a,b,c,... run from 1 to m.

The formulae (2.1) and (2.5) show us that $D\Gamma = (L^{i}_{jk}(x,y), \tilde{L}^{a}_{bk}(x,y), \tilde{C}^{i}_{ja}(x,y), C^{a}_{bc}(x,y))$ are the local coefficients of the d-connection D with respect to the local frame $(\frac{\delta}{\delta x^{i}}, \frac{\partial}{\partial y^{a}})$.

If a change of local coordinates on E is performed, using the law of transformation under it of the local frame $(\frac{\delta}{\delta x^{i}}, \frac{\partial}{\partial y^{a}})$ and the formulae (2.1) and (2.8), after a long but not difficult computation, we obtain the following transformation laws of the local coefficients of a d-connection:

$$(2.9)\quad L^{i'}_{j'k'} = \frac{\partial x^{i'}}{\partial x^{i}}\frac{\partial x^{j}}{\partial x^{j'}}\frac{\partial x^{k}}{\partial x^{k'}}L^{i}_{jk} + \frac{\partial x^{i'}}{\partial x^{k}}\frac{\partial^{2}x^{k}}{\partial x^{j'}\partial x^{k'}},$$

$$(2.10)\quad \tilde{L}^{a'}_{b'k'} = M^{a'}_{a}M^{b}_{b'}\frac{\partial x^{k}}{\partial x^{k'}}\tilde{L}^{a}_{bk} + M^{a'}_{c}\frac{\partial M^{c}_{b'}}{\partial x^{k'}},$$

(2.11) $\tilde{C}^{i'}_{j'c'} = \frac{\partial x^{i'}}{\partial x^{i}} \frac{\partial x^{j}}{\partial x^{j'}} M^{c}_{c'} \tilde{C}^{i}_{jc}$,

(2.12) $C^{a'}_{b'c'} = M^{a'}_{a} M^{b}_{b'} M^{c}_{c'} C^{a}_{bc}$.

We shall now prove:

Theorem 2.1. *If on every open set $p^{-1}(U)$ the real functions $D\Gamma = (L^{i}_{jk}, \tilde{L}^{a}_{bk}, \tilde{C}^{i}_{jc}, C^{a}_{bc})$ are given such that on intersections of these open sets the formulae (2.9)-(2.12) are satisfied, then there exists an unique d-connection D on E whose local coefficients with respect to the adapted local frame $(\frac{\delta}{\delta x^{i}}, \frac{\partial}{\partial y^{a}})$ are the functions $D\Gamma$. Moreover, the functions $(L^{i}_{jk}, \tilde{L}^{a}_{bk})$ locally define the operator of h-covariant derivation associated to D and the functions $(\tilde{C}^{i}_{jc}, C^{a}_{bc})$ locally define the operator of v-covariant derivation associated to D.*

Proof. If $X = X^{k}\frac{\delta}{\delta x^{k}} + \tilde{X}^{c}\frac{\partial}{\partial y^{c}}$ and $Y = Y^{j}\frac{\delta}{\delta x^{j}} + \tilde{Y}^{b}\frac{\partial}{\partial y^{b}}$, we define an operator $D: \chi(E) \times \chi(E) \rightarrow \chi(E)$, $(X,Y) \rightarrow D_{X}Y$ as follows:

(2.13) $D_{X}hY = [X^{k}(\frac{\delta Y^{i}}{\delta x^{k}} + L^{i}_{jk}Y^{j}) + \tilde{X}^{c}(\frac{\partial Y^{i}}{\partial y^{c}} + \tilde{C}^{i}_{jc}Y^{j})]\frac{\delta}{\delta x^{i}}$.

(2.14) $D_{X}vY = [X^{k}(\frac{\delta \tilde{Y}^{a}}{\delta x^{k}} + \tilde{L}^{a}_{bk}\tilde{Y}^{b}) + \tilde{X}^{c}(\frac{\partial \tilde{Y}^{a}}{\partial y^{c}} + C^{a}_{bc}\tilde{Y}^{b})]\frac{\partial}{\partial y^{a}}$.

(2.15) $D_{X}Y = D_{X}hY + D_{X}vY$.

By (2.9)-(2.12), the vector fields $D_{X}hY$ and $D_{X}vY$ are well defined. From (2.13)-(2.15) we deduce by a direct checking that D defines a d-connection on E i.e. it satisfies the usual properties of an operator of covariant derivation (Koszul's axioms) as well as, for instance, (i) from the Theorem 1.1. By (2.13), (2.14) we get

$$(2.16)\quad \begin{array}{ll} D_{\frac{\delta}{\delta x^{k}}}\frac{\delta}{\delta x^{j}} = L^{i}_{jk}\frac{\delta}{\delta x^{i}}, & D_{\frac{\delta}{\delta x^{k}}}\frac{\partial}{\partial y^{b}} = \tilde{L}^{a}_{bk}\frac{\partial}{\partial y^{a}}, \\ D_{\frac{\partial}{\partial y^{b}}}\frac{\delta}{\delta x^{j}} = \tilde{C}^{i}_{jb}\frac{\delta}{\delta x^{i}}, & D_{\frac{\partial}{\partial y^{c}}}\frac{\partial}{\partial y^{b}} = C^{a}_{bc}\frac{\partial}{\partial y^{a}}, \end{array}$$

from which we infer the uniqueness of D by contradiction as well as the second statement of our theorem if we set $D_{\frac{\delta}{\delta x^k}} f = \frac{\delta f}{\delta x^k}$ and $D_{\frac{\partial}{\partial y^a}} f = \frac{\partial f}{\partial y^a}$ for every $f \in \mathcal{F}(E)$, q.e.d.

We note that (2.13) and (2.14) define in fact the Otsuki connections on E associated to h and v, respectively. For the theory of the Otsuki connections we refer to T. Otsuki [122].

We conclude this section with the following theorem of existence of d-connections on E.

Theorem 2.2. *Let $\xi = (E,p,M)$ be a vector bundle whose base manifold is paracompact and N be a non-linear connection in ξ. Then there exists a d-connection on E.*

Proof. Let $(p^{-1}(U),h)$ be a local chart on E such that (U,φ) is a local chart on M. Since M is paracompact, there exists a linear connection on M of local coefficients, say $\Gamma^i_{jk}(x)$. Let $N^a_i(x,y)$ be the local coefficients of the non-linear connection N. We set $L^i_{jk}(x,y) = \Gamma^i_{jk}(x)$ and $\tilde{L}^a_{bi}(x,y) = \frac{\partial N^a_i(x,y)}{\partial y^b}$. Thus we obtain two sets of functions which transform, under a change of coordinates on E, by (2.9) and (2.10), respectively. It results that $D\Gamma = (\Gamma^i_{jk}(x), \frac{\partial N^a_i(x,y)}{\partial y^b}, 0, 0)$ defines a d-connection on E, q.e.d.

A d-connection $D\Gamma = (L^i_{jk}(x,y), \frac{\partial N^a_k(x,y)}{\partial y^b}, 0, C^a_{bc})$ is called a *connection of Berwald type on E.*

§3. Torsion and Curvature of d-Connections

Let $\xi = (E,p,M)$ be a vector bundle endowed with a non-linear connection N. Since a d-connection D is a particular linear connection on E, its torsion is given by

(3.1) $T(X,Y) = D_X Y - D_Y X - [X,Y], \quad X,Y \in \chi(M).$

By using the projectors h and v associated to N the following decomposition of T is derived

(3.1)′ $T(X,Y) = T(hX,hY) + T(hX,vY) + T(vX,hY) + T(vX,vY).$

Every vector field from the right side of $(3.1)'$ has an horizontal component and a vertical one. Taking into account the skewsymmetry of the mapping T and the equation $h[vX,vY] = 0$ we deduce

Theorem 3.1. *The torsion of a d-connection is completely determined by the following d-tensor fields:*

$$
(3.2)\quad
\begin{aligned}
hT(hX,hY) &= \overset{h}{D}_X hY - \overset{h}{D}_Y hX - h[hX,hY],\\
vT(hX,hY) &= -v[hX,hY],\\
hT(hX,vY) &= -\overset{v}{D}_Y hX - h[hX,vY],\\
vT(hX,vY) &= \overset{h}{D}_X vY - v[hX,vY],\\
vT(vX,vY) &= \overset{v}{D}_X vY - \overset{v}{D}_Y vX - v[vX,vY],\quad X,Y \in \chi\ (E).
\end{aligned}
$$

Corrolary 3.1. *The following properties hold good:*

a) $vT(hX,hY)=0$ *if and only if the horizontal distribution is integrable.*

b) $hT(hX,vY)=0$ *if and only if* $v\overset{h}{D}_X vY = v[hX,vY]$, $X,Y \in \chi(E)$.

We shall say that $hT(hX,hY)$ is $h(hh)$- torsion of D, that $vT(hX,hY)$ is $v(hh)$-torsion of D and so on.

In order to get the local form of the d-tensor fields which determine the torsion of D, called simple the torsions of D, we note the following equation

$$
(3.3)\quad [\frac{\delta}{\delta x^j}, \frac{\delta}{\delta x^k}] = R^a{}_{jk}\frac{\partial}{\partial y^a}.
$$

where $R^a{}_{jk} = -\Omega^a_{jk}$ given by (5.9), Ch. II is the integrability tensor of the horizontal distribution. It can be written as

$$
(3.4)\quad R^a{}_{jk} = \frac{\delta N^a_j}{\delta x^k} - \frac{\delta N^a_k}{\delta x^j}.
$$

Also we have

$$
(3.5)\quad [\frac{\delta}{\delta x^j}, \frac{\partial}{\partial y^b}] = \frac{\partial N^a_j}{\partial y^b}\frac{\partial}{\partial y^a}.
$$

Introducing the notations

$$\begin{aligned}
&hT(\frac{\delta}{\delta x^k},\frac{\delta}{\delta x^j})=T^i{}_{jk}\frac{\delta}{\delta x^i},\quad vT(\frac{\delta}{\delta x^k},\frac{\delta}{\delta x^j})=\tilde{T}^a{}_{jk}\frac{\partial}{\partial y^a},\\
(3.6)\quad &hT(\frac{\partial}{\partial y^b},\frac{\delta}{\delta x^j})=\tilde{P}^i{}_{jb}\frac{\delta}{\delta x^i},\quad vT(\frac{\partial}{\partial y^b},\frac{\delta}{\delta x^j})=P^a{}_{jb}\frac{\partial}{\partial y^a},\\
&vT(\frac{\partial}{\partial y^c},\frac{\partial}{\partial y^b})=S^a{}_{bc}\frac{\partial}{\partial y^a},
\end{aligned}$$

by the Theorem 3.1 easily get

Theorem 3.2. *The local components, in the frame* $(\frac{\delta}{\delta x^i},\frac{\partial}{\partial y^a})$ *of the torsions of a d-connection* $D\Gamma=(L^i_{jk},\tilde{L}^a_{bk},\tilde{C}^i_{ja},C^a_{bc})$ *introduced in (3.6) have the following expressions:*

$$\begin{aligned}
&T^i{}_{jk}=L^i_{jk}-L^i_{kj},\quad \tilde{T}^a{}_{jk}=R^a{}_{jk},\quad \tilde{P}^i{}_{jb}=\tilde{C}^i_{jb}\\
(3.7)\quad &P^a{}_{jb}=\frac{\partial N^a_j}{\partial y^b}-\tilde{L}^a_{bj},\quad S^a{}_{bc}=C^a_{bc}-C^a_{cb}.
\end{aligned}$$

Corollary 3.2. *A d-connection D whose local coefficients are* $D\Gamma=(L^j_{jk},\tilde{L}^a_{bk},\tilde{C}^i_{jb},C^a_{bc})$ *is without torsions if and only if:*

$$(3.8)\quad L^i_{jk}=L^i_{kj},\ R^a{}_{jk}=0,\ \tilde{C}^i{}_{jb}=0,\ \frac{\partial N^a_j}{\partial y^b}=\tilde{L}^a_{bj},\ C^a{}_{bc}=C^a{}_{cb}.$$

In the following we shall deal with the curvature of a d-connection D. This is given by

$$(3.9)\quad R(X,Y)Z=D_XD_YZ-D_YD_XZ-D_{[X,Y]}Z,\ X,Y,Z\in\chi\ (E).$$

We prove

Theorem 3.3. *The curvature tensor field R of the d-connection D has the properties*

$$\begin{aligned}
(3.10)\quad &vR(X,Y)hZ=0,\ hR(X,Y)vZ=0,\\
&R(X,Y)Z=hR(X,Y)hZ+vR(X,Y)vZ,\ X,Y,Z\in\chi\ (E).
\end{aligned}$$

Proof. Since D preserves by parallelism the vertical and horizontal distributions, by (3.9) the operator R(X,Y) carries horizontal vector fields to horizontals and vertical vector fields to verticals. Thus the first two equations from (3.10) hold. The next one is an easy consequence of the first two.

By Theorem 3.3 and the equation $R(X,Y) = -R(X,Y)$, $X,Y \in \chi(E)$ we get

Theorem 3.4. *The curvature of a d-connection D on the total space E of a vector bundle is completely determined by the following six d-tensor fields:*

$$
(3.11)\quad
\begin{aligned}
R(hX,hY)hZ &= \overset{h}{D}_X\overset{h}{D}_Y hZ - \overset{h}{D}_Y\overset{h}{D}_X hZ - \overset{h}{D}_{[hX,hY]}hZ - \overset{v}{D}_{[hX,hY]}hZ,\\
R(hX,hY)vZ &= \overset{h}{D}_X\overset{h}{D}_Y vZ - \overset{h}{D}_Y\overset{h}{D}_X vZ - \overset{h}{D}_{[hX,hY]}vZ - \overset{v}{D}_{[hX,hY]}vZ,\\
R(vX,hY)hZ &= \overset{v}{D}_X\overset{h}{D}_Y hZ - \overset{h}{D}_Y\overset{v}{D}_X hZ - \overset{h}{D}_{[vX,hY]}hZ - \overset{v}{D}_{[vX,hY]}hZ,\\
R(vX,hY)vZ &= \overset{v}{D}_X\overset{h}{D}_Y vZ - \overset{h}{D}_Y\overset{v}{D}_X vZ - \overset{h}{D}_{[vX,hY]}vZ - \overset{v}{D}_{[vX,hY]}vZ,\\
R(vX,vY)hZ &= \overset{v}{D}_X\overset{v}{D}_Y hZ - \overset{v}{D}_Y\overset{v}{D}_X hZ - \overset{v}{D}_{[vX,vY]}hZ,\\
R(vX,vY)vZ &= \overset{v}{D}_X\overset{v}{D}_Y vZ - \overset{v}{D}_Y\overset{v}{D}_X vZ - \overset{v}{D}_{[vX,vY]}vZ.
\end{aligned}
$$

We introduce the local components of these d-tensor fields as follows:

$$
(3.12)\quad
\begin{aligned}
&R\left(\frac{\delta}{\delta x^k},\frac{\delta}{\delta x^j}\right)\frac{\delta}{\delta x^h} = R_{h\ jk}^{\ i}\frac{\delta}{\delta x^i}, \quad R\left(\frac{\delta}{\delta x^k},\frac{\delta}{\delta x^j}\right)\frac{\partial}{\partial y^b} = \tilde{R}_{b\ jk}^{\ a}\frac{\partial}{\partial y^a},\\
&R\left(\frac{\partial}{\partial y^c},\frac{\delta}{\delta x^k}\right)\frac{\delta}{\delta x^j} = \tilde{P}_{j\ kc}^{\ i}\frac{\delta}{\delta x^i}, \quad R\left(\frac{\partial}{\partial y^c},\frac{\delta}{\delta x^k}\right)\frac{\partial}{\partial y^b} = P_{b\ kc}^{\ a}\frac{\partial}{\partial y^a},\\
&R\left(\frac{\partial}{\partial y^c},\frac{\partial}{\partial y^b}\right)\frac{\delta}{\delta x^j} = \tilde{S}_{j\ bc}^{\ i}\frac{\delta}{\delta x^i}, \quad R\left(\frac{\partial}{\partial y^d},\frac{\partial}{\partial y^c}\right)\frac{\partial}{\partial y^b} = S_{b\ cd}^{\ a}\frac{\partial}{\partial y^a}.
\end{aligned}
$$

By a direct computation, using (3.3) and (3.5) we get

Theorem 3.5. *The local components of the curvature of a d-connection D on E whose local coefficients are* $D\Gamma = (L^i_{jk}, \tilde{L}^a_{bk}, \tilde{C}^i_{jb}, C^a_{bc})$ *are as follows:*

$$
(3.13)\quad
\begin{aligned}
R_{h\ jk}^{\ i} &= \frac{\delta L_{hj}^{i}}{\delta x^{k}} - \frac{\delta L_{hk}^{i}}{\delta x^{j}} + L_{hj}^{m} L_{mk}^{i} - L_{hk}^{m} L_{mj}^{i} + \tilde{C}_{ha}^{i} R^{a}_{\ jk},\\
\tilde{R}_{b\ jk}^{\ a} &= \frac{\delta \tilde{L}_{bj}^{a}}{\delta x^{k}} - \frac{\delta \tilde{L}_{bk}^{a}}{\delta x^{j}} + \tilde{L}_{bj}^{c} \tilde{L}_{ck}^{a} - \tilde{L}_{bk}^{c} \tilde{L}_{cj}^{a} + C_{bc}^{a} R^{c}_{\ jk},\\
\tilde{P}_{j\ ka}^{\ i} &= \frac{\partial L_{jk}^{i}}{\partial y^{a}} - \tilde{C}_{ja|k}^{i} + \tilde{C}_{jb}^{i} P^{b}_{\ ka},\\
P_{b\ kc}^{\ a} &= \frac{\partial \tilde{L}_{bk}^{a}}{\partial y^{c}} - C_{bc|k}^{a} + C_{bd}^{a} P^{d}_{\ kc},\\
\tilde{S}_{j\ bc}^{\ i} &= \frac{\partial \tilde{C}_{jb}^{i}}{\partial y^{c}} - \frac{\partial \tilde{C}_{jc}^{i}}{\partial y^{b}} + \tilde{C}_{jb}^{h} \tilde{C}_{hc}^{i} - \tilde{C}_{jc}^{h} \tilde{C}_{hb}^{i},\\
S_{b\ cd}^{\ a} &= \frac{\partial C_{bc}^{a}}{\partial y^{d}} - \frac{\partial C_{bd}^{a}}{\partial y^{c}} + C_{bc}^{e} C_{ed}^{a} - C_{bd}^{e} C_{ec}^{a}.
\end{aligned}
$$

Here small bar denotes, as usual, the h-covariant derivative.

The torsions and curvatures of a d-connection are not independent. These satisfy, as we shall now prove, fifteen Bianchi identities. As it is well known the torsion T and curvature R of every linear connection D on E satisfy the following Bianchi identities:

$$
(3.14)\quad
\begin{aligned}
&\sum [(D_X T)(Y,Z) - R(X,Y)Z + T(T(X,Y),Z)] = 0,\\
&\sum [(D_X R)(U,Y,Z) + R(T(X,Y),Z)U] = 0,
\end{aligned}
$$

where $\sum$ means cyclic sum over X,Y,Z.

If D is a d-connection, then by (3.10) and

$$
(3.15)\quad v(D_X R)(U,Y,hZ) = 0,\ \ h(D_X R)(U,Y,vZ) = 0,
$$

the identities (3.14) become

$$
(3.16)\quad
\begin{aligned}
&\sum [h(D_X T)(Y,Z) - hR(X,Y)Z + hT(hT(X,Y),Z) + hT(vT(X,Y),Z)] = 0,\\
&\sum [v(D_X T)(Y,Z) - vR(X,Y)Z + vT(hT(X,Y),Z) + vT(vT(X,Y),Z)] = 0,\\
&\sum [h(D_X R)(U,Y,Z) + hR(hT(X,Y),Z)U + hR(vT(X,Y),Z)U] = 0,\\
&\sum [v(D_X R)(U,Y,Z) + vR(hT(X,Y),Z)U + vR(vT(X,Y),Z)U] = 0.
\end{aligned}
$$

For applications, the form of these identities in the frame $(\frac{\delta}{\delta x^{i}}, \frac{\partial}{\partial y^{a}})$ is needed. In order to get it we shall insert in (3.16) the vector fields X,Y,Z as in the following table

X	$\frac{\delta}{\delta x^j}$	$\frac{\delta}{\delta x^j}$	$\frac{\partial}{\partial y^d}$	$\frac{\partial}{\partial y^d}$
Y	$\frac{\delta}{\delta x^k}$	$\frac{\delta}{\delta x^k}$	$\frac{\partial}{\partial y^c}$	$\frac{\partial}{\partial y^c}$
Z	$\frac{\delta}{\delta x^l}$	$\frac{\delta}{\delta x^l}$	$\frac{\partial}{\partial y^b}$	$\frac{\partial}{\partial y^b}$

and U will be taken successively equal to $\frac{\delta}{\delta x^h}$ and $\frac{\partial}{\partial y^a}$.
Then taking into account (3.6), (3.7) and (3.12) one obtains the local form of the Bianchi identities:

$$(3.17)_1 \quad \begin{cases} \overset{0}{\sum} [T^i{}_{jk|h} + T^m{}_{jk} T^i{}_{hm} + R^a{}_{jk} \tilde{C}^i{}_{ha} - R_j{}^i{}_{kh}] = 0 \\ \overset{0}{\sum} [R^a{}_{jk|h} + T^m{}_{jk} R^a{}_{hm} + R^b{}_{jk} P^a{}_{hb}] = 0 \end{cases}$$

$$(3.17)_2 \quad \begin{cases} \tilde{C}^i_{jb|k} - \tilde{C}^i_{kb|j} - T^i{}_{jk}|_b + \tilde{C}^m_{jb} T^i{}_{km} - \tilde{C}^m_{kb} T^i{}_{jm} + \tilde{C}^i_{mb} T^i{}_{jm} + T^m{}_{jk} \tilde{C}^i_{mb} + \\ + P^d{}_{jb} \tilde{C}^i_{kd} - P^d{}_{kb} C^i_{jd} + \tilde{P}^i_{j\ kb} - \tilde{P}^i_{k\ jb} = 0, \\ P^a{}_{jb|k} - P^a{}_{kb|j} - R^a{}_{jk}|_b + \tilde{C}^m_{jb} R^a{}_{km} - \tilde{C}^m_{kb} R^a{}_{jm} + T^m{}_{jk} P^a{}_{mb} + \\ + P^d{}_{jb} P^a{}_{kd} - P^d{}_{kb} P^a{}_{jd} - R^d{}_{jk} S^a{}_{bd} + \tilde{R}^a_{b\ jk} = 0, \end{cases}$$

$$(3.18) \quad \begin{cases} \tilde{C}^i_{jb}|_c - \tilde{C}^i_{jc}|_b + \tilde{C}^m_{jc} \tilde{C}^i_{mb} - \tilde{C}^m_{jb} \tilde{C}^i_{mc} + S^d{}_{bc} \tilde{C}^i_{jd} - \tilde{S}^i_{j\ bc} = 0, \\ P^a{}_{jb}|_c - P^a{}_{jc}|_b + S^a{}_{bc|j} + \tilde{C}^m_{jc} P^a{}_{mb} - \tilde{C}^m_{jb} P^a{}_{mc} + P^d{}_{jb} S^a{}_{cd} - \\ - P^d{}_{jc} S^a{}_{bd} + S^d{}_{bc} P^a{}_{jd} + P^a_{b\ jc} - P^a_{c\ jb} = 0, \end{cases}$$

$$(3.19)\quad \overset{0}{\sum}[S^{a}{}_{bc}|_{d}+S^{f}{}_{bc}S^{a}{}_{df}-S_{b}^{a}{}_{cd}]=0,$$

$$(3.20)_1\begin{cases}\overset{0}{\sum}[R_{k}^{i}{}_{hj\,||\,l}-T^{m}{}_{hj}R_{k}^{i}{}_{lm}-R^{a}{}_{hj}\tilde{P}_{k}^{i}{}_{la}]=0,\\ \overset{0}{\sum}[\tilde{R}_{d}^{a}{}_{hj\,||\,l}-T^{m}{}_{hj}\tilde{R}_{d}^{a}{}_{lm}-R^{c}{}_{hj}P_{d}^{a}{}_{lc}]=0,\end{cases}$$

$$(3.20)_2\begin{cases}\tilde{P}_{k}^{i}{}_{jd\,|\,l}-\tilde{P}_{k}^{i}{}_{ld\,|\,j}+R_{k}^{i}{}_{lj}|_{d}-\tilde{C}_{jd}^{m}R_{k}^{i}{}_{lm}+\tilde{C}_{ld}^{m}R_{k}^{i}{}_{jm}-T^{m}{}_{jl}\tilde{P}_{k}^{i}{}_{md}-\\ -P^{a}{}_{jd}\tilde{P}_{k}^{i}{}_{la}+P^{a}{}_{ld}\tilde{P}_{k}^{i}{}_{ja}-R^{a}{}_{jl}\tilde{S}_{k}^{i}{}_{ad}=0,\\ P_{c}^{a}{}_{jd\,|\,l}-P_{c}^{a}{}_{ld\,|\,j}+\tilde{R}_{c}^{a}{}_{lj}|_{d}-\tilde{C}_{jd}^{m}\tilde{R}_{c}^{a}{}_{lm}+\tilde{C}_{ld}^{m}R_{c}^{a}{}_{jm}-T^{m}{}_{jl}P_{c}^{a}{}_{md}-\\ -P^{f}{}_{jd}P_{c}^{a}{}_{lf}+P^{f}{}_{ld}P_{c}^{a}{}_{jf}-R^{f}{}_{jl}S_{c}^{a}{}_{fd}=0,\end{cases}$$

$$(3.20)_3\begin{cases}\tilde{P}_{k}^{i}{}_{jd}|_{c}-\tilde{P}_{k}^{i}{}_{jc}|_{d}+S_{k}^{i}{}_{dc\,|\,j}-\tilde{C}^{m}{}_{jc}\tilde{P}_{k}^{i}{}_{md}+\tilde{C}^{m}{}_{jd}\tilde{P}_{k}^{i}{}_{mc}+P^{a}{}_{jc}\tilde{S}_{k}^{i}{}_{da}-P^{a}{}_{jd}\tilde{S}_{k}^{i}{}_{ca}+S^{a}{}_{cd}\tilde{P}_{k}^{i}{}_{ja}=0,\\ P_{b}^{a}{}_{jd}|_{c}-P_{b}^{a}{}_{jc}|_{d}+S_{b}^{a}{}_{cd\,|\,j}-\tilde{C}^{m}{}_{jc}P_{b}^{a}{}_{md}+\tilde{C}^{m}{}_{jd}P_{b}^{a}{}_{mc}+P^{f}{}_{jc}S_{bdf}^{a}-P^{f}{}_{jd}S_{b}^{a}{}_{cf}+S^{f}{}_{cd}P_{b}^{a}{}_{jf}=0,\end{cases}$$

$$(3.20)_4\begin{cases}\overset{0}{\sum}[S_{k}^{i}{}_{bc}|_{d}-S^{a}{}_{bc}\tilde{S}_{k}^{i}{}_{da}]=0,\\ \overset{0}{\sum}[S_{f}^{a}{}_{bc}|_{d}-S^{c}{}_{bc}S_{f}^{a}{}_{dc}]=0,\end{cases}$$

where $\overset{0}{\sum}$ means cyclic sum over (l,k,j) in $(3.17)_1$, over (h,j,l) in $(3.20)_1$ and over (b,c,d) in (3.19) and $(3.20)_4$.

These identities become simpler if $T^{i}{}_{jk}=0$ and $S^{a}{}_{bc}=0$, a case that we shall meet later.

We shall conclude this section by deriving the Ricci identities of a d-connection on E. We shall establish such identities for vector fields only, although these may be written for every tensor field.

A simple arrangement of (3.11) gives us the following commutation formulae known as the Ricci identities:

$$
(3.21)\quad \begin{cases}
D^h_X D^h_Y hZ - D^h_Y D^h_X hZ = R(hX,hY)hZ + D^h_{[hX,hY]}hZ + D^v_{[hX,hY]}hZ, \\
D^v_X D^h_Y hZ - D^h_Y D^v_X hZ = R(vX,hY)hZ + D^h_{[vX,hY]}hZ + D^v_{[vX,hY]}hZ, \\
D^v_X D^v_Y hZ - D^v_Y D^v_X hZ = R(vX,vY)hZ + D^v_{[vX,vY]}hZ,
\end{cases}
$$

$$
(3.22)\quad \begin{cases}
D^h_X D^h_Y vZ - D^h_Y D^h_X vZ = R(hX,hY)vZ + D^h_{[hX,hY]}vZ + D^v_{[hX,hY]}vZ, \\
D^v_X D^h_Y vZ - D^h_Y D^v_X vZ = R(vX,hY)vZ + D^h_{[vX,hY]}vZ + D^v_{[vX,hY]}vZ, \\
D^v_X D^v_Y vZ \; - \; D^v_Y D^v_X Z = R(vX,vY)vZ + D^v_{[vX,vY]}vZ.
\end{cases}
$$

Then taking into account the local form of the h- and v- covariant derivatives, as well as (3.6), (3.7) and (3.12), by a direct calculation one proves

Theorem 3.6. *If* $X = X^i(x,y)\dfrac{\delta}{\delta x^i} + \tilde{X}^a(x,y)\dfrac{\partial}{\partial y^a}$ *is a vector field on E and D is a d-connection on E, then the following Ricci identities hold good*

$$
(3.23)\quad \begin{cases}
X^i_{\ |k|l} - X^i_{\ |l|k} = R_h{}^i{}_{kl} X^h - T^h{}_{kl} X^i_{\ |h} - R^a{}_{kl} X^i|_a, \\
X^i_{\ |k}|_d - X^i|_{d|k} = \tilde{P}_h{}^i{}_{kd} X^h - C^h_{\ kd} X^i_{\ |h} - P^a{}_{kd} X^i|_a, \\
X^i|_b|_c - X^i|_c|_b = \tilde{S}_h{}^i{}_{bc} X^h - S^a{}_{bc} X^i|_a
\end{cases}
$$

and

$$
\text{3.24)}\quad \begin{cases}
\tilde{X}^a_{\ |k|l} - \tilde{X}_{|l|k} = R_b{}^a{}_{kl} \tilde{X}^b - T^h{}_{kl} \tilde{X}^a_{\ |h} - R^b{}_{kl} \tilde{X}^a|_b, \\
\tilde{X}^a_{\ |k}|_b - \tilde{X}^a|_{b|k} = P_c{}^a{}_{kb} \tilde{X}^c - C^h_{\ kb} \tilde{X}^a_{\ |h} - P^d{}_{kb} \tilde{X}^a|_d, \\
\tilde{X}^a|_b|_c - \tilde{X}|_c|_b = S_d{}^a{}_{bc} \tilde{X}^d - S^d{}_{bc} \tilde{X}^a|_d.
\end{cases}
$$

§4. Structure Equations of a d-Connection

In the treating of some problems from geometry of the total space E of a vector bundle $\xi = (E,p,M)$, the computations can be shortened if instead of h- and v- covariant derivatives the covariant differentiation associated with a d-connection D on E is used.

Let $c:[0,1]\to E$, $t\to c(t)$ be a curve on E and $\frac{dc}{dt}$ be the tangent vector field along it. This vector field decomposes in the frame $(\frac{\delta}{\delta x^i}, \frac{\partial}{\partial y^a})$ as follows

$$(4.1)\quad \frac{dc}{dt}=\frac{dx^i}{dt}\frac{\delta}{\delta x^i}+\frac{\delta y^a}{dt}\frac{\partial}{\partial y^a}.$$

A vector field $D_{dc}Y$, where $dc=dx^i\frac{\delta}{\delta x^i}+\delta y^a\frac{\partial}{\partial y^a}$ will be denoted by DY and will be called the *covariant differentiation* of $Y\in\chi(E)$.
If $Y=hY=Y^i\frac{\delta}{\delta x^i}$, then by (2.1) and (2.5) we get

$$(4.2)\quad DhY=(DY^i)\frac{\delta}{\delta x^i},$$

where

$$(4.2)'\quad DY^i=Y^i_{|j}dx^j+Y^i|_a\delta y^a.$$

The last equation can be written as

$$(4.3)\quad DY^i=dY^i+\omega^i_jY^j,$$

where

$$(4.3)'\quad \omega^i_j=L^i_{jk}dx^k+C^i_{ja}\delta y^a.$$

Analogously, if $Y=vY=\tilde{Y}^a\frac{\partial}{\partial y^a}$ then

$$(4.4)\quad DvY=(D\tilde{Y}^a)\frac{\partial}{\partial y^a},$$

where

$$(4.4)'\quad D\tilde{Y}^a=\tilde{Y}^a_{|j}dx^j+\tilde{Y}^a|_b\delta y^b$$

or, equivalently,

(4.5) $D\tilde{Y}^a = d\tilde{Y} + \tilde{\omega}^a_b \tilde{Y}^b$,

where

(4.5)′ $\tilde{\omega}^a_b = \tilde{L}^a_{bj} dx^j + C^a_{bc} \delta y^c$.

Thus we have introduced the so called *connection 1-forms* ω^i_j and $\tilde{\omega}^a_b$. We extend the covariant differentiation to every d-tensor field as in the following example: if $t = t^i_a \frac{\delta}{\delta x^i} \otimes \delta y^a$ then

(4.6) $Dt = (Dt^i_a) \frac{\delta}{\delta x^i} \otimes \delta y^a$,

where

(4.6)′ $Dt^i_a = t^i_{a|j} dx^j + t^i_a|_b \delta y^b = dt^i_a + \omega^i_j t^j_a - \tilde{\omega}^b_a t^i_b$.

Now we prove

Theorem 4.1. *Let $\omega^i_j, \tilde{\omega}^a_b$ be the connection 1-forms of a d-connection D on E. Then the following structure equations for D hold:*

(4.7)
$$d(dx^i) - dx^h \wedge \omega^i_h = -\Omega^i, \quad d(\delta y^a) - \delta y^b \wedge \tilde{\omega}^a_b = -\tilde{\Omega}^a,$$

(4.8)
$$d\omega^i_j - \omega^h_j \wedge \omega^i_h = -\Omega^i_j, \quad d\tilde{\omega}^a_b - \tilde{\omega}^c_b \wedge \tilde{\omega}^a_c = -\tilde{\Omega}^a_b,$$

in which the torsion 2-forms Ω^i and $\tilde{\Omega}^a$ are given by

(4.9)
$$\Omega^i = \frac{1}{2} T^i{}_{jk} dx^j \wedge dx^k + \tilde{C}^i_{jk} dx^j \wedge \delta y^c,$$
$$\tilde{\Omega}^a = \frac{1}{2} R^a{}_{jk} dx^j \wedge dx^k + P^a{}_{jc} dx^j \wedge \delta y^c + \frac{1}{2} S^a{}_{bc} \delta y^b \wedge \delta y^c$$

and the curvature 2-forms Ω^i_j and $\tilde{\Omega}^a_b$ are given by

$$(4.10)\quad \begin{aligned} \Omega^i_j &= \frac{1}{2}R_j{}^i{}_{kh}dx^k\wedge dx^h + \tilde{P}^i_{j\ kc}dx^k\wedge\delta y^c + \frac{1}{2}\tilde{S}^i_{j\ bc}\delta y^b\wedge\delta y^c, \\ \tilde{\Omega}^a_b &= \frac{1}{2}\tilde{R}^a_{b\ kh}dx^k\wedge dx^h + P^a_{b\ kc}dx^k\wedge\delta y^c + \frac{1}{2}S^a_{b\ cd}\delta y^c\wedge\delta y^d. \end{aligned}$$

Proof. The first equation (4.7), with Ω^i given by the first equation (4.9) results by exterior multiplying of both sides of (4.3)$'$ by dx^j. By exterior multiplying of both sides of (4.5)$'$ by δy^a and noticing that

$$d(\delta y^a) = -\frac{1}{2}R^a{}_{jk}dx^j\wedge dx^k - \frac{\partial N^a_j}{\partial y^b}dx^j\wedge\delta y^b$$

one obtains the second equation (4.7) with $\tilde{\Omega}^a$ given by the second equation (4.9). Then exterior differentiating (4.3)$'$ and (4.5)$'$ and using them again one obtains (4.8) with the curvature 2-forms Ω^i_j and $\tilde{\Omega}^a_b$ given by (4.10).

Exterior differentiating the equations (4.7) and (4.8) and using them again, the Bianchi identities are recovered.

Returning to the covariant differentiation DY of a vector field $Y \in \chi$ (E) we shall say that Y is parallel along curve c on E if DY = 0. If $Y = Y^i\frac{\delta}{\delta x^i} + \tilde{Y}^a\frac{\partial}{\partial y^a}$ then we have

$$(4.11)\quad DY = (DY^i)\frac{\delta}{\delta x^i} + (D\tilde{Y}^a)\frac{\partial}{\partial y^a}.$$

By (4.11), (4.3) and (4.4)$'$ one gets

Theorem 4.2. *A vector field* $Y = Y^i\frac{\delta}{\delta x^i} + \tilde{Y}^a\frac{\partial}{\partial y^a}$ *on E is parallel along every curve on E if and only if*

$$(4.12)\quad Y^i_{|j} = 0,\ Y^i|_a = 0,\ \tilde{Y}^a_{|j} = 0,\ \tilde{Y}^a|_b = 0.$$

A d-connection D on E is said to be with absolute parallelism of vectors if the differential consequences of the system (4.12) are all identities.

If the differential consequences of the system (4.12) are written, by using the

Ricci identities, we directly obtain

Theorem 4.3. *A d-connection D on E is with absolute parallelism of vectors if and only if its six curvature d-tensor fields vanish identically.*

§5. Metric Structures on the Total Space of a Vector Bundle

Let $\xi = (E,p,M)$ be a vector bundle over a connected and paracompact manifold M.

Definition 5.1. *A metric structure on the total space E of the vector bundle ξ is a symmetric covariant tensor field G of the type (0,2) which is non degenerate at each point $u \in E$ and of constant signature on E. If G is positive definite we say it defines a Riemannian structure on E.*

The restriction of a Riemannian structure G on E to the vertical distribution on E is completely determined by

$$(5.1) \quad h_{ab} = G(\frac{\partial}{\partial y^a}, \frac{\partial}{\partial y^b}), \quad a,b = 1,2,\ldots,m.$$

Performing a change of coordinates on E it comes out that the functions h_{ab} may be viewed as the components of a d-tensor field on E. Assuming that rank $(h_{ab}) = m$, let (h^{cd}) be the inverse of the matrix (h_{ab}).

We now prove:

Theorem 5.1. *Let E be the total space of a vector bundle $\xi = (E,p,M)$. Assuming E is endowed with a metric structure G, there exists an unique non-linear connection N in ξ such that*

$$(5.2) \quad G(hX, vY) = 0, \quad X, Y \in \chi(E),$$

where h and v are the projectors associated with N.

Proof. Since M is paracompact there exists at least a non-linear connection in ξ. It is given by the so called horizontal distribution which is spanned by the local vector fields $\frac{\delta}{\delta x^i} = \frac{\partial}{\partial x^i} - N_i^a \frac{\partial}{\partial y^a}$. We shall determine N_i^a such that (5.2) holds. It is clear that (5.2) is equivalent to $G(\frac{\delta}{\delta x^i}, \frac{\partial}{\partial y^a}) = 0$ or $G_{ia} - N_i^b h_{ab} = 0$, where $G_{ia} = G(\frac{\partial}{\partial x^i}, \frac{\partial}{\partial y^a})$, which gives $N_i^b = h^{ba} G_{ia}$. Thus the coefficients (N_i^a) are uniquely determined and the proof is finished.

In this chapter we shall use only this non-linear connection in ξ. It produces the following decomposition of G:

(5.3) $G(X,Y) = G(hX,hY) + G(vX,vY), \quad X,Y \in \chi(E)$.

With the other words, G decomposes in a sum of two d-tensor fields: hG of type $\begin{pmatrix} 0 & 0 \\ 2 & 0 \end{pmatrix}$ defined by $hG(X,Y) = G(hX,hY)$ and vG of type $\begin{pmatrix} 0 & 0 \\ 0 & 2 \end{pmatrix}$ defined by $vG(X,Y) = G(vX,vY)$.

Locally, these d-tensor fields can be written as

(5.4) $hG = g_{ij} dx^i \otimes dx^j, \quad vG = h_{ab} \delta y^a \otimes \delta y^b$,

where

(5.5) $g_{ij} = G(\frac{\delta}{\delta x^i}, \frac{\delta}{\delta x^j})$, rank $(g_{ij}) = n$.

Thus the decomposition (5.3) looks locally as follows:

(5.6) $G = g_{ij} dx^i \otimes dx^j + h_{ab} \delta y^a \otimes \delta y^b$.

Definition 5.2. *A d-connection D on E endowed with a metric structure G is said to be a metric d-connection if $D_X G = 0$ for every $X \in \chi(E)$.*

Proposition 5.1. *If a linear connection D on E has the properties:*

a) D preserves by parallelism the vertical distribution $u \to V_u E$, $u \in E$,

b) $D_X G = 0$, $X \in \chi(E)$,

then it is a metric d-connection.

Proof. It is enough to prove that D preserves by parallelism the horizontal distribution $u \to H_u E$. Using a) and (5.2) in

$$(D_X G)(hY,vZ) = XG(hY,vZ) - G(D_X hY,vZ) - G(hY,D_X vZ) = 0$$

one gets $G(D_X hY, vZ) = 0$ for every $Z \in \chi(E)$. Thus, by (5.2) $D_X hY$ is an horizontal vector field, q.e.d.

Proposition 5.2. *A d-connection on E endowed with a metric structure G is a metric d-connection if and only if*

$$(5.7)\quad \overset{h}{D}_X hG=0,\ \overset{h}{D}_X vG=0,\ \overset{v}{D}_X hG=0\ \overset{v}{D}_X vG=0,\ X\in\chi(E).$$

Proof. The equation $D_X G = 0$ implies $\overset{h}{D}_X G = 0$ and $\overset{v}{D}_X G = 0$.

By (5.3) we have

$$(*)\qquad (\overset{h}{D}_X hG)(Y,Z) + (\overset{h}{D}_X vG)(Y,Z) = (\overset{h}{D}_X G)(Y,Z) = 0$$

$$(**)\qquad (\overset{v}{D}_X hG)(Y,Z) + (\overset{v}{D}_X vG)(Y,Z) = (\overset{v}{D}_X G)(Y,Z) = 0,\ X,Y,Z \in \chi(E).$$

Taking in (*) $Y = hY, Z = hZ$ one gets $\overset{h}{D}_X hG = 0$ and then taking $Y = vY$, $Z = vZ$ one obtains $\overset{h}{D}_X vG = 0$. Similary, putting in (**) $Y = hY$, $Z = hZ$ one obtains $\overset{v}{D}_X hG = 0$ and then $Y = vY$, $Z = vZ$ one gets $\overset{v}{D}_X vG = 0$. Conversely, using (5.7) in (*) one results $\overset{h}{D}_X G = 0$ and then by (**) one deduces $\overset{v}{D}_X G = 0$. From these two equations it follows $D_X G = 0$.

We shall now discuss the existence of metric d-connections on E. First we prove:

Theorem 5.2. *If $\mathring{D}$ is a fixed d- connection on E, then the d-connection given by the following formulae is metric with respect to G.*

$$(5.8)\quad \begin{aligned}
&2hG(\overset{h}{D}_X hY,Z) = hX(hG)(Y,Z) + hY(hG)(Z,X) - hZ(hG)(Y,Z) - \\
&-hG(X,[hY,hZ]) + hG((Y,[hZ,hX]) + hG(Z,[hX,hY]),\\
&\overset{h}{D}_X vY = \overset{\circ h}{D}_X vY + B(vY,hX),\ \text{such that}\\
&\qquad 2vG(B(vY,hX),Z) = (\overset{\circ h}{D}_X vG)(vY,vZ),\\
&\overset{v}{D}_X hY = \overset{\circ h}{D}_X hY + \tilde{B}(hY,vX),\ \text{such that}\\
&\qquad 2hG(\tilde{B}(hY,vX),Z) = (\mathring{D}_X hG)(hY,hZ),\\
&2vG(\overset{v}{D}_X Y,Z) = vX(vG)(Y,Z) + vY(vG)(Z,X) - vZ(vG)(X,Y) - \\
&-vG(X,[vY,vZ]) + (vG)(Y,[vZ,vX]) + (vG)(Z,[vX,vY]).
\end{aligned}$$

Proof. It is obvious that the formulae (5.8) uniquely determine D^h_X and D^h_Y hence they uniquely determine a d-connection D on E. By a direct computation one checks that D^h_X and D^h_Y verify (5.7). Thus D is a metric d-connection. We note that the h(hh)- and v(vv)-torsions of D vanish.

Next we have:

Theorem 5.3. *Let G be a metric structure on E. There exist metric d-connections on E which depend on G, only. One of them is given by (5.8), in which*

$$\text{(5.9)}\quad \dot{D}^h_X vY = v[hX, vY],\ \dot{D}^v_X hY = h[vX, hY],\ X, Y \in \chi(E).$$

Proof. It is evident that $D^h_X hY$ and $D^v_X vY$ given by the first and the fourth equations from (5.8) depend on G only. If we choose the d-connection $\dot{D}_X$ such that $v\dot{T}(hX, vY) = h\dot{T}(hX, vY) = 0$ then the equations (5.9) hold and by the second and the third equations from (5.8) $D^h_X vY$ and $D^v_X hY$ depend on G, only.

Now we shall express a metric d-connection and related results in terms of local coordinate systems.

As we have seen, a metric structure G uniquely determines a non-linear connection N and then G takes the local form (5.6), where the basis $(dx^i, \delta y^a)$, was used.

Translating the Proposition 5.2 in local coordinates one obtains:

Proposition 5.3. *A d-connection on E is a metric connection if and only if*

$$\text{(5.10)}\quad g_{ij|k} = 0,\ g_{ij}|_a = 0,\ h_{ab|k} = 0,\ h_{ab}|_c = 0.$$

If we proceed similarly with the Theorem 5.3 we deduce:

Theorem 5.4. *If the total space E of the vector bundle ξ is endowed with the metric structure G given by (5.6) then there exists on E a metric d-connection, depending only on G, whose h(hh)- and v(vv) - torsions vanish.*

Its local coefficients $\overset{c}{D\Gamma} = (\overset{c}{L}{}^i_{jk}, \overset{c}{\tilde{L}}{}^a_{bk}, \overset{c}{\tilde{C}}{}^i_{ja}, \overset{c}{C}{}^a_{bc})$ are as follows:

$$(5.11)\quad \begin{aligned} \overset{c}{L}{}^{i}_{jk} &= \frac{1}{2} g^{ih}\left(\frac{\delta g_{hj}}{\delta x^{k}} + \frac{\delta g_{hk}}{\delta x^{j}} - \frac{\delta g_{jk}}{\delta x^{h}}\right), \\ \overset{c}{L}{}^{a}_{bi} &= \frac{\partial N^{a}_{i}}{\partial y^{b}} + \frac{1}{2} h^{ac}\left(\frac{\delta h_{bc}}{\delta x^{i}} - \frac{\partial N^{d}_{i}}{\partial y^{b}} h_{dc} - \frac{\partial N^{d}_{i}}{\partial y^{c}} h_{db}\right), \\ \overset{c}{\tilde{C}}{}^{i}_{jc} &= \frac{1}{2} g^{ik} \frac{\partial g_{jk}}{\partial y^{c}}, \\ \overset{c}{C}{}^{a}_{bc} &= \frac{1}{2} h^{ad}\left(\frac{\partial h_{db}}{\partial y^{c}} + \frac{\partial h_{dc}}{\partial y^{b}} - \frac{\partial h_{bc}}{\partial y^{d}}\right). \end{aligned}$$

Definition 5.2. *The metric d-connection given by (5.11) will be called the canonical d-connection associated with G.*

Let $D\overset{\circ}{\Gamma} = (\overset{\circ}{L}{}^{i}_{jk}, \overset{\circ}{\tilde{L}}{}^{a}_{bk}, \overset{\circ}{\tilde{C}}{}^{i}_{ja}, \overset{\circ}{C}{}^{a}_{bc})$ be a d-connection on E which is endowed with a metric structure G. If we denote by $\overset{\circ}{|}$ and $\overset{\circ}{|}$ the h- and v-covariant derivations of $D\overset{\circ}{\Gamma}$, then by a direct calculation one checks that the d-connection whose local coefficients are given by

$$(5.12)\quad \begin{aligned} L^{i}_{jk} &= \overset{\circ}{L}{}^{i}_{jk} + \frac{1}{2} g^{ih} g_{hj\overset{\circ}{|}k}, \quad \tilde{L}^{a}_{bk} = \overset{\circ}{\tilde{L}}{}^{a}_{bk} + \frac{1}{2} h^{ac} h_{cb\overset{\circ}{|}k}, \\ \tilde{C}^{i}_{jc} &= \overset{\circ}{\tilde{C}}{}^{i}_{jc} + \frac{1}{2} g^{ih} g_{jh}\overset{\circ}{|}_{c}, \quad C^{a}_{bc} = \overset{\circ}{C}{}^{a}_{bc} + \frac{1}{2} h^{ad} h_{db}\overset{\circ}{|}_{c}, \end{aligned}$$

is a metrical d-connection.

This method of metrisation of a d-connection is due to A. Kawaguchi, [77]. We call it the Kawaguchi metrisation process.

Let us associate to G the following operators of Obata type:

$$(5.13)\quad \begin{aligned} O^{ih}_{jk} &= \frac{1}{2}(\delta^{i}_{j}\delta^{h}_{k} - g_{jk}g^{ih}), \quad O^{*ih}_{jk} = \frac{1}{2}(\delta^{i}_{j}\delta^{h}_{k} + g_{jk}g^{ih}), \\ O^{ab}_{cd} &= \frac{1}{2}(\delta^{a}_{c}\delta^{b}_{d} - h_{cd}h^{ab}), \quad O^{*ab}_{cd} = \frac{1}{2}(\delta^{a}_{c}\delta^{b}_{d} + h_{cd}h^{ab}). \end{aligned}$$

Theorem 5.5. *The set of all metric d-connections with respect to G on the total space E of a bundle ξ is given by*

$$(5.14)\quad \begin{aligned} &L^i_{jk} = \overset{c}{L}{}^i_{jk} + O^{di}_{jh} X^h_{dk}, \quad \tilde{L}^a_{bk} = \overset{c}{\tilde{L}}{}^a_{bk} + O^{ca}_{bd} Y^d_{ck}, \\ &\tilde{C}^i_{jc} = \overset{c}{\tilde{C}}{}^i_{jc} + O^{*hi}_{jk} \tilde{X}^k_{hc}, \quad C^a_{bc} = \overset{c}{C}{}^a_{bc} + O^{*ea}_{bd} \tilde{Y}^d_{ec}, \end{aligned}$$

where $(\overset{c}{L}{}^i_{jk}, \overset{c}{\tilde{L}}{}^a_{bk}, \overset{c}{\tilde{C}}{}^i_{jc}, \overset{c}{C}{}^a_{bc})$ is the canonical d-connection (5.11) and X^i_{jk}, $\tilde{X}^i_{jc}$, Y^a_{bk}, $\tilde{Y}^a_{bc}$ are arbitrary d-tensor fields.

Proof. See R. Miron [103].

We have proved in the above the existence of metric d-connections whose h(hh)- and v(vv)-torsions vanish. But there are certain problems, especially related to the applications of the geometry of E in the Theory of Relativity, in which metrical d-connections with h(hh)- and v(vv)-torsions prescribed are nedeed. In the following we show that such metric d-connections do exist.

Definition 5.3. *A d-connection D on E is hv-metric with respect to a metric structure G if*

$$(5.15)\quad D^h_X hG = 0, \quad D^v_X vG = 0, \quad X \in \chi(E).$$

An easy computation in local coordinates leads to

Proposition 5.4. *A d-connection* $D\Gamma = (L^i_{jk}, \tilde{L}^a_{bk}, \tilde{C}^i_{ja}, C^a_{bc})$ *is hv-metric with respect to* $G = g_{ij} dx^i \otimes dx^j + h_{ab} \delta y^a \otimes \delta y^b$ *if and only if*

$$(5.16)\quad g_{ij|k} = 0, \quad h_{ab}|_c = 0.$$

Let us consider a d-connection of the Berwald type

$$\overset{b}{D\Gamma} = (L^i_{jk}, \frac{\partial N^a_k}{\partial y^b}, 0, C^a_{bc}), \text{ where}$$

$$(5.17)\quad \begin{aligned} L^i_{jk} &= \frac{1}{2} g^{ih} \left(\frac{\delta g_{jh}}{\delta x^k} + \frac{\delta g_{kh}}{\delta x^j} - \frac{\delta g_{jk}}{\delta x^h} \right), \\ C^a_{bc} &= \frac{1}{2} h^{ad} \left(\frac{\partial h_{bd}}{\partial y^c} + \frac{\partial h_{cd}}{\partial y^b} - \frac{\partial h_{bc}}{\partial y^d} \right). \end{aligned}$$

It is easy to check this connection is hv-metric. We shall prove:

Theorem 5.6. *There exists an unique hv-metric d-connection of the Berwald type with h(hh) and v(vv)-torsions prescribed.*

Proof. Let us fix the Berwald connection $\overset{b}{D\Gamma}$ introduced in the above. Then, by the general theory of connections, every other d-connection of the Berwald type, is of the form $(L^i_{jk}+\tau^i_{jk}, \dfrac{\partial N^a_k}{\partial y^b}, 0, C^a_{bc}+\tilde{\tau}^a_{bc})$, where τ^i_{jk} and $\tilde{\tau}^a_{bc}$ are arbitrary d-tensor fields. Let be given T^i_{jk} and S^a_{bc} two d-tensor fields which are skewsymmetric with respect to the covariance indices. We shall determine the d-tensor fields τ^i_{jk} and $\tilde{\tau}^a_{bc}$ such that the Berwald type connection of general form given in the above to be hv-metric and to have T^i_{jk} and S^a_{bc} as h(hh)- and v(vv)-torsions, respectively. These conditions show us that τ^i_{jk} and $\tilde{\tau}^a_{bc}$ must satisfy the following systems of equations

$$(5.18)\quad \begin{cases} \tau^i_{kj}-\tau^i_{jk}=T^i{}_{kj} \\ \tau^h_{ik}g_{hj}+\tau^h_{jk}g_{ih}=0 \end{cases} \qquad (5.19)\quad \begin{cases} \tilde{\tau}^a_{bc}-\tilde{\tau}^a_{cb}=S^a{}_{bc} \\ \tilde{\tau}^d_{bc}h_{ad}+\tilde{\tau}^d_{ac}h_{bd}=0. \end{cases}$$

If in the second equation (5.18) we cyclicly permute the indices i,j,k then we add the equations such obtained and take into account the first equation (5.18) we obtain

$$(5.20)\quad \tau^i_{jk}=\frac{1}{2}g^{ih}(g_{kl}T^l{}_{hj}+g_{jl}T^l{}_{hk}-g_{hl}T^l{}_{jk}).$$

If we similarly proceed with the equations (5.19) we deduce

$$(5.21)\quad \tilde{\tau}^a_{bc}=\frac{1}{2}h^{ad}(h_{be}S^e{}_{bc}+h_{ce}S^e{}_{db}-h_{de}S^e{}_{bc}).$$

Consequently τ^i_{jk} and $\tilde{\tau}^a_{bc}$ are uniquely determined, q.e.d.

From (5.12) we see directly that the Kawaguchi metrisation process leaves unchanged the h(hh)- and v(vv)-torsions. Thus we have:

Theorem 5.7. *Let E be the total space of a vector bundle ξ endowed with a metric structure G. There exists on E a metric d-connection completely determined by G whose h(hh)- and v(vv)-torsions are prescribed. It is obtained from the hv-metric Berwald type connection given by Theorem 5.6 via the Kawaguchi metrization process and has the following local coefficients*

$$L^{\,i}_{jk} = \frac{1}{2} g^{ih} \left(\frac{\delta g_{jh}}{\delta x^k} + \frac{\delta g_{kh}}{\delta x^j} - \frac{\delta g_{jk}}{\delta x^h} \right) + \tau^{\,i}_{jk} ,$$

$$(5.22) \quad \tilde{L}^{a}_{bk} = \overset{c}{\tilde{L}}{}^{a}_{bk} , \quad \tilde{C}^{i}_{jc} = \overset{c}{\tilde{C}}{}^{i}_{jc} ,$$

$$C^{a}_{bc} = \frac{1}{2} h^{ad} \left(\frac{\partial h_{db}}{\partial y^c} + \frac{\partial h_{dc}}{\partial y^b} - \frac{\partial h_{bc}}{\partial y^d} \right) + \tilde{\tau}^{\,a}_{bc} ,$$

where $\tau^{\,i}_{jk}$ and $\tilde{\tau}^{\,a}_{bc}$ *are given by (5.20) and (5.21), respectively.*

It is well known there exists an unique linear connection on E which is metric with respect to G and symmetric, that is, it has no torsions. This is called the Levi-Civita connection of G. Note it is not a d-connection of G on E. Further, we have:

Theorem 5.8. *The canonical metric d-connection (5.11) coincides with the Levi-Civita connection of the metric structure G if and only if:*

a) The horizontal distribution $u \to H_u E$, $u \in E$, *is integrable.*
b) G is constant on the fibres of the vertical bundle (VE, τ_{VE}, E).
c) The v(hv)-torsion vanishes.

Proof. Because of uniqueness of the Levi-Civita connection of G it follows that the metric d-connection (5.11) coincides with it if and only if this has no torsions, that is, a), b) and c) hold good.

We shall give the local coefficients of the Levi-Civita connection of G in the adapted frame $\left(\frac{\delta}{\delta x^i}, \frac{\partial}{\partial y^a}\right)$. These coefficients will be expressed by using the local coefficients of the canonical metrical d-connection $\overset{c}{D\Gamma}$ from (5.11).

If we denote by ∇ the Levi-Civita connection of G, then by a well known fact about the difference of two linear connections, we can write:

$$(5.23) \quad \nabla_X = \overset{c}{D}_X + \tau_X , \quad X \in \chi(E),$$

where τ_X is a tensor field of the type (1.1) on E. Taking into account that the linear connections ∇ and $\overset{c}{D}$ are metric with respect to G and ∇ is without torsion the following system of equations for the determining of τ_X is obtained:

$$
\begin{aligned}
&G(\tau(Y,X),Z)+G(Y,\tau(Z,X))=0,\\
&\tau(X,Y)-\tau(Y,X)=\overset{c}{T}(X,Y), \quad X,Y\in\chi(E),
\end{aligned} \tag{5.24}
$$

where we have set $\tau_X(Y)=\tau(Y,X)$ and $\overset{c}{T}$ is the torsion of $\overset{c}{D}$.

In the adapted frame $\left(\frac{\delta}{\delta x^i},\frac{\partial}{\partial y^a}\right)$ the Levi-Civita connection looks as follows:

$$
\begin{aligned}
&\nabla_{\frac{\delta}{\delta x^k}}\frac{\delta}{\delta x^j}=L^i_{jk}\frac{\delta}{\delta x^i}+\tilde{L}^a_{jk}\frac{\partial}{\partial y^a};\quad \nabla_{\frac{\delta}{\delta x^k}}\frac{\partial}{\partial y^b}=\tilde{K}^i_{bk}\frac{\delta}{\delta x^i}+K^a_{bk}\frac{\partial}{\partial y^a};\\
&\nabla_{\frac{\partial}{\partial y^b}}\frac{\delta}{\delta x^j}=M^i_{jb}\frac{\delta}{\delta x^i}+\tilde{M}^a_{jb}\frac{\partial}{\partial y^a};\quad \nabla_{\frac{\partial}{\partial y^c}}\frac{\partial}{\partial y^b}=\tilde{C}^i_{bc}\frac{\delta}{\delta x^i}+C^a_{bc}\frac{\partial}{\partial y^a}.
\end{aligned} \tag{5.25}
$$

Writing the system of equations (5.24) in the adapted frame $\left(\frac{\delta}{\delta x^i},\frac{\partial}{\partial y^a}\right)$ one gets a system of equations which allows us to determine the local components of τ_X. Inserting these local components in the local form of the equation (5.23) one obtains:

Theorem 5.9. *The local coefficients of the Levi-Civita connection ∇ of the metric structure G on the total space E of a vector bundle ξ are given as follows:*

$$
\begin{aligned}
&L^i_{jk}=\overset{c}{L}{}^i_{jk},\ \tilde{L}^a_{jk}=-\overset{c}{\tilde{C}}{}^i_{jb}g_{ik}h^{ab}-\frac{1}{2}\overset{c}{R}{}^a{}_{jk},\\
&\tilde{K}^i_{bk}=\frac{1}{2}R^c{}_{hk}h_{cb}g^{hi}-O^{ih}_{lk}\overset{c}{\tilde{C}}{}^l_{hb},\ K^a_{bk}=\overset{c}{\tilde{L}}{}^a_{bk}-\overset{*}{O}{}^{ad}_{eb}\overset{c}{P}{}^e{}_{kd},\\
&M^i_{kb}=\overset{c}{\tilde{C}}{}^i_{kb}+\frac{1}{2}\overset{c}{R}{}^d{}_{bk}h_{db}g^{li}+O^{ih}_{lk}\overset{c}{\tilde{C}}{}^l_{hb},\ \tilde{M}^a_{jb}=O^{ad}_{eb}\overset{c}{P}{}^e{}_{kd},\\
&\tilde{C}^i_{ab}=\frac{1}{2}(\overset{c}{P}{}^d{}_{jb}h_{da}+\overset{c}{P}{}^d{}_{ja}h_{db})g^{ij},\ C^a_{bc}=\overset{c}{C}{}^a_{bc}.
\end{aligned} \tag{5.26}
$$

In the end of this section we shall introduce two d-tensor fields on E which will play an important role in some applications to Physics of our theory.

First we notice there exists on E a remarkable vertical vector field $C=y^a\frac{\partial}{\partial y^a}$ which will be called the *Liouville vector field.*

Let D be a d-connection on E.

Definition 5.4. *A map $D:\chi(E)\to\chi(E)$ given by $D(X)=D^h_XC$ is called h-*

deflection tensor field of $\mathbf{D}$ *and a map* $\mathbf{d}:\chi(E)\to\chi(E)$ *given by* $\mathbf{d}(X)=D^v_X C$ *is called v-deflection tensor field of* $\mathbf{D}$.

Locally, if the adapted frame $(\frac{\delta}{\delta x^i}, \frac{\partial}{\partial y^a})$ is used, we have

$$(5.27)\quad \begin{cases} D(\frac{\delta}{\delta x^i}) = D^a_i \frac{\partial}{\partial y^a}, & \text{where } D^a_i = y^a_{|i}, \\ D(\frac{\partial}{\partial y^a}) = 0, \end{cases}$$

$$(5.28)\quad \begin{cases} d(\frac{\delta}{\delta x^i}) = 0, \\ d(\frac{\partial}{\partial y^a}) = d^b_a \frac{\partial}{\partial y^b}, & \text{where } d^b_a = y^b|_a. \end{cases}$$

Applying the Ricci identities (3.24) to the Liouville vector field $\mathbf{C}$ one immediately gets:

Theorem 5.10. *The local components* D^a_i *and* d^a_b *of the h- and v-deflection tensor fields of a d-connection satisfy the following identities:*

$$(5.29)\quad \begin{aligned} &D^a_{k|h} - D^a_{h|k} = \tilde{R}_b{}^a{}_{kh} y^b - T^l{}_{kh} D^a_l - R^b{}_{kh} d^a_b, \\ &D^a_k|_b - d^a_{b|k} = P_c{}^a{}_{kb} y^c - \tilde{C}^h_{kb} D^a_h - P^c{}_{kb} d^a_c, \\ &d^a_b|_c - d^a_c|_b = S_d{}^a{}_{bc} y^d - S^e{}_{bc} d^a_e. \end{aligned}$$

Now let us take D a metrical connection with respect to

$$G = g_{ij}(x,y)dx^i \otimes dx^j + h_{ab}(x,y)\delta y^a \otimes \delta y^b$$

and let us put

$$(5.30)\quad D_{ai} = h_{ac} D^c_i, \quad d_{ab} = h_{ac} d^c_b.$$

Then $h_{ab|k}=0$, $h_{ab}|_c=0$ together with (5.29) yield

$$\begin{aligned} & D_{ak|h} - D_{ah|k} = \tilde{R}_{abkh} y^b - T^i{}_{kh} D_{ai} - R^b{}_{kl} d_{ab}, \\ (5.31)\quad & D_{ak|b} - d_{ab|k} = P_{ackb} y^c - \tilde{C}^h_{kb} D_{ah} - P^e{}_{kb} d_{ae}, \\ & d_{ab|c} - d_{ac|b} = S_{adbc} y^d - S^e{}_{bc} d_{ae}. \end{aligned}$$

Definition 5.5. *A d-connection D on E is said to be of Cartan type if its h-deflection tensor field vanishes and its v-deflection tensor field is identity tensor field.*

From the local characterization of h- and v-deflection tensor fields it follows:

Theorem 5.11. *A d-connection* $D\Gamma = (L^i_{jk},\ \tilde{L}^a_{bk},\ \tilde{C}^i_{ja},\ C^a_{bc})$ *is of Cartan type if and only if*

$$(5.32)\quad \tilde{L}^a_{bk} y^b = N^a_k, \quad C^a_{eb} y^e = 0,$$

hold.

Proof. We have $y^a_{|k} = L^a_{bk} y^b - N^a_k$ and $y^a|_b = \delta^a_b + C^a{}_{eb} y^e$.
It is clear that D is of Cartan type if and only if $y^a_{|k} = 0$ and $y^a|_b = \delta^a_b$ i.e. if and only if (5.32) holds.

By (5.29) one gets

Theorem 5.12. *If a d-connection on E is of Cartan type the following identities involving the torsions and curvatures hold*

$$(5.33)\quad R^a{}_{kb} = \tilde{R}^a_{b\ kh} y^b, \quad P^a{}_{kb} = P^a_{c\ kb} y^c, \quad S^a{}_{bc} = S^a_{d\ bc} y^d.$$

Chapter IV

Geometrical Theory of Embeddings of Vector Bundles

In Chapter I the morphisms of vector bundles were introduced and some algebraic and topological considerations on them were made. In particular, the so-called embeddings of vector bundles were defined. The introducing of the non-linear connections in Ch. II and, then, the metrical structures in Ch. III, together with the related concepts, has much enriched the geometric structure of a vector bundle and of its total space. In this circumstance the geometry of the embeddings of vector bundles becomes much more interesting. In this Chapter such a geometric theory will be developed. As we shall see later this theory offers a very convenient framework for study of the subspaces in Lagrange or Finsler spaces.

The moving frame method, as well as tensorial and exterior calculus will be used. Explicit forms of the basic equations are given in local coordinates.

In §1 we recall the notion of embedding of vector bundles and we give its local form. Also, the geometric structures which will be taken into consideration are recalled. Then, in §2, a moving frame is constructed. The non-linear connection and a relative covariant derivation are discussed in §3. The Gauss-Weingarten formulae are established in §4 and, in §5, the Gauss-Codazzi equations are derived.

§1. Embeddings of Vector Bundles

Let $\xi=(E,p,M)$ be a vector bundle. We assume its total space E is endowed with a Riemannian or pseudo-Riemannian structure G and we consider the non-linear connection N defined by the distribution on E which is orthogonal to the vertical distribution V with respect to G. Also we consider on E the canonical metrical d-connection D found in §5 of Ch. III.

As usual (x^i,y^a), $i=1,2,\ldots,n$, $a=1,2,\ldots,m$ will denote the local coordinates on E and $(\frac{\delta}{\delta x^i},\frac{\partial}{\partial y^a})$, where $\frac{\delta}{\delta x^i}=\frac{\partial}{\partial x^i}-N_i^a\frac{\partial}{\partial y^a}$ is the frame adapted to the non-linear connection N and to vertical distribution V. Its dual is $(dx^i,\delta y^a)$, where $\delta y^a=dy^a+N_i^a dx^i$. In order to avoid very long formulae, the exterior calculus in handling D, as it was presented in §4 of Ch. III, will be used. The metric d-tensor fields g_{ij} and h_{ab} will be used for rising or lowering indices. For instance, we shall set

(1.1) $\Omega_{ij} = \Omega_i^h g_{hj}$, $\tilde{\Omega}_{ab} = \tilde{\Omega}_a^c h_{cb}$,

where Ω_i^j and $\tilde{\Omega}_a^b$ are the curvature 2-forms of D. Since the d-connection D is metrical we have

(1.2) $\Omega_{ij} = -\Omega_{ji}$, $\tilde{\Omega}_{ab} = -\tilde{\Omega}_{ba}$.

Now let $\xi' = (E', p', M')$ be another vector bundle with dim M $= n'$ and $\dim E' = n' + m'$, $1 \leq n' < n$, $1 \leq m' < m$.

Definition 1.1. *A pair (f, f') is said to be an embedding of the vector bundle $\xi' = (E', p', M')$ in $\xi = (E, p, M)$ if $f: E \to E'$ and $f': M' \to M$ are embeddings of manifolds, $p' \circ f = f' \circ p$ and the mapping $f_{x'}: E'_{x'} \to E_{f'(x')}$ is a monomorphism for every $x' \in M'$.*

Since all considerations in the following have a local character sometimes we shall identify M' to $f'(M')$ and E' to $f(E')$ and we shall think $\xi' = (E', p', M')$ as a vector subbundle of $\tilde{\xi} = (E, p_{|M'}, M')$.

Let (u^α, v^A) be the local coordinates on E'. From now on $\alpha, \beta, \gamma, \ldots$ range from 1 to n' and A,B,C,... range from 1 to m'. The embedding (f, f') is locally represented in the form

$$(1.3)\quad \begin{aligned} & x^i = x^i(u^1, \ldots, u^{n'}), \ \mathrm{rank}\left(\frac{\partial x^i}{\partial u^\alpha}\right) = n', \\ & y^a = B_A^a(u) v^A, \ \mathrm{rank}(B_A^a(u)) = m'. \end{aligned}$$

As an example, let $f': M' \to M$ be an embedding and let (TM', τ', M') and (TM, τ, M) be the tangent bundle over M' and M, respectively. Then (f'^T, f'),, where $f'^T: TM' \to TM$ is the tangent map to f' is an embedding of these bundles and (1.3) takes the form

$$(1.4)\quad \begin{aligned} & x^i = x^i(u^1, \ldots, u^{n'}), \quad \mathrm{rank}\left(\frac{\partial x^i}{\partial u^\alpha}\right) = n', \\ & y^i = \frac{\partial x^i}{\partial u^\alpha} v^\alpha. \end{aligned}$$

The last equations are basic in the theory of subspaces in Lagrange or Finsler spaces (R. Miron [106], M. Matsumoto [99], M. Anastasiei [9], T. Sakaguchi [136] etc.).

§2. Moving Frame on E′ in E

Looking at (1.3) let us set

$$(2.1)\quad \begin{aligned} &B^i_\alpha = \frac{\partial x^i}{\partial u^\alpha}, \quad B^i_{\alpha\beta} = \frac{\partial^2 x^i}{\partial u^\alpha \partial u^\beta}, \\ &B^a_{A\alpha} = \frac{\partial B^a_A}{\partial u^\alpha}, \quad B^a_{0\alpha} = v^A B^a_{A\alpha}, \end{aligned}$$

and let us note that B^i_α $(\alpha = 1,\ldots,n')$ and B^a_A $(A = 1,2,\ldots,m')$ are $n'+m'$ linear independent locally d-vector fields which are tangent to E′.

We complete this system of d-vector fields with n-n′ d-vector fields $B^i_{\bar\alpha}$, $(\bar\alpha,\ \bar\beta = 1,2,\ldots,n-n')$ given by the equations

$$(2.2)\quad g_{ij}B^i_\alpha B^j_{\bar\beta} = 0, \quad g_{ij}B^i_{\bar\alpha}B^j_{\bar\beta} = \delta_{\bar\alpha\bar\beta}$$

as well as with m-m′ d-vectors fields $B^a_{\bar A}(u,v)$, $(\bar A, \bar B, = 1,\ldots,m-m')$ given by the equations

$$(2.2)'\quad h_{ab}B^a_A B^b_{\bar B} = 0, \quad h_{ab}B^a_{\bar A}B^b_{\bar B} = \delta_{\bar A\bar B}\,.$$

It is obvious that $B^i_{\bar\alpha}(u,v)$ are linearly independent since they are mutually orthogonal. Moreover, they are all of length 1 and are normal to the space spanned by $(B^i_\alpha(u))$ $\alpha = 1,2,\ldots,n$, in the horizontal distribution associated to N. Analogously, $B^a_{\bar A}$ are linear independent of length 1, mutually orthogonal and are normal to the space spanned by $(B^a_A(u))$, $A = 1,2,\ldots,m'$ in the vertical distribution on E.

Consequently,

$$(2.3)\quad \mathfrak{R} = \{(u,v), B^i_\alpha(u), B^a_A(u), B^i_{\bar\alpha}(u,v), B^a_{\bar A}(u,v)\}$$

is a natural moving frame along E′ as a submanifold in E.

Now we can represent in $\mathfrak{R}$ the geometric objects on E restricted to E′. For instance, the components of the tensor G are given in $\mathfrak{R}$ as follows

$$(2.4)\quad g_{\alpha\beta} = g_{ij}B^i_\alpha B^j_\beta, \quad h_{AB} = h_{ab}B^a_A B^b_B\,.$$

The construction of $\mathfrak{R}$ depends on coordinates on E and E$'$ as well as on the choice of the orthogonal frames $(B^i_{\bar\alpha})$ and $(B^a_{\bar A})$. So that we are led to consider as mixed geometric objects on E$'$ those which are so with respect to the change of coordinates $(x^i,y^a)\to(\bar x^i,\bar y^a)$ on E, $(u^\alpha,v^A)\to(\bar u^\alpha,\bar v^A)$ on E$'$ as well as to the orthogonal transformations

$$(B^i_{\bar\alpha})\to(\bar B^i_{\bar\alpha}),\ (B^a_{\bar A})\to(\bar B^a_{\bar A}).$$

For example, $B^i_\alpha(u)$, $B^a_A(u)$, $B^i_{\bar\alpha}(u,v)$, $B^a_{\bar A}(u,v)$ g_{ij}, h_{ab}, $g_{\alpha\beta}$, h_{AB}, $\delta_{\bar\alpha\bar\beta}$, $\delta_{\bar A\bar B}$ are mixed geometric objects (tensors) on E$'$.

Let be

$$(2.4)'\quad \mathfrak{R}^* = \{(u,v);\, B^\alpha_i(u,v),\, B^A_a(u,v),\, B^{\bar\alpha}_i(u,v),\, B^{\bar A}_\alpha(u,v)\}$$

the frame which is dual to $\mathfrak{R}$, that is, the following equations hold

$$(2.5)\quad \begin{aligned} &B^i_\beta B^\alpha_i = \delta^\alpha_\beta,\ B^i_\beta B^{\bar\alpha}_i = 0,\ B^\alpha_i B^i_{\bar\beta} = 0,\ B^{\bar\alpha}_i B^i_{\bar\beta} = \delta^{\bar\alpha}_{\bar\beta},\\ &B^a_C B^A_a = \delta^A_C,\ B^a_C B^{\bar A}_a = 0,\ B^A_a B^a_{\bar B} = 0,\ B^{\bar A}_a B^a_{\bar B} = \delta^{\bar A}_{\bar B}. \end{aligned}$$

$$(2.5)'\quad \begin{aligned} &B^i_\alpha B^\alpha_j + B^i_{\bar\alpha} B^{\bar\alpha}_j = \delta^i_j,\\ &B^a_A B^A_b + B^a_{\bar A} B^{\bar A}_b = \delta^a_b. \end{aligned}$$

By (2.5) and (2.5)$'$ it follows that

$$(2.6)\quad p^i_j = B^i_\alpha B^\alpha_j,\quad q^i_j = B^i_{\bar\alpha} B^{\bar\alpha}_j$$

are supplementary projectors on the tangent space spanned by (B^i_α) and on normal space spanned by $(B^i_{\bar\alpha}(u,v))$, respectively, as well as that

$$(2.6)'\quad \tilde p^a_b = B^a_A B^A_b,\quad \tilde q^a_b = B^a_{\bar A} B^{\bar A}_b,$$

are supplementary projectors on the tangent space spanned by $(B^a_A(u))$ and on the normal space spanned by $(B^a_{\bar A}(u,v))$, respectively.

It is now obvious that we can represent g_{ij} and h_{ab} in $\mathfrak{R}$ under the following form

$$(2.7)\quad g_{ij} = g_{\alpha\beta} B^\alpha_i B^\beta_j + \delta_{\bar\alpha\bar\beta} B^{\bar\alpha}_i B^{\bar\beta}_j;\quad h_{ab} = h_{AB} B^A_a B^B_b + \delta_{\bar A\bar B} B^{\bar A}_a B^{\bar B}_b.$$

Then their reciprocals take the form

$$(2.7)' \quad g^{ij} = g^{\alpha\beta} B^i_\alpha B^j_\beta + \delta^{\bar{\alpha}\bar{\beta}} B^i_{\bar{\alpha}} B^j_{\bar{\beta}}; \quad h^{ab} = h^{AB} B^a_A B^b_B + \delta^{\bar{A}\bar{B}} B^a_{\bar{A}} B^b_{\bar{B}}.$$

§3. Induced Non-linear Connections. Relative Covariant Derivation

We shall see now that the non-linear connection N determines a non-linear connection N on E′, when it is conveniently projected on the tangent space to E′. Observing that in order to give N it is enough to give the coframe $(du^\alpha, \delta v^A)$, where $\delta v^A = dv^A + \tilde{N}^A_\alpha(u,v) du^\alpha$ we shall set

$$(3.1) \quad \delta v^A = B^A_a \delta y^a.$$

Thus $\tilde{N}$ is uniquely determined. It will be called the *induced non-linear connection.* By (3.1) we get

Proposition 3.1. *The local coefficients of the induced non-linear connection* $\tilde{N}$ *are as follows*

$$(3.2) \quad \tilde{N}^A_\alpha = B^A_a (B^a_{o\alpha} + N^a_i B^i_\alpha).$$

Next, a direct calculation leads to

Theorem 3.1. *The coframe* $(dx^i, \delta y^a)$ *is represented in* $\mathfrak{R}$ *under the form*

$$(3.3) \quad dx^i = B^i_\alpha du^\alpha, \quad \delta y^a = B^a_A \delta v^A + B^a_{\bar{A}} H^{\bar{A}}_\alpha du^\alpha$$

in which

$$(3.4) \quad H^{\bar{A}}_\alpha = B^{\bar{A}}_a (B^a_{o\alpha} + N^a_i B^i_\alpha)$$

is a mixed tensor field.

By (3.3) one yields

Proposition 3.2. *The coframe* $(du^\alpha, \delta v^\alpha)$ *is mapped to the coframe* $(B^i_\alpha du^\alpha, B^a_A \delta v^A)$ *if and only if* $H^{\bar{A}}_\alpha = 0$.

Now we shall construct an operator of relative covariant derivation ∇ in the algebra of mixed tensor fields. It is clear that ∇ will be known if its action on functions and on the vector fields of the following form

(3.5) $X^i(x(u),y(u,v)), X^a(x(u),y(u,v)), X^\alpha(u,v), X^A(u,v), X^{\bar{\alpha}}(u,v), X^{\bar{A}}(u,v)$

is known.

We shall use a convention that indices which are different in their nature denote also a difference between the geometric objects to which they are assigned. For instance X^α is different from X^i, as well as from $X^{\bar{\alpha}}$ and so on.

Now we shall indicate how ∇ acts on the vector fields from (3.5).

Definition 3.1. *We call a coupling of the metrical d-connection D to the induced non-linear connection* $\tilde{N}$ *along* E' *the operator* $\tilde{D}$ *defined by D when it is taken modulo (3.3), that is*

(3.6) $\tilde{D}X^i = DX^i$ (*modulo* (3.3)), $\tilde{D}X^a = DX^a$ (*modulo* (3.3)).

Then, let

(3.7) $\tilde{D}X^i = dX^i + X^j \tilde{\omega}^i_j$, $\tilde{D}X^a = dX^a + X^b \tilde{\omega}^a_b$,

where $\tilde{\omega}^i_j$ and $\tilde{\omega}^a_b$ are the connection 1-forms of $\tilde{D}$.

We have

Theorem 3.2. *The connection 1-forms of* $\tilde{D}$ *are given by*

(3.7)' $\tilde{\omega}^i_j = L^i_{j\alpha} du^\alpha + C^i_{jA} \delta v^A$, $\tilde{\omega}^a_b = L^a_{b\alpha} du^\alpha + C^a_{bA} \delta v^A$,

where the local coefficients of $\tilde{D}$ *are as follows*

$$(3.8) \quad \begin{aligned} L^i_{j\gamma} &= L^i_{jk} B^k_\gamma + \tilde{C}^i_{jd} B^d_{\bar{A}} H^{\bar{A}}_\gamma, \quad C^i_{jD} = C^i_{jd} B^d_D, \\ L^a_{b\gamma} &= \tilde{L}^a_{bk} B^k_\gamma + C^a_{bd} B^d_{\bar{A}} H^{\bar{A}}_\gamma, \quad C^a_{bD} = C^a_{bc} B^c_D. \end{aligned}$$

Corollary 3.1. *The coupling* $\tilde{D}$ *of the metrical d-connection D to* $\tilde{N}$ *depends on* *the metric G and the embedding (f, f') only.*

It is clear that DX^i and DX^a from (3.7) can be written in the form

$$(3.6)' \quad \tilde{D}X^i = X^i_{|\alpha} du^\alpha + X^i|_A \delta v^A, \quad \tilde{D}X^a = X^a_{|\alpha} du^\alpha + X^a|_A \delta v^A$$

in which $X^i_{|\alpha}, X^i|_A, X^a_{|\alpha}, X^a|_A$ can be easily written down.

Definition 3.2. *We call the induced tangent connection on E' by the metrical d-connection D the operator of covariant derivation D^T given by*

$$(3.9) \quad \begin{aligned} D^T X^\alpha &= B^\alpha_i \tilde{D}X^i \quad \text{for } X^i = B^i_\alpha X^\alpha, \\ D^T X^A &= B^A_\alpha \tilde{D}X^a \quad \text{for } X^a = B^a_A X^a. \end{aligned}$$

Next we shall set

$$(3.10) \quad D^T X^\alpha = dX^\alpha + X^\beta \omega^\alpha_\beta, \quad D^T X^A = dX^A + X^B \omega^A_B,$$

where ω^α_β and ω^A_B are the connection 1-forms of D^T.

We have

Theorem 3.3. *The connection 1-forms of the induced tangent connection are given by*

$$(3.11) \quad \omega^\alpha_\beta = L^\alpha_{\beta\gamma} du^\gamma + C^\alpha_{\beta D} \delta v^D, \quad \omega^A_B = L^A_{B\gamma} du^\gamma + C^A_{BD} \delta v^D,$$

where

$$(3.12) \quad \begin{aligned} L^\alpha_{\beta\gamma} &= B^\alpha_i (B^i_{\beta\gamma} + B^j_\beta L^i_{j\gamma}), \quad C^\alpha_{\beta D} = B^\alpha_i B^j_\beta C^i_{jD}, \\ L^A_{B\gamma} &= B^A_a (B^a_{B\gamma} + B^c_B L^a_{c\gamma}), \quad C^A_{BD} = B^A_a B^c_B C^a_{cD}. \end{aligned}$$

As in the case of $\tilde{D}$ we may write

$$(3.10)' \quad D^T = X^\alpha_{|\gamma} du^\gamma + X^\alpha|_D \delta v^D, \quad D^T X^A = X^A_{|\gamma} du^\gamma + X^A|_d \delta v^D$$

in which $X^\alpha_{|\gamma}$, $X^\alpha|_D$, $X^A_{|\gamma}$ and $X^A|_D$ can be easily written down.

By (3.11) we get

Corollary 3.2. *The induced tangent connection D^T depends on the metric G and the embedding (f,f'), only.*

Finally, we introduce

Definition 3.3. *We call the induced normal connection by the metrical d-connection D, the operator of covariant derivation $D^{\perp}$ given by*

$$(3.13)\quad \begin{aligned} &D^{\perp}X^{\bar\alpha} = B_i^{\bar\alpha}\tilde{D}X^i \text{ for } X^i = B_{\bar\alpha}^{i}X^{\bar\alpha}, \\ &D^{\perp}X^{\bar A} = B_a^{\bar A}\tilde{D}X^a \text{ for } X^a = B_{\bar A}^{a}X^{\bar A}. \end{aligned}$$

As before, we set

$$(3.14)\quad D^{\perp}X^{\bar\alpha} = dX^{\bar\alpha} + X^{\bar\beta}\omega_{\bar\beta}^{\bar\alpha};\ D^{\perp}X^{\bar A} = dX^{\bar A} + X^{\bar B}\omega_{\bar B}^{\bar A}$$

in which $\omega_{\bar\beta}^{\bar\alpha}$, $\omega_{\bar B}^{\bar A}$ are the connection 1-forms of the induced normal connection $D^{\perp}$.

By (3.13) and (3.14) we deduce

Theorem 3.4. *The connection 1-forms of the induced normal connection $D^{\perp}$ are as follows*

$$(3.15)\quad \omega_{\bar\beta}^{\bar\alpha} = L_{\bar\beta\gamma}^{\bar\alpha}du^{\gamma} + C_{\bar\beta D}^{\bar\alpha}\delta v^D,\ \omega_{\bar B}^{\bar A} = L_{\bar B\gamma}^{\bar A}du^{\gamma} + C_{\bar BD}^{\bar A}\delta v^D,$$

where

$$(3.16)\quad \begin{aligned} &L_{\bar\beta\gamma}^{\bar\alpha} = B_i^{\bar\alpha}\Big(\frac{\delta B_{\bar\beta}^{i}}{\delta u^{\gamma}} + B_{\bar\beta}^{j}L_{j\gamma}^{i}\Big);\quad C_{\bar\beta D}^{\bar\alpha} = B_i^{\bar\alpha}\Big(\frac{\partial B_{\bar\beta}^{i}}{\partial v^D} + B_{\bar\beta}^{j}C_{jD}^{i}\Big), \\ &L_{\bar B\gamma}^{\bar A} = B_a^{\bar A}\Big(\frac{\delta B_{\bar B}^{a}}{\delta u^{\gamma}} + B_{\bar B}^{b}L_{b\gamma}^{a}\Big);\quad C_{\bar BD}^{\bar A} = B_a^{\bar A}\Big(\frac{\partial B_{\bar B}^{a}}{\partial v^D} + B_{\bar B}^{b}C_{bD}^{a}\Big). \end{aligned}$$

And in this case we may write

$$(3.14)'\quad D^{\perp}X^{\bar\alpha} = X_{|\gamma}^{\bar\alpha} + X^{\bar\alpha}|_D\delta v^D,\ D^{\perp}X^{\bar A} = X_{|\gamma}^{\bar A}du^{\gamma} + X^{\bar A}|_D\delta v^D.$$

Also, we have

Corollary 3.3. *The induced normal connection $D^{\perp}$ depends on the metric G and the embedding (f,f'), only.*

Now we are in a position to define the relative covariant derivation ∇ announced before.

Definition 3.4. *We call relative or mixed covariant derivation in the algebra of mixed tensor fields an operator* ∇ *such that* $\nabla f = df$, $f \in \mathcal{F}(E')$ *and*

$$\nabla X^i = \tilde{D}X^i,\ \nabla X^a = \tilde{D}X^a,\ \nabla X^\alpha = D^T X^\alpha,\ \nabla X^A = \tilde{D}X^A,\ \nabla X^{\bar{\alpha}} = D^\perp X^{\bar{\alpha}},\ \nabla X^{\bar{A}} = D^\perp X^{\bar{A}}.$$

The 1-forms $\tilde{\omega}^i_j, \tilde{\omega}^a_b; \omega^\alpha_\beta, \omega^A_B$ *and* $\omega^{\bar{\alpha}}_{\bar{\beta}}, \omega^{\bar{A}}_{\bar{B}}$ *will be called the connection 1-forms of* ∇.

Clearly there exists a relative covariant derivation, it is unique and depends only on the metric G and the embedding (f, f′).

The torsions of ∇ are as follows

$$(3.17)\quad \begin{aligned} &T^\alpha{}_{\beta\gamma} = L^\alpha_{\beta\gamma} - L^\alpha_{\gamma\beta};\ C^\alpha_{\beta D};\ R^A{}_{\beta\gamma} = \frac{\delta \tilde{N}^A_\beta}{\delta u^\gamma} - \frac{\delta \tilde{N}^A_\gamma}{\delta u^\beta}, \\ &P^A{}_{\gamma D} = \frac{\partial \tilde{N}^A_\gamma}{\partial v^D} - L^A_{D\gamma};\ S^A{}_{BC} = C^A_{BC} - C^A_{CB}, \end{aligned}$$

while its curvatures, written in a compact form, are given by

$$(3.18)\quad \begin{aligned} R^{\underline{A}}_{\underline{B}|\gamma\delta} &= \frac{\delta L^{\underline{A}}_{\underline{B}\gamma}}{\delta u^\delta} - \frac{\delta L^{\underline{A}}_{\underline{B}\delta}}{\delta u^\gamma} + L^{\underline{C}}_{\underline{B}\gamma} L^{\underline{A}}_{\underline{C}\delta} - L^{\underline{C}}_{\underline{B}\delta} L^{\underline{A}}_{\underline{C}\gamma} + C^{\underline{A}}_{\underline{B}D} R^D{}_{\gamma\delta}, \\ P^{\underline{A}}_{\underline{B}|\gamma D} &= \frac{\partial L^{\underline{A}}_{\underline{B}\gamma}}{\partial v^D} - \frac{\delta C^{\underline{A}}_{\underline{B}D}}{\delta u^\gamma} + L^{\underline{C}}_{\underline{B}\gamma} C^{\underline{A}}_{\underline{C}D} - C^{\underline{C}}_{\underline{B}D} L^{\underline{A}}_{\underline{C}\gamma} - C^{\underline{A}}_{\underline{B}E} \frac{\partial \tilde{N}^E_\gamma}{\partial v^D}, \\ S^{\underline{A}}_{\underline{B}|CD} &= \frac{\partial C^{\underline{A}}_{\underline{B}C}}{\partial v^D} - \frac{\partial C^{\underline{A}}_{\underline{B}D}}{\partial v^C} + C^{\underline{C}}_{\underline{B}C} C^{\underline{A}}_{\underline{C}D} - C^{\underline{C}}_{\underline{B}D} C^{\underline{A}}_{\underline{C}C}, \end{aligned}$$

in which $(\underline{A}, \underline{B}, \underline{C}) \in \{(i,j,k), (a,b,c), (\alpha,\beta,\gamma), (A,B,C), (\bar{\alpha},\bar{\beta},\bar{\gamma}), (\bar{A},\bar{B},\bar{F})\}$

Then it easily to prove

Theorem 3.5. *The structure equations of* ∇ *are*

$$(i)\quad \begin{aligned} &d(du^\alpha) - du^\beta \wedge \omega^\alpha_\beta = -\Omega^\alpha;\ d(\delta v^A) - \delta v^C \wedge \omega^A_C = -\Omega^A, \\ &d\omega^\alpha_\beta - \omega^\gamma_\beta \wedge \omega^\alpha_\gamma = -\Omega^\alpha_\beta;\ d\tilde{\omega}^a_b - \tilde{\omega}^c_b \wedge \tilde{\omega}^a_c = -\tilde{\Omega}^a_b, \end{aligned}$$

$$\text{(ii)}\quad \begin{aligned} &d\tilde{\omega}^i_j-\tilde{\omega}^k_j\wedge\tilde{\omega}^i_k=-\tilde{\Omega}^i_j\,;\ d\omega^A_B-\omega^C_B\wedge\omega^A_C=-\Omega^A_B\,,\\ &d\omega^{\bar\alpha}_{\bar\beta}-\omega^{\bar\gamma}_{\bar\beta}\wedge\omega^{\bar\alpha}_{\bar\gamma}=-\Omega^{\bar\alpha}_{\bar\beta}\,;\ d\omega^{\bar A}_{\bar B}-\omega^{\bar C}_{\bar B}\wedge\omega^{\bar A}_{\bar C}=-\Omega^{\bar A}_{\bar B}\,, \end{aligned}$$

in which Ω^α, Ω^A *are the torsion 2-forms of* ∇ *given by*

$$\text{(3.19)}\quad \begin{aligned} \Omega^\alpha &= \frac{1}{2}T^\alpha_{\beta\gamma}du^\beta\wedge du^\gamma+C^\alpha_{\beta C}du^\beta\wedge\delta v^C\\ \Omega^A &= \frac{1}{2}R^A_{\beta\gamma}du^\beta\wedge du^\gamma+P^A_{\beta C}du^\beta\wedge\delta v^C+\frac{1}{2}S^A_{BC}\,\delta v^B\wedge\delta v^C \end{aligned}$$

and $\tilde{\Omega}^i_j,\ \tilde{\Omega}^a_b,\ldots$ *are the curvature 2-forms of* ∇ *given by*

$$\text{(3.19)}'\quad \Omega^{\underline{A}}_{\underline{B}}=\frac{1}{2}R_{\underline{B}}{}^{\underline{A}}{}_{\gamma\Delta}du^\gamma\wedge du^\delta+P_{\underline{B}}{}^{\underline{A}}{}_{\gamma D}du^\gamma\wedge\delta v^D+\frac{1}{2}S_{\underline{B}}{}^{\underline{A}}{}_{CD}\delta v^C\wedge\delta v^D$$

with $(\underline{A},\underline{B})\in\{(\alpha,\beta),(A,B);(i,j),(a,b);(\bar\alpha,\bar\beta),(\bar A,\bar B)\}$.

We shall adopt, as in §1, the notations

$$\text{(3.20)}\quad \begin{aligned} &\tilde{\Omega}_{ij}=\tilde{\Omega}^k_i g_{kj},\ \tilde{\Omega}_{ab}=\tilde{\Omega}^c_a h_{cb},\ \Omega_{\alpha\beta}=\Omega^\gamma_\alpha g_{\gamma\beta}\,,\\ &\Omega_{AB}=\Omega^C_A\, h_{CB}\,,\ \Omega_{\bar\alpha\bar\beta}=\Omega^{\bar\gamma}_{\bar\alpha}\,\delta_{\bar\gamma\bar\beta}\,,\ \Omega_{\bar A\bar B}=\Omega^{\bar C}_{\bar A}\,\delta_{\bar C\bar B}\,, \end{aligned}$$

which will be useful for the writing of the fundamental equations of the embedding (f, f').

As a direct consequence of the Theorem 3.2 we notice

Proposition 3.3. The following equations hold good

$$\text{(3.21)}\quad \nabla g_{ij}=0,\quad \nabla h_{ab}=0.$$

§4. The Gauss-Weingarten Formulae

As usual in the theory of the embeddings we are interested in finding the moving equations of the moving frame $\mathfrak{R}$ given in (2.3) along E'. These equations, called also Gauss-Weingarten formulae, are obtained when the relative covariant derivatives of the vector fields from $\mathfrak{R}$ are expressed again in the frame $\mathfrak{R}$.

Thus we have

Theorem 4.1. *The following Gauss-Weingarten formulae hold good*

$$(4.1)\quad \nabla B^i_\alpha = B^i_{\bar\beta}\Pi^{\bar\beta}_\alpha,\quad \nabla B^a_A = B^a_{\bar C}\Pi^{\bar C}_A,$$

$$(4.2)\quad \nabla B^i_{\bar\alpha} = -B^i_\beta \Pi^\beta_{\bar\alpha},\ \nabla B^a_{\bar A} = -B^a_B\,\Pi^B_{\bar A},$$

where

$$(4.3)\quad \begin{array}{l} \Pi^{\bar\beta}_\alpha = H_\alpha{}^{\bar\beta}{}_\gamma du^\gamma + H_\alpha{}^{\bar\beta}{}_D \delta v^D,\ \Pi^\beta_{\bar\alpha} = g^{\beta\gamma}\delta_{\bar\alpha\bar\beta}\Pi^{\bar\beta}_\gamma, \\ \Pi^{\bar C}_A = K_A{}^{\bar C}{}_\gamma du^\gamma + K_A{}^{\bar C}{}_D \delta v^D,\ \Pi^A_{\bar C} = h^{AB}\,\delta_{\bar C\bar D}\,\Pi^{\bar D}_B \end{array}$$

and

$$(4.4)\quad \begin{array}{l} H_\alpha{}^{\bar\beta}{}_\gamma = B^{\bar\beta}_k (B^k_{\alpha\gamma} + B^h_\alpha L^k_{h\gamma}),\ H_\alpha{}^{\bar\beta}{}_D = B^{\bar\beta}_k B^h_\alpha C^k_{hD}, \\ K_A{}^{\bar B}{}_\gamma = B^{\bar B}_a (B^a_{A\gamma} + B^C_A L^a_{C\gamma}),\ K_A{}^{\bar B}{}_D = B_c{}^{\bar B} B^d_A C^c_{dD}. \end{array}$$

Proof. From $\nabla B^i_\alpha = B^i_{\alpha|\gamma} du^\gamma + B^i_\alpha|_D\ \delta v^D =$

$$= \left(\frac{\delta B^i_\alpha}{\delta u^\gamma} + B^j_\alpha L^i_{j\gamma} - B^i_\beta L^\beta_{\alpha\gamma}\right) du^\gamma + \left(\frac{\partial B^i_\alpha}{\partial v^D} + B^j_\alpha C^i_{jD} - B^i_\beta C^\beta_{\alpha D}\right)\delta v^D,$$

taking into account the Theorem 3.4 we find

$$\nabla B^i_\alpha = B^i_{\bar\beta}(H_\alpha{}^{\bar\beta}{}_\gamma du^\gamma + H_\alpha{}^{\bar\beta}{}_D \delta v^D) = B^i_{\bar\beta}\Pi^{\bar\beta}_\alpha$$

that is, the first formula (4.1). One proceeds analogously in order to find the second formula (4.1). Now, by applying ∇ to the both sides of the equations $g_{ij}B^i_\alpha B^j_{\bar\beta} = 0$ and $h_{ab}B^a_A B^b_{\bar B} = 0$ and taking into account the Proposition 3.3 one obtains (4.2).

The Theorem 4.1 shows us that

$$(4.5)\quad H_\alpha{}^{\bar\beta}{}_\gamma,\ H_\alpha{}^{\bar\beta}{}_D,\ K_A{}^{\bar\beta}{}_\gamma,\ K_A{}^{\bar B}{}_D$$

are mixed tensor fields. They will be called the second fundamental tensor fields of the embedding (f, f').

We now prove

Proposition 4.1. *The following equations hold good*

(4.6) $\nabla g_{\alpha\beta}=0,\ \nabla h_{AB}=0.$

Proof. Covariantly deriving, with respect to ∇ the identities (2.4) and by using the Gauss formulae (4.1) one obtains (4.6).

Proposition 4.2. *We have*

(4.7) $\Omega_{\alpha\beta}=-\Omega_{\beta\alpha},\ \Omega_{AB}=-\Omega_{BA}.$

Proof. These equations are direct consequences of (4.6).

Applying ∇ to $y^a=B^a_A(u)v^A$ and taking into account (4.1) one deduces

(4.8) $\nabla y^a = B^a_{\bar{B}}\Pi^{\bar{B}}_A v^A + B^a_A \nabla v^A.$

By (4.8) we easily obtain

Proposition 4.3. *If $\nabla v^A=0$ then ∇y^a is normal to E' in the vertical distribution $u \to V_u E$ on E.*

Proposition 4.4. *If $K_A{}^{\bar{\beta}}{}_\gamma=0,\ K_A{}^{\bar{B}}{}_D=0$ then $\nabla y^a=B^a_A\nabla v^A$, that is, ∇y^a is tangent to E'.*

Now we shall give a geometrical meaning for the vanishing of the second fundamental tensor fields of the embedding (f, f'). To this aim let $c: I \to E'$ $(I \subseteq \mathbf{R})$ be a smooth curve given in a local chart on E' by the equations

(4.9) $u^\alpha=u^\alpha(t),\ \ v^A=v^A(t),\ t\in I$

and $X^i=B^i_\alpha(u)X^\alpha(u,v)$ be an horizontal vector field along of c which is tangent to E'. By (4.1), along of curve c we have

$$(4.10)\quad \frac{\nabla X^i}{dt}=B^i_{\bar{\beta}}\frac{\Pi^{\bar{\beta}}_\alpha}{dt}+B^i_\alpha\frac{\nabla X^\alpha}{dt}.$$

Definition 4.1. *We say that the vector bundle ξ' is h-totally geodesic if for every* *curve c on E' and for every horizontal vector field X^i along of c and tangent to E', the vector field $\frac{\nabla X^i}{dt}$ is tangent to E', too.*

From (4.10) we immediately get

Theorem 4.2. *The vector bundle ξ' is h-totally geodesic if and only if the second fundamental tensors $H_{\alpha}{}^{\bar\beta}{}_{\gamma}$ and $H_{\alpha}{}^{\bar\beta}{}_{D}$ identically vanish.*

Also, we may formulate

Definition 4.2. *The vector bundle ξ' is said to be v-totally geodesic if for every curve c on E' and every vertical vector field X^a along of c and tangent to E', the vector field $\frac{\nabla X^a}{dt}$ is tangent to E', too.*

Just as before we can prove

Theorem 4.3. *The vector bundle ξ' is v-totally geodesic if and only if the second fundamental tensors $K_{A}{}^{\bar B}{}_{\alpha}$ and $K_{A}{}^{\bar B}{}_{D}$ identically vanish.*

§5. The Gauss-Codazzi Equations

The Gauss-Codazzi equations of the vector subbundle ξ' embedded in the vector bundle ξ are nothing but the integrability conditions of the systems of equations (4.1) and (4.2). We shall analyze these conditions by using the exterior calculus. Thus we shall prove

Theorem 5.1. *The Gauss-Codazzi equations of the embedding (f,f') are as follows:*

$$\begin{aligned} &B^i_\alpha B^j_\beta \tilde\Omega_{ij} - \Omega_{\alpha\beta} = \Pi_{\beta\bar\varphi}\wedge\Pi^{\bar\varphi}_\alpha\,;\ B^a_A B^b_B \tilde\Omega_{ab} - \Omega_{AB} = \Pi_{B\bar C}\wedge\Pi^{\bar C}_A\,,\\ &B^i_{\bar\alpha} B^j_{\bar\beta} \tilde\Omega_{ij} - \Omega_{\bar\alpha\bar\beta} = \Pi_{\varphi\bar\beta}\wedge\Pi^{\varphi}_{\bar\alpha}\,;\ B^a_{\bar A} B^b_{\bar B} \tilde\Omega_{ab} - \Omega_{\bar A\bar B} = \Pi_{C\bar B}\wedge\Pi^{C}_{\bar A}\,,\end{aligned} \tag{5.1}$$

$$\begin{aligned} &-B^i_\alpha B^j_{\bar\beta}\tilde\Omega_{ij} = \delta_{\bar\beta\bar\varphi}(d\Pi^{\bar\varphi}_\alpha + \Pi^{\bar\varphi}_\gamma\wedge\omega^\gamma_\alpha - \Pi^{\bar\gamma}_\alpha\wedge\omega^{\bar\varphi}_{\bar\gamma})\,,\\ &-B^a_A B^b_{\bar B}\tilde\Omega_{ab} = \delta_{\bar B\bar C}(d\Pi^{\bar C}_A + \Pi^{\bar C}_D\wedge\omega^D_A - \Pi^{\bar D}_A\wedge\omega^{\bar C}_{\bar D})\,.\end{aligned} \tag{5.2}$$

Proof. The first equation (4.1) can be written in the form

$$dB^{\ i}_{\alpha} + B^{\ k}_{\alpha}\tilde{\omega}^{\ i}_{k} - B^{\ i}_{\beta}\omega^{\beta}_{\alpha} = B^{\ i}_{\bar{\beta}}\Pi^{\bar{\beta}}_{\alpha}.$$

Exterior differentiating the both sides of this equation and using (4.1) and (4.2) one gets

$$-B^{\ h}_{\alpha}\tilde{\Omega}^{\ i}_{h} + B^{\ i}_{\beta}\tilde{\Omega}^{\beta}_{\alpha} = B^{\ i}_{\varphi}\Pi^{\bar{\beta}}_{\alpha}\wedge\Pi^{\varphi}_{\bar{\beta}} + B^{\ i}_{\bar{\varphi}}(d\Pi^{\bar{\varphi}}_{\alpha} + \Pi^{\bar{\varphi}}_{\gamma}\wedge\omega^{\gamma}_{\alpha} - \Pi^{\bar{\gamma}}_{\alpha}\wedge\omega^{\bar{\varphi}}_{\bar{\gamma}})$$

from which (5.1) and (5.2) are easily deduced.

In order to obtain all fundamental equations of the embedding (f, f') we must find the relations between the torsion 2-forms Ω^i, Ω^a of the metrical d-connection D and the torsion 2-forms Ω^α, Ω^A of ∇. Also, we must add to the Pfaff system of equations (4.1), (4.2) the system of equations $dx^i = B^i_\alpha(u)du^\alpha$, $y^a = B^a_A(u)v^A$.

Therefore, we obtain

Theorem 5.2. *The fundamental equations of embedding (f,f') are (5.1), (5.2), (4.8) as well as*

$$(5.3)\qquad \begin{aligned} &d(dx^i) - dx^j\wedge\omega^i_j = -\Omega^i \quad (\textit{modulo } (3.3))\\ &d(\delta y^a) - \delta y^b\wedge\omega^a_b = -\Omega^a \quad (\textit{modulo } (3.3)). \end{aligned}$$

We end here our very general theory on the embeddings of vector bundles. We underline the importance of this theory for applications especially in the particular cases: a) $n' = m'$, b) $n' = n$, $m' = 1$, c) $m' = m + 1$.

The case a) will be examined in detail when subspaces in Finsler or Lagrange spaces are studied in a subsequent chapter.

Chapter V

Einstein Equations

Several physical theories need mathematical patterns which allow the study of physical quantities depending on two sets of variables (x^i, y^a). Here (x^i), $i=1,2,\dots,n$ represents a point and (y^a), $a=1,2,\dots,m$, represents in general a vector which can be regarded as a set of internal variables associated to each point. The cases occurring most frequently are $m=n=4$, as for instance the bilocal field theory of Yukawa, Particle Physics, Theory of Relativity, High Energy Cosmic Ray Physics, Relativistic Optics and others. The case $m<n=4$ occurs in some Unitary Theories.

One of these mathematical patterns could be the geometric theory of the total space of a fibre bundle over a smooth manifold, where (x^i) are the coordinates of the base manifold and (y^a) are those of the fibres. In particular, vector bundles can be used.

We shall argue this point constructing a set of Einstein equations on the total space E of a vector bundle over a smooth manifold.

These equations were suggested by researches due to Y. Takano [148,149], H. Ishikawa [72,73], S. Ikeda [67,68] and were correctly established, for the first time by R. Miron [107,108].

Thus in the first section we shall consider a metric structure G on the total space E of a vector bundle (E,p,M), the metric d-connection D having h(hh) and v(vv)-torsions prescribed, which was showed in §5 of Chapter III and we shall write the Einstein equation associated to it. Expressing this equation in the frame adapted to the non-linear connection on E defined by G, a set of four equations, called also the Einstein equations since they are similar to the classical Einstein equations, is obtained.

In the second section this set of the Einstein equations is transformed in a more suitable form, which allows an easier handling of the conservation laws.

Last section is devoted to some particular cases. These concern both the metric structure G and the dimensions of the base manifold and fibres.

§1. Einstein Equations

The Einstein equation associated to the metric d-connection D with respect to G and having h(hh) - and v(vv) - torsions prescribed is of the form

$$(1.1)\quad \mathrm{Ric}(D) - \frac{1}{2}\mathrm{Sc}(D)G = \kappa \mathfrak{T},$$

where Ric(D) and Sc(D) are the Ricci tensor field and the scalar curvature of D, respectively, κ is constant and $\mathfrak{T}$ is the energy-momentum tensor.

Now we shall express (1.1) in the frame $(\frac{\delta}{\delta x^i}, \frac{\partial}{\partial y^a})$ adapted to the non-linear connection N defined by G and to the vertical distribution V on E. In order to do this we shall set

$$(X_\alpha) = (\frac{\delta}{\delta x^i}, \frac{\partial}{\partial y^a}),\ X_i = \frac{\delta}{\delta x^i},\ X_a = \frac{\partial}{\partial y^a},\ \alpha = 1,2,\ldots,n+m.$$

In the following the Greek indices $\alpha, \beta, \gamma, \ldots$ run from 1 to $n+m$.
We have

$$(1.2)\ D_{X_\gamma} X_\beta = \Gamma^\alpha_{\beta\gamma} X_\alpha,$$

$$(1.3)\ \mathbf{T}^\alpha{}_{\beta\gamma} = \Gamma^\alpha_{\beta\gamma} - \Gamma_{\gamma\beta} + W^\alpha_{\beta\gamma},$$

where the anholonomic coefficients $W^\alpha_{\beta\gamma}$ are defined by

$$(1.3)'\ [X_\alpha, X_\beta] = W^\gamma_{\alpha\beta} X_\gamma.$$

Next we have

$$(1.4)\ \mathbf{R}_\beta{}^\alpha{}_{\gamma\delta} = X_\delta \Gamma^\alpha_{\beta\gamma} - X_\gamma \Gamma^\alpha_{\beta\delta} + \Gamma^\varphi_{\beta\gamma}\Gamma^\alpha_{\varphi\delta} - \Gamma^\varphi_{\beta\delta}\Gamma^\alpha_{\varphi\gamma} + \Gamma^\alpha_{\beta\varphi} W^\varphi_{\gamma\delta}.$$

The formulae (1.3) and (1.4) give the local components with respect to the frame (X_α) of the torsion **T** and the curvature **R** of D introduced by

$$(1.5)\ \mathbf{T}(X_\gamma, X_\beta) = T^\alpha_{\beta\gamma} X_\alpha,\quad R(X_\delta, X_\gamma) X_\beta = R_\beta{}^\alpha{}_{\gamma\delta} X_\alpha.$$

Consequently, the Bianchi identities in the frame (X_α) look as follows

$$(1.6)\ \sum (D_{X_\alpha} R_\delta{}^\varphi{}_{\beta\gamma} + R_\delta{}^\varphi{}_{\alpha\psi} T^\psi{}_{\beta\gamma}) = 0,$$

$$(1.6)'\ \sum (D_{X_\alpha} T^\delta{}_{\beta\gamma} + T^\varphi{}_{\alpha\beta} T^\delta{}_{\gamma\varphi} - R_\alpha{}^\delta{}_{\beta\gamma}) = 0,$$

where $\sum$ means cyclic sum over α, β, γ.
We shall separate in the previous formulae the components with respect to

$X_i = \frac{\delta}{\delta x^i}$ and to $X_a = \frac{\partial}{\partial y^a}$. Thus we obtain

Proposition 1.1. *In the frame* (X_α) *the metric d-connection D has the following coefficients:*

$$(1.7)\quad \begin{aligned} &\Gamma^i_{jk} = L^i_{jk},\ \Gamma^i_{ja} = \tilde{C}^i_{ja},\ \Gamma^i_{aj} = 0,\ \Gamma^i_{ab} = 0, \\ &\Gamma^a_{jk} = 0,\ \Gamma^a_{jb} = 0,\ \Gamma^a_{bk} = \tilde{L}^a_{bk},\ \Gamma^a_{bc} = C^a_{bc}. \end{aligned}$$

Proposition 1.2. *The anholonomic coefficients are as follows*

$$(1.8)\quad \begin{aligned} &W^k_{ij} = 0,\ W^k_{aj} = 0,\ W^k_{ia} = 0,\ W^k_{ab} = 0, \\ &W^a_{ij} = R^a{}_{ij},\ W^b_{ai} = -\frac{\partial N^b_i}{\partial y^a},\ W^b_{ia} = \frac{\partial N^b_i}{\partial y^a},\ W^c_{ab} = 0. \end{aligned}$$

Proposition 1.3. *In the frame* (X_α) *the torsion T of D has following components*

$$(1.9)\quad \begin{aligned} &T^i{}_{jk} = T^i{}_{jk},\ T^i{}_{ja} = \tilde{C}^i{}_{ja},\ T^i{}_{aj} = -\tilde{C}^i{}_{ja},\ T^i{}_{ab} = 0, \\ &T^a{}_{ij} = R^a{}_{ij},\ T^a{}_{ib} = -P^a{}_{bi},\ T^a{}_{bi} = P^a{}_{bi},\ T^a{}_{bc} = S^a{}_{bc}. \end{aligned}$$

Proposition 1.4. *The curvature* **R** *of D has in the frame* (X_α) *the components:*

$$(1.10)\quad \begin{aligned} &R_j{}^i{}_{kl} = R_j{}^i{}_{kl},\ R_b{}^i{}_{kl} = 0,\ R_j{}^a{}_{kl} = 0,\ R_b{}^a{}_{kl} = \tilde{R}_b{}^a{}_{kl}, \\ &R_j{}^i{}_{kd} = P_j{}^i{}_{kd},\ R_b{}^i{}_{kd} = 0,\ R_j{}^a{}_{kd} = 0,\ R_b{}^a{}_{kd} = P_b{}^a{}_{kd}, \\ &R_j{}^i{}_{dk} = -P_j{}^i{}_{kd},\ R_b{}^i{}_{dk} = 0,\ R_j{}^a{}_{dk} = 0,\ R_b{}^a{}_{dk} = -P_b{}^a{}_{kd}, \\ &R_j{}^i{}_{cd} = S_j{}^i{}_{cd},\ R_b{}^i{}_{cd} = 0,\ R_j{}^a{}_{cd} = 0,\ R_b{}^a{}_{cd} = S_b{}^a{}_{cd}. \end{aligned}$$

Proposition 1.5. *The components of the Ricci tensor Ric(D) in the frame*(X_α) *are as follows:*

$$(1.11)\quad \begin{aligned} &Ric(D)(\frac{\delta}{\delta x^i}, \frac{\delta}{\delta x^j}) = R_i{}^k{}_{jk} = R_{ij},\ Ric(D)(\frac{\delta}{\delta x^i}, \frac{\partial}{\partial y^a}) = -\overset{2}{P}_{ia} \\ &Ric(D)(\frac{\partial}{\partial y^a}, \frac{\delta}{\delta x^i}) = \overset{1}{P}_{ai},\ Ric(D)(\frac{\partial}{\partial y^a}, \frac{\partial}{\partial y^b}) = S_a{}^c{}_{bc} = S_{ab}, \end{aligned}$$

where we have set $\overset{1}{P}_{ai} = \tilde{P}_a{}^b{}_{ib}$, $\overset{2}{P}_{ia} = P_i{}^k{}_{ka}$.

Proposition 1.6. *The scalar curvature Sc(D) is given by*

$$(1.12)\quad Sc(D) = R + S,$$

where $R = g^{ij} R_{ij}$, $S = h^{ab} S_{ab}$.

The equation (1.1) written in the frame (X_α) looks as follows

$$(1.13)\quad \mathrm{Ric}(D)(X_\alpha, X_\beta) - \frac{1}{2}\mathrm{Sc}(D)\, G(X_\alpha, X_\beta) = \kappa \mathfrak{I}_{\alpha\beta}.$$

By the previous propositions we get

Theorem 1.1. *The Einstein equations associated to the metric d-connection D are as follows*

$$(1.14)\quad \begin{aligned} & R_{ij} - \frac{1}{2}(R+S) g_{ij} = \kappa \mathfrak{I}_{ij}, \quad \overset{1}{P}_{ai} = \kappa \overset{1}{\mathfrak{I}}_{ai}, \\ & S_{ab} - \frac{1}{2}(S+R) h_{ab} = \kappa \tilde{\mathfrak{I}}_{ab}, \quad \overset{2}{P}_{ia} = -\kappa \overset{2}{\mathfrak{I}}_{ia}, \end{aligned}$$

where $\mathfrak{I}_{ij}$, $\tilde{\mathfrak{I}}_{ab}$, $\overset{1}{\mathfrak{I}}_{ai}$, $,\overset{2}{\mathfrak{I}}_{ia}$ *are the components in the adapted frame of the energy-momentum tensor field.*

The system (1.14) has $(n+m)^2$ equations while the unknown functions N_i^a, g_{ij}, h_{ab}, $T^i{}_{jk}$, $S^a{}_{bc}$ are in number of

$$nm + n(n+1)/2 + m(m+1)/2 + n^2(n-1)/2 + m^2(m-1)/2.$$

This number is greater then the number of the equations if $n \geq 3$ and $m \geq 1$.

If we take into account the form (1.13) of the equation (1.1) we deduce the following form of the conservation law

$$(1.15)\quad D_{X_\alpha}\left(\mathrm{Ric}(D)^\alpha_\beta - \frac{1}{2}\mathrm{Sc}(D)\delta^\alpha_\beta\right) = 0,$$

where δ^α_β is the Kronecker symbol and $\mathrm{Ric}(D)^\alpha_\beta = G^{\alpha\gamma}\mathrm{Ric}(D)(X_\gamma, X_\beta)$.

If we separate (1.15) on components with respect to X_i and X_a we get

Theorem 1.2. *In the adapted frame* (X_α) *the conservation law takes the form*

$$(1.16)\qquad \begin{aligned} &[R^i_j - \frac{1}{2}(R+S)\delta^i_j]_{|i} + \overset{1}{P}{}^{a}{}_{i}\,|_a = 0,\\ &[S^a_b - \frac{1}{2}(R+S)\delta^a_b]\,|_a - \overset{2}{P}{}^{i}{}_{b|i} = 0,\end{aligned}$$

where

$$\overset{1}{P}{}^{a}{}_{j} = h^{ab}\overset{1}{P}_{bj},\ \overset{2}{P}{}^{i}{}_{b} = g^{ij}\overset{2}{P}_{jb},\ R^i_j = g^{ik}R_{kj},\ S^a_b = h^{ac}S_{cb}.$$

Of course, the conservation law (1.15) implies the vanishing of the divergence $D_{X_\alpha}\mathfrak{T}^\alpha_\beta$ of the energy-momentum tensor. Thus we have

Theorem 1.3. *In the frame* (X_α) *the divergence of the energy-momentum tensor is as follows*

$$(1.17)\quad \mathfrak{T}^i_{j|i} + \overset{1}{\mathfrak{T}}{}^{a}{}_{j}|_a = \frac{1}{\kappa}U_i = 0,\ \ \overset{2}{\mathfrak{T}}{}^{i}{}_{a|i} + \tilde{\mathfrak{T}}^{a}{}_{b}|_a = \frac{1}{\kappa}U_a = 0,$$

where

$$U_\alpha = \frac{1}{2}(G^{\beta\delta}R_{\delta}{}^{\gamma}{}_{\varphi\beta}T^{\varphi}{}_{\alpha\gamma} - G^{\beta\delta}R_{\delta}{}^{\gamma}{}_{\varphi\alpha}T^{\varphi}{}_{\varphi\gamma} + R^{\beta}_{\varphi}T^{\varphi}{}_{\beta\alpha}).$$

It is clear that if D is without torsions, i.e. D coincides to the Levi-Civita connection of G, then the d-covector U_α vanishes, hence the conservation laws are identically verified. In general, the conservation laws (1.16) do not satisfy $U_\alpha = 0$. We shall see later several important particular cases when $U_\alpha = 0$ holds good.

§2. Einstein Equations in the Case m=1

Now we shall discuss the case when the dimension of fibres of the vector bundle $\xi = (E,p,M)$ is one.

In this case the local coordinates on E are $(x^i, y^1), i=1,2,\ldots,n$. It is convenient to denote y^1 by x^0. The transformations $(x^i, x^0) \to (x^{i'}, x^{0'})$ of the local coordinates on the total space E are of the form:

$$(2.1)\quad \begin{aligned} & x^{i'} = x^{i'}(x^1,\dots,x^n),\ \operatorname{rank}\left(\frac{\partial x^{i'}}{\partial x^i}\right) = n\,, \\ & x^{0'} = f(x^1,\dots,x^n)x^0,\ f(x^1,\dots,x^n) \neq 0\,. \end{aligned}$$

These formulae show us that E may be thought of as a general framework for a projective theory of physical fields (cf. M.A. Tonnelat, [157], p.233).

A non-linear connection in ξ is defined by a set of functions$(N_k(x^1,\dots,x^n,x^0))$ given on every domain of local chart on E such that under a change of coordinates (2.1) the vector fields $\frac{\delta}{\delta x^k} = \frac{\partial}{\partial x^k} - N_k\frac{\partial}{\partial x^0}$ transform as follows: $\frac{\delta}{\delta x^k} = \frac{\partial x^{k'}}{\partial x^k}\frac{\delta}{\delta x^{k'}}$. A d-connection D takes in this case, with respect to the adapted frame $(\frac{\delta}{\delta x^i},\frac{\partial}{\partial x^0})$ the form

$$(2.2)\quad \begin{aligned} & D_{\frac{\delta}{\delta x^k}}\frac{\delta}{\delta x^j} = L^i_{jk}\frac{\delta}{\delta x^i},\quad D_{\frac{\delta}{\delta x^k}}\frac{\partial}{\partial x^0} = L_k\frac{\partial}{\partial x^0}, \\ & D_{\frac{\partial}{\partial x^0}}\frac{\delta}{\delta x^j} = C^i_j\frac{\delta}{\delta x^i},\quad D_{\frac{\partial}{\partial x^0}}\frac{\partial}{\partial x^0} = C\frac{\partial}{\partial x^0}\,. \end{aligned}$$

Thus D is completely determined by its local coefficients $D\Gamma = (L^i_{jk}, L_k, C^i_j, C)$. The torsions of $D\Gamma$ are as follows

$$(2.3)\quad T^i{}_{jk} = L^i_{jk} - L^i_{kj},\ \tilde{R}_{kj} = \frac{\delta N_k}{\delta x^j} - \frac{\delta N_j}{\delta x^k},\ \tilde{P}_k = \frac{\partial N_k}{\partial x^0} - L_k,\ C\,.$$

The d-connection $D\Gamma$ has the following four curvatures

$$(2.4)\quad \begin{aligned} & R_j{}^i{}_{kh} = \frac{\delta L^i_{jk}}{\delta x^k} - \frac{\delta L^i_{jh}}{\delta x^k} + L^l_{jk}L^i_{lh} - L^l_{jh}L^i_{lk} + C^i_j\tilde{R}_{kh}, \\ & R_{kh} = \frac{\delta L_k}{\delta x^h} - \frac{\delta L_h}{\delta x^k} + C\tilde{R}_{kh}, \\ & P_j{}^i{}_{k0} = \frac{\partial L^i_{jk}}{\partial x^0} - \frac{\partial C^i_j}{\partial x^k} + \frac{\partial (N_k C^i_j)}{\partial x^0} + C^i_h L^h_{jk} - C^h_j L^i_{hk}, \\ & P_{0k} = \frac{\partial L_k}{\partial x^0} - \frac{\partial C}{\partial x^k} + \frac{\partial (CN_k)}{\partial x^0}. \end{aligned}$$

Now we consider on E the metric structure given by

$$G = g_{ij}(x,x^0)dx^i \otimes dx^j + g_{00}(x,x^0)\delta x^0 \otimes \delta x^0,$$

where $\delta x^0 = dx^0 + N_k dx^k$, $\mathrm{rank}(g_{ij}) = n$ and $g_{00}(x, x^0) \neq 0$.

By the general theory presented in the above it follows there exists a metrical d-connection on E with h(hh)-torsion prescribed. This has, in the adapted frame $(\frac{\delta}{\delta x^i}, \frac{\partial}{\partial x^0})$ the local coefficients

$$\text{(2.5)}\quad \begin{aligned} L^i_{jk} &= \frac{1}{2} g^{ih}\left(\frac{\delta g_{hk}}{\delta x^j} + \frac{\delta g_{hj}}{\delta x^k} - \frac{\delta g_{jk}}{\delta x^h} + g_{hr} T^r{}_{jk} - g_{jr} T^r{}_{hk} + g_{kr} T^r{}_{jk}\right), \\ L_k &= \frac{1}{2} g_{00}^{-1} \frac{\delta g_{00}}{\delta x^0}, \quad C^i_j = \frac{1}{2} g^{ik} \frac{\partial g_{kj}}{\partial x^0}, \quad C = \frac{1}{2} g_{00}^{-1} \frac{\partial g_{00}}{\partial x^0}. \end{aligned}$$

Next we have

Theorem 2.1. *The Einstein equations of the metrical d-connection (2.5) on the total space of a vector bundle whose fibres are 1-dimensional are as follows*

$$\text{(2.6)}\quad \begin{aligned} &R_{ij} - \frac{1}{2} R g_{ij} = \kappa \mathfrak{T}_{ij}, \\ &\overset{1}{P}_{0i} = \kappa \mathfrak{T}_{0i}, \quad \overset{2}{P}_i = -\kappa \overset{2}{\mathfrak{T}}_{i0}, \quad R g_{00} = -2 \tilde{\mathfrak{T}}_{00}, \end{aligned}$$

where $\overset{2}{P}_i = P_i{}^k{}_{k0}$ *and* $\mathfrak{T}_{ij}, \overset{1}{\mathfrak{T}}_{0i}, \overset{2}{\mathfrak{T}}_{i0}, \tilde{\mathfrak{T}}_{00}$ *are the components of the energy-momentum tensor field.*

We note that in vacuum, that is, if the energy-momentum tensor field vanishes, these equations reduce to

$$\text{(2.7)}\quad R_{ij} = 0, \quad P_i{}^k{}_{k0} = 0, \quad P_{0k} = 0.$$

Taking into account (2.6), by a direct calculation we find

Theorem 2.2. *The conservation laws are given by*

$$\text{(2.8)}\quad \begin{aligned} &\left(R^i_j - \frac{1}{2} R \delta^i_j\right)_{|i} + \left(g_{00}^{-1} P_{0j}\right)|_0 = 0, \\ &\frac{1}{2}\frac{\partial R}{\partial x^0} + g^{ij} \overset{2}{P}_{j|i} = 0. \end{aligned}$$

Finally, we remark that if $(M, g_{ij}(x))$, $i,j = \overline{1,4}$ is a Lorentz space and we consider the metric structure

$$G = g_{ij}(x)dx^i \otimes dx^j + g_{00}(x^1,\dots,x^4,x^0)dx^0 \otimes dx^0$$

on the total space of a vector bundle over M, whose fibres are 1-dimensional, the first equation (2.6) coincides with the Einstein equations for the space $(M, g_{ij}(x))$, if we assume $T^i_{jk} = 0$ in $(2.5)_1$ since in this case $N_k = 0$ and $C^i_j = 0$.

§3. Another Form of the Einstein Equations

An essential feature of the Einstein equations (1.14) is the mixing of the R_{ij}-component and S_{ab}-component which is explicitly represented by the presence of S in $(1.14)_1$ and of R in $(1.14)_2$. Although various physical meanings of this feature may be found (cf. S. Ikeda [67], [68]) we shall give a new form of the Einstein equations in which this mixing is removed. This new form leads to simpler conservation laws that can be verified more easily. Also this new form is more feasible for certain applications.

We preserve the framework from the first section and we prove

Theorem 3.1. *For $n>2$, $m>2$ the Einstein equations (1.14) are equivalent to the following equations*

$$(3.1)\quad \begin{aligned} &R_{ij} - \frac{1}{2}Rg_{ij} = \kappa \overset{H}{\mathfrak{T}}_{ij}, \quad \overset{1}{P}_{ai} = \overset{1}{\mathfrak{T}}_{ai},\\ &S_{ab} - \frac{1}{2}Sh_{ab} = \kappa \overset{V}{\mathfrak{T}}_{ab}, \quad \overset{2}{P}_{ia} = -\kappa \overset{2}{\mathfrak{T}}_{ia}. \end{aligned}$$

Proof. First, we shall show that the equations (1.14) imply (3.1). Multiplying the first equation (1.14) by g^{ij} and the second by h^{ab} and putting $\mathfrak{T} = g^{ij}\mathfrak{T}_{ij}$, $\tilde{\mathfrak{T}} = h^{ab}\tilde{\mathfrak{T}}_{ab}$ one obtains a system of equations in R and S which can be solved to give

$$R = \frac{\kappa}{2-m-n}[(2-m)\mathfrak{T} + n\tilde{\mathfrak{T}}], \quad S = \frac{\kappa}{2-m-n}[m\mathfrak{T} + (2-n)\tilde{\mathfrak{T}}].$$

Now introducing

$$\begin{aligned} \overset{H}{\mathfrak{T}}_{ij} &= \mathfrak{T}_{ij} + \frac{1}{2(2-m-n)}[m\mathfrak{T} + (2-n)\tilde{\mathfrak{T}}]g_{ij},\\ \overset{V}{\mathfrak{T}}_{ab} &= \tilde{\mathfrak{T}}_{ab} + \frac{1}{2(2-m-n)}[(2-m)\mathfrak{T} + n\tilde{\mathfrak{T}}]h_{ab} \end{aligned}$$

one obtains the system (3.1).

Conversely, multiplying the first equation (3.1) by g^{ij} and the second by h^{ab} one obtains

$$R = \frac{2\kappa}{2-n}\overset{H}{\mathfrak{T}}, \quad S = \frac{2\kappa}{2-m}\overset{V}{\mathfrak{T}},$$

where $\overset{H}{\mathfrak{T}} = g^{ij}\overset{H}{\mathfrak{T}}_{ij}$, $\overset{V}{\mathfrak{T}} = h^{ab}\overset{V}{\mathfrak{T}}_{ab}$.

From these, one easily deduces (1.14).

From now we assume $n>2$, $m \geq 1, m \neq 2$.

In order to derive the conservation laws corresponding to the new form (3.1) of the Einstein equations we shall use the following Bianchi identities for the canonical metrical d-connection D:

$$(3.2) \quad \sum_{(bcd)} S_{b\ cd}^{\ a} = 0,$$

$$(3.3) \quad \sum_{(bcd)} S_{e\ bc}^{\ a}|_d = 0,$$

$$(3.4) \quad \sum_{(j,k,s)} (R_{j\ ks}^{\ h} - \tilde{C}_{ja}^{h} R^{a}_{\ ks}) = 0,$$

$$(3.5) \quad \sum_{(j,k,s)} (R_{i\ jk|s}^{\ h} - \tilde{P}_{i\ ja}^{\ h} R^{a}_{\ ks}) = 0,$$

where $\sum$ denotes the cyclic sum over indicated indices.

Now we shall compute the divergence of the Einstein tensor fields

$$\overset{V}{E}_{ab} = S_{ab} - \frac{1}{2} S h_{ab}, \quad \overset{H}{E}_{ij} = R_{ij} - \frac{1}{2} R g_{ij}.$$

First we notice that $S_{abcd} + S_{bacd} = 0$ and this implies $S_{a\ bc}^{\ a} = 0$.

Contracting $a=b$ in (3.2), the symmetry of the Ricci tensor follows. Next contracting $a=c$ in (3.3) one gets

$$S_{ed}|_b - S_{eb}|_d + S_{e\ bd}^{\ a}|_a = 0.$$

Rising an index one obtains

$$S^{e}{}_{d}|_{b}-S^{e}{}_{b}|_{d}+S^{ea}{}_{bd}|_{a}=0.$$

Now contracting e and b and noticing that $S^{a}_{a}=S$, we have

$$S^{b}{}_{d}|_{b}-S|_{d}+S^{ea}{}_{ed}|_{a}=0$$

or

(3.6) $\overset{V}{E}{}^{b}{}_{d}|_{b}=0.$

For h=s the Bianchi identity (3.4) leads to the following form of the skew symmetric part of the h-Ricci tensor field R_{ij}:

(3.7) $R_{jk}-R_{kj}=\sum_{(jkh)}\tilde{C}^{h}_{ja}R^{a}{}_{kh}.$

Contracting h=k in (3.5) and rising the first index one gets

$$R^{i}_{j|k}-R^{i}_{k|j}+R^{ih}{}_{kj|h}=-(\tilde{P}^{ih}{}_{ja}R^{a}{}_{hk}+\overset{2}{P}{}^{i}_{a}R^{a}{}_{kj}+\tilde{P}^{ih}{}_{ka}R^{a}{}_{jk}).$$

Noticing that $R^{i}_{i}=R$ one obtains for i=k

$$R^{i}_{j|i}-R_{|j}+R^{i}_{j|i}=-(\tilde{P}^{ih}{}_{ja}R^{a}{}_{hi}+\overset{2}{P}{}^{i}_{a}R^{a}{}_{ij}+\overset{2}{P}{}^{i}_{a}R^{a}{}_{ji}).$$

or

(3.8) $\overset{H}{E}{}^{i}_{j|i}=-\frac{1}{2}(\tilde{P}^{ih}{}_{ja}R^{a}{}_{hi}+2\overset{2}{P}{}^{i}_{a}R^{a}{}_{ij}).$

From (3.6) and (3.8) we get

Theorem 3.2. *The components* $\overset{H}{\mathfrak{T}}_{ij}$ *and* $\overset{V}{\mathfrak{T}}_{ab}$ *of the energy-momentum tensor field must satisfy the following equations*

$$(3.9)\quad \overset{H}{\mathfrak{I}}{}^{i}_{j|i} = -\frac{1}{2\kappa}(\tilde{P}^{ih}{}_{ja}R^{a}{}_{hi} + 2R^{a}{}_{ij}\overset{2}{P}{}^{i}_{a}), \quad \overset{V}{\mathfrak{I}}{}^{a}_{b}|_{a} = 0.$$

Theorem 3.3. *In each of the following cases:*
a) The horizontal distribution N is integrable,
b) The curvature $\tilde{P}^{i}_{j\,ka}$ vanishes,
the equations (3.9) take the form

$$(3.10)\quad \overset{H}{\mathfrak{I}}{}^{i}_{j|i} = o, \quad \overset{V}{\mathfrak{I}}{}^{a}_{b}|_{a} = 0.$$

The equations (3.9) may be thought of as partial conservation laws. In fact, if we take into account the conservation laws (1.17) we have

Theorem 3.4. *The conservation laws associated to the new form of the Einstein equations (3.1) are as follows:*

$$(3.11)\quad \begin{aligned} &\overset{H}{\mathfrak{I}}{}^{i}_{j|i} + \frac{1}{m-2}\overset{V}{S}{}_{|j} + \overset{1}{\mathfrak{I}}{}^{a}_{j}|_{a} = 0, \\ &\overset{V}{\mathfrak{I}}{}^{a}_{b}|_{a} + \frac{1}{n-2}\overset{V}{R}|_{b} + \overset{2}{\mathfrak{I}}{}^{i}_{a|i} = 0. \end{aligned}$$

Proof. By (1.14) and (3.1) we have $\overset{H}{\mathfrak{I}}_{ij} - \frac{1}{2\kappa}Sg_{ij} = \mathfrak{I}_{ij}$ from which one easily deduces the first equation (3.11). The second equation (3.11) is analogously proved.

§4. Einstein Equations for Some Particular Metrics on E

Let us recapitulate the ingredients which were used for deriving the Einstein equations. First we have the metric tensor field G by means of which a non-linear connection is derived. Then in the formulae which define the metrical d-connection D, the tensor fields $T^{i}{}_{jk}$ and $S^{a}{}_{bc}$ are assumed to be given a priori. Therefore, the known or given degrees of freedom are in number

$$\frac{(n+m)(n+m+1)}{2} + \frac{n^2(n-1)}{2} + \frac{m^2(m-1)}{2} = \frac{1}{2}(m^3+n^3+m+n+2mn).$$

On the other hand, the necessary degrees of freedom to solve (1.14) or (3.1) are in number $(m+n)^2$. When $n=4$, the latter is less then the former if $m \geq 1$. So that we can solve (1.14) or (3.1) in principle.

When $T^{i}{}_{jk}=0$ and $S^{a}{}_{bc}=0$ the given degrees of freedom $\frac{1}{2}(n+m)(n+m+1)$ is less then $(m+n)^2$, hence this case is improper. If the torsion $T^{i}{}_{jk}$ ($S^{a}{}_{bc}$) is only neglected the known degrees of freedom become more then $(n+m)^2$ when $n=4$ only if $m \geq 5$. These cases are improper for unitary theories. Therefore, the appearance of both torsions $T^{i}{}_{jk}$ and $S^{a}{}_{bc}$ is supported.

Let us notice that the given degrees of freedom remain unchanged in number if instead to give the components of G with respect to natural frame we give a priori a non-linear connection on E of local coefficients (N_i^a) and then we give G in the form

$$(4.1)\quad G = g_{ij}(x,y)dx^i \otimes dx^j + h_{ab}(x,y)\delta y^a \otimes \delta y^b,$$

where as usual $(dx^i, \delta y^a)$ $(\delta y^a = dy^a + N_i^a dx^i)$ is the dual frame to the adapted frame $(\frac{\delta}{\delta x^i}, \frac{\partial}{\partial y^a})$. The metrical d-connection D remains unchanged.

This change of point of view on G allows us to consider easier some particular metric tensor fields on E.

Definition 4.1. *We say that the metric tensor field G given by (4.1) is h-Riemannian (v-Riemannian) if the functions $g_{ij}(x,y)$ ($h_{ab}(x,y)$) do not depend on y.*

A glance at the canonical metrical d-connection $D\overset{c}{\Gamma}$ gives us

Proposition 4.1. *The metric tensor field G is h-Riemannian if and only if $\overset{c}{C}{}^{i}_{ja}=0$.*

Proposition 4.2. *The metric tensor field G is v-Riemannian if and only if $\overset{c}{C}{}^{a}_{bd}=0$.*

These propositions imply

Proposition 4.3. *The metric tensor field G is h- and v-Riemannian if and only if $\overset{c}{C}{}^{i}_{ja}=0$, $\overset{c}{C}{}^{a}_{bd}=0$.*

An example of h- and v-Riemannian metric tensor field on E is obtained as follows.

If the base manifold M of the vector bundle $\xi=(E,p,M)$ is paracompact, then there exists a Riemannian metric on M as well as a Riemannian metric in the vector bundle (E,p,M). The former can be given as $g_{ij}dx^i \otimes dx^j$ while the latter is nothing but a smooth assignment to each point $x \in M$ of an inner product in the fibre $E_x = p^{-1}(x)$. Its local components in a natural frame in E_x are of the form $(h_{ab}(x))$. Thus these two Riemannian metrics provide by means of (4.1) an h- and v-Riemannian metric tensor field on E.

Let us consider an h- and v-Riemannian metric G on E and let D be the metrical d-connection derived from it having h(hh)-torsion $T^i{}_{jk}$ and v(vv)-torsion $S^a{}_{bc}$ given a priori. Suppose these torsions depend on x only. Then the Einstein equations (3.1) take the form

$$
(4.2)\quad \begin{aligned}
&R_{ij}(x) - \frac{1}{2}Rg_{ij}(x) = \kappa \overset{H}{\mathfrak{T}}_{ij}(x), \quad \overset{1}{P}_{ai}(x,y) = \kappa \overset{1}{\mathfrak{T}}_{ai}(x,y),\\
&S_{ab}(x) - \frac{1}{2}Sh_{ab}(x) = \kappa \overset{V}{\mathfrak{T}}_{ab}(x), \quad \overset{2}{P}_{ia} = \kappa \overset{2}{\mathfrak{T}}_{ia} = 0.
\end{aligned}
$$

For the case when the Einstein d-tensor fields $R_{ij} - \frac{1}{2}Rg_{ij}$ and $S_{ab} - \frac{1}{2}Sh_{ab}$ from (4.2) are divergence less, i.e. they are conserved we refer to O. Amici and B. Casciaro [6]. Although these are theoretically interesting we do not insist on them since S. Ikeda in [68] asserts that the cases when G is h- and v-Riemannian "seems to have no essential physical meaning". In the same paper S. Ikeda finds as much more significant from the point of view of physicists the case when G is h-Riemannian and v-locally Minkowski.

Definition 4.2. *We shall say that the metric G given by (4.1) is v-locally Minkowski if for every point $u \in E$ there exists a local chart around it on E such that on its domain h_{ab} depends on y only.*

Thus if G is h-Riemannian and v-locally Minkowski (see R. Miron, S. Watanabe and S. Ikeda, [115]) for every point $u \in E$ there exists a local chart on E around it such that G has the form

$$
(4.3)\quad G = g_{ij}(x)dx^i \otimes dx^j + h_{ab}(y)\delta y^a \otimes \delta y^b.
$$

In this case if, moreover, we assume that $T^i{}_{jk}$ depends on x only and $S^a{}_{bc}$ depends on y only, the Einstein equations take the following form

$$(4.4)\quad \begin{aligned} &R_{ij}(x) - \frac{1}{2}R(x)g_{ij}(x) = \kappa \overset{H}{\mathfrak{T}}_{ij}(x),\ \overset{1}{P}_{ai}(x,y) = \kappa \overset{1}{\mathfrak{T}}_{ai}(x,y),\\ &S_{ab}(y) - \frac{1}{2}S(y)h_{ab}(y) = \kappa \overset{V}{\mathfrak{T}}_{ab}(y),\ \overset{2}{P}_{ia}(x,y) = \kappa \overset{2}{\mathfrak{T}}_{ia}(x,y) = 0. \end{aligned}$$

The most interesting feature of this situation is a very weak interaction between (x)-field spanned by the points (x) and the internal (y)-field spanned by vectors (y).

Now we shall give an example of h-Riemannian and v-locally Minkowski metric due to M. Anastasiei [11].

Let $\xi=(E,p,M)$ be a vector bundle. Suppose that its fibre type, say V, is endowed with a norm $\|\ \|: V \to \mathbb{R}_+$, i.e. V is a Minkowski space. If $v=v^a e_a$, where (e_a) is a basis of V, we set $\|v\| = f(v^1,..,v^m) = f(v^a)$ and suppose that f is differentiable at least of class C^3 for $v \neq 0$. The set

$$\{T/T \in GL(m,\mathbb{R}),\ \|Tv\| = \|v\|,\ v \in V\}$$

is a Lie group. Let H be a subgroup of it.

Definition 4.3. *A vector bundles $\xi=(E,p,M)$ is said to be a $\{V,H\}$-bundle if there exists a bundle atlas $\{(U_\alpha,\varphi_\alpha,V)\}$ such that the mappings $\varphi_{\beta,x} \circ \varphi^{-1}_{\alpha,x}$ belong to H for every $x \in U_\alpha \cap U_\beta \neq \Phi$. We also say that the vector bundle ξ admits an H-structure.*

The local fibres of a $\{V,H\}$-bundle are isomorphic and isometric each to others.

Using the bundle atlas $(U_\alpha,\varphi_\alpha,V)$ it comes out that the norm f defines a function on E, say F, such that in any atlas on E the matrix

$$(4.5)\quad h_{ab}(x,y) = \left(\frac{1}{2}\frac{\partial^2 F^2}{\partial y^a \partial y^b}\right)$$

is non-singular and the quadratic form $h_{ab}\eta^a\eta^b$ $(\eta \in \mathbb{R}^m)$ is positive definite.

If we consider on E the atlas induced by the bundle atlas $(U_\alpha,\varphi_\alpha,V)$ then F and (h_{ab}) depend on y only.

Now let $g_{ij}(x)$ be the local coefficients of a Lorentz metric on the base manifold M. It is clear that G given by (4.1) with this $g_{ij}(x)$ and $h_{ab}(y)$ from (4.5) is an h-Riemannian and v-locally Minkowski metric on E.

Chapter VI

Generalized Einstein-Yang Mills Equations

The methods from the differential geometry of fibre bundles were used in the gauge fields theory for a long time. Usually, the relationships between the gauge fields theory and the theory of fibre bundles implied bundles whose fibres were some Lie groups. In the last years G.S. Asanov ([19], [20]) made various attempts to reach a higher step of gauge generality taking as fibres more general geometrical spaces. He also used a very general notion of gauge transformation. Since till now we have developed a geometrical theory of vector bundles we consider that is useful the reader be informed on such a general gauge theory at least at the level of vector bundles. Here are the main purposes of the chapter. First we treat a very general gauge transformation in §1. Some particular cases are pointed out. Then, in §2, a gauge covariant derivative is introduced. Metrical gauge d-connections are discussed in §3. A way to establish a set of generalized Einstein-Yang Mills equations is pointed out in §4.

§1. Gauge Transformations

Let $\xi = (E,p,M)$ be a vector bundle of rank m i.e. the dimension of its type fibre is m. Recall that E has a structure of n + m - dimensional manifold.

Definition 1.1. *A gauge transformation of the vector bundle ξ is a pair (f,f_0) of difeomorfisms $f: E\to E$ and $f_0: M\to M$ such that the following diagram is commutative.*

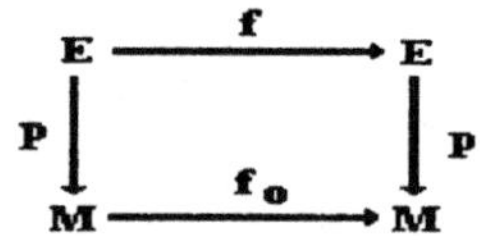

Let (U,ψ) be a local chart on the base manifold M and $(p^{-1}(U),\varphi,\mathbb{R}^m)$ a vectorial chart associated to it. We assume, restricting U if necessary, that $f_0(U)\subset U$. If $x^1,\dots,x^n$ are local coordinates on U, then the mapping $\psi\circ f_0\circ\psi^{-1}$ has the form $\tilde{x}^i = X^i(x^1,\dots,x^n)$ with rank $\left(\dfrac{\partial X^i}{\partial x^k}\right) = n$.

The local chart on E induced by (U,ψ) and $(p^{-1}(U),\varphi,\mathbb{R}^m)$ is $(p^{-1}(U),h)$, where

$h: p^{-1}(U) \to \mathbf{R}^n \times \mathbf{R}^m$, $h(u) = (\psi(p(u)), \varphi_{p(u)}(u))$. The mapping $h \circ f \circ h^{-1}$ which locally represents f and maps $\psi(U) \times \mathbf{R}^m$ to $\psi(f_0(U)) \times \mathbf{R}^m$ is of the form $(x^i, y^a) \to (X^i(x^1, \dots, x^n), Y^a(x^1, \dots, x^n, y^1, \dots, y^n))$ because of the condition $p \circ f = f_0 \circ p$.

Thus a gauge transformation of vector bundle ξ may be given locally as follows

$$\tag{1.1} \begin{aligned} &\tilde{x}^i = X^i(x^1, \dots, x^n), \text{ rank } \left(\frac{\partial X^i}{\partial x^k}\right) = n, \\ &\tilde{y}^a = Y^a(x^1, \dots, x^n, y^1, \dots, y^m), \text{ rank } \left(\frac{\partial Y^a}{\partial y^b}\right) = m. \end{aligned}$$

Recall that a change of local coordinates $(x,y) \to (x', y')$ on E is of the form

$$\tag{1.2} \begin{aligned} &x'^i = x'^i(x^1, \dots, x^n), \text{ rank } \left(\frac{\partial x'^i}{\partial x^k}\right) = n, \\ &y'^i = M^a_b(x) y^b, \text{ rank } (M^a_b(x)) = m. \end{aligned}$$

Notice a small change of notations in (1.2) if one compares to (3.3) in Ch.I. This will help us in this chapter in which almost all considerations will be presented in local coordinates.

Since f and f_0 are globally defined the following compatibility conditions of (1.1) and (1.2) must be satisfied.

$$\tag{1.3} \begin{aligned} &\tilde{x}'^i(X^k(x^1, \dots, x^n)) = X'^i(x'^k(x^1, \dots, x^n)), \\ &M^a_b(X^k(x^1, \dots, x^n)) Y^b(x,y) = Y'^a(x'^k(x), M^b_c(x) y^c), \end{aligned}$$

where X'^i and Y'^a represent the gauge transformation (f, f_0) in a local chart $(p^{-1}(U'), (x'^i, y'^a))$ with $U \cap U' \neq \emptyset$.

We shall denote $X^i_k = \dfrac{\partial X^i}{\partial x^k}$, $Y^a_b = \dfrac{\partial Y^a}{\partial y^b}$ and by $\bar{X}^i_k$ and $\bar{Y}^a_b$ the inverses of these matrices, respectively.

Let $\left(\dfrac{\partial}{\partial x^i}, \dfrac{\partial}{\partial y^a}\right)$ be the natural basis of the tangent space $T_u E$ $(u \in E)$ with respect to the local coordinates (x^i, y^a). These local vector fields transform, under (1.2) as follows

$$(1.4)\quad \begin{aligned} &\frac{\partial}{\partial x^i} = \frac{\partial x'^k}{\partial x^i}\frac{\partial}{\partial x'^k} + \frac{\partial M^a_b}{\partial x^i} y^b \frac{\partial}{\partial y'^a}, \\ &\frac{\partial}{\partial y^a} = M^b_a(x)\frac{\partial}{\partial y'^b}. \end{aligned}$$

Then if we regard (1.1) as a change of coordinates we find the following formulae

$$(1.5)\quad \begin{aligned} &\frac{\partial}{\partial x^i} = X^k_i \frac{\partial}{\partial \tilde{x}^k} + \frac{\partial Y^a}{\partial x^i}\frac{\partial}{\partial \tilde{y}^a}, \\ &\frac{\partial}{\partial y^a} = Y^b_a(x,y)\frac{\partial}{\partial \tilde{y}^b}, \end{aligned}$$

which relate $(\frac{\partial}{\partial x^i}, \frac{\partial}{\partial y^a})$ to $\frac{\partial}{\partial \tilde{x}^i}$ and $\frac{\partial}{\partial \tilde{y}^a}$.

Now we shall come back for a moment to the so-called d-tensor fields on E introduced in Ch.III. They were defined there in connection with a non-linear connection on E. But it is interesting to notice they may be defined without any non-linear connection on E as a set of real functions $W^{\dots i\dots a\dots}_{\dots j\dots b\dots}$ on $p^{-1}(U)$ which under (1.2) transforms by (1.6) in Ch.III. In the notation used here, (1.6) from Ch.III reads as follows

$$(1.6)\quad W'^{\dots h\dots c\dots}_{\dots k\dots d\dots}(x', y') = \frac{\partial x'^h}{\partial x^i}\dots M^c_a \dots \frac{\partial x^j}{\partial x'^k}\dots \bar{M}^b_d \dots W^{\dots i\dots a\dots}_{\dots j\dots b\dots}(x,y),$$

where $(\frac{\partial x^k}{\partial x'^j})$ is the inverse of the matrix $(\frac{\partial x'^i}{\partial x^j})$ and $(\bar{M}^b_a)$ is the inverse of the matrix (M^a_b).

Definition 1.2. *A d-tensor field given by its local components $W^{\dots i\dots a\dots}_{\dots j\dots b\dots}$ is said to be a gauge d-tensor field if the tensorial law operates also under(1.1), that is,*

$$(1.7)\quad \tilde{W}^{\dots h\dots c\dots}_{\dots k\dots d\dots}(\tilde{x},\tilde{y}) = X^h_i \dots Y^c_a \dots \bar{X}^j_k \dots \bar{Y}^b_d \dots W^{\dots i\dots a}_{\dots j\dots b\dots}.$$

Notice that by (1.4) and (1.5), $\frac{\partial}{\partial y^a}$ are gauge d-vector fields, while $\frac{\partial}{\partial x^i}$ are not so. Also, if $L: E \to R$ is a gauge scalar, that is, $L'(x', y') = L(x, y)$ and $\tilde{L}(\tilde{x},\tilde{y}) = L(x,y)$ then $\frac{\partial L}{\partial y^a}$ is a gauge d-tensor field of type (0,1). Deriving again with respect to y^a we obtain a d-tensor field of type (0,2) only, not a gauge d-tensor field. Thus we need a gauge covariant derivative. We shall look for it in the next section.

Now we note some *particular gauge transformations:*

$$\tilde{x}^i = x^i, \quad \tilde{y}^a = Y^a(x,y), \ \text{rank}\left(\frac{\partial Y^a}{\partial y^b}\right) = m. \tag{1.8}$$

Under the gauge transformations (1.8) the points of the base manifold M are not shifted. In this case $X^i_j = \delta^i_j$ and by $(1.3)_2$ the entities $Y^a(x,y)$ define a d-vector field.

$$\tilde{x}^i = X^i(x), \quad \tilde{y}^a = Y^a_b(x) y^b. \tag{1.9}$$

Such a gauge transformation is nothing but an automorphism of the vector bundle ξ (the mapping f restricted to each fibre is linear).

Notice if the gauge transformation (1.9) is used then the derivative with respect to y^a preserves the gauge nature of a gauge field.

Of course, the following combination of (1.8) and (1.9) may be considered

$$\tilde{x}^i = x^i, \quad \tilde{y}^a = Y^a_b(x) y^b. \tag{1.10}$$

In this case by $(1.3)_2$, $Y^a_b(x,y)$ must be a d-tensor field of type (1.1).

Particularly, the following gauge transformation could be useful

$$\tilde{x}^i = X^i(x), \quad \tilde{y}^a = \varphi(x) \delta^a_b y^b, \tag{1.11}$$

where φ is a real mapping on M.

Homotheties of ξ are particular gauge transformations:

$$\tilde{x}^i = x^i, \quad \tilde{y}^a = \lambda y^a, \ \lambda \in \mathbf{R}, \ \lambda > 0. \tag{1.12}$$

Let us denote by Diff(M) and Diff(E) the group of diffeomorphisms of M and E, respectively. It is easy to see that the composition of two gauge transformations is again a gauge transformation as well as that the inverse of a gauge transformation is a gauge transformation.

Thus we have

Proposition 1.1. *The set of all gauge transformations has a structure of group with respect to the composition of mappings.*

The group of the gauge transformations will be denoted by **G**. It is obviously a subgroup of Diff(M)xDiff(E).

This group is very large. One may consider various subgroups of it which preserve certain geometrical objects in order to find particular gauge theories.

Remark 1.1. We defined the gauge transformations as global transformations. In many problems only local gauge transformations are used.

§2. Gauge Covariant Derivatives

Let us remember that in the presence of a non-linear connection in ξ we were able to define two algorithms called h- and v- covariant derivation, both transforming a d-tensor field of type (p,q) in a new d-tensor field of type (p,q+1). In other words, the algebra of d-tensor fields is closed with respect to these algorithms. Now we shall try to modify these algoritms such that they preserve closed the algebra of gauge d-tensor fields.

Let N be a non-linear connection in ξ, of the local coefficients (N_i^a). These coefficients transform under (1.2) as follows (cf. (2.4) in Ch.II)

$$(2.1)\quad N'^{a}_{i}(x',y')\frac{\partial x'^{i}}{\partial x^{k}} = M^{a}_{b}(x)N^{b}_{k}(x,y) - \frac{\partial M^{a'}_{a}(x)}{\partial x^{k}}y^{a}.$$

If we introduce the local vector fields

$$(2.2)\quad \frac{\delta}{\delta x^{i}} = \frac{\partial}{\partial x^{i}} - N^{a}_{i}(x,y)\frac{\partial}{\partial y^{a}},$$

by (1.4) and taking into account (2.1) we get

$$(2.3)\quad \frac{\delta}{\delta x^{i}} = \frac{\partial x'^{k}}{\partial x^{i}}\frac{\delta}{\delta x'^{k}}.$$

Thus $(\frac{\delta}{\delta x^{i}})$ are d- vector fields. Now it is easy to see that if L is a scalar then $\frac{\delta L}{\delta x^{i}}$ is a d-covector field. We wish $\frac{\delta L}{\delta x^{i}}$ to be also a gauge d- covector field, that is, we postulate

(2.4) $$\frac{\delta}{\delta x^i} = X_i^k \frac{\delta}{\delta \tilde{x}^k},$$

where as it is obvious, we have to set

(2.5) $$\frac{\delta}{\delta \tilde{x}^k} = \frac{\partial}{\partial \tilde{x}^k} - \tilde{N}_k^a(\tilde{x},\tilde{y})\frac{\partial}{\partial \tilde{y}^a}.$$

Here $\tilde{N}_i^a$ is just N_i^a in the point $(\tilde{x},\tilde{y})$.

By (1.5) it follows that (N_i^a) must satisfy beside (2.1) the following condition

(2.6) $$\tilde{N}_i^a(\tilde{x},\tilde{y})X_k^i(x) = Y_b^a(x,y)N_k^b(x,y) - \frac{\partial Y^a(x,y)}{\partial x^k}.$$

Definition 2.1. *A set of functions (N_i^a) satisfying (2.1) and (2.6) will be called a gauge non-linear connection in the vector bundle ξ or on E.*

As we have seen before the algorithms of h- and v- covariant derivation are based on a d-connection on E. Let $D\Gamma = (L_{jk}^i, K_{bk}^a, M_{jc}^i, C_{bc}^a)$ be a d-connection on E given by its local coefficients. These transform under (1.2) as follows (cf. (2.9)-(2.12) in Ch.III):

(2.7) $$\begin{aligned}
L'^i_{jk}(x',y') &= \frac{\partial x'^i}{\partial x^s}\frac{\partial x^r}{\partial x'^j}\frac{\partial x^h}{\partial x'^k}L^s_{rh}(x,y) - \frac{\partial x^r}{\partial x'^j}\frac{\partial x^s}{\partial x'^k}\frac{\partial^2 x'^i}{\partial x^r \partial x^s},\\
K'^a_{bk}(x',y') &= M^a_c\bar{M}^d_b\frac{\partial x^h}{\partial x'^k}K^c_{dh} - \bar{M}^c_b\frac{\partial x'^j}{\partial x^k}\frac{\partial M^a_c}{\partial x^j},\\
M'^i_{ja}(x',y') &= \frac{\partial x'^i}{\partial x^h}\frac{\partial x^k}{\partial x'^j}\bar{M}^b_a M^h_{kb},\\
C'^a_{bc}(x',y') &= M^a_d\bar{M}^e_b\bar{M}^f_c C^d_{ef}.
\end{aligned}$$

Notice that M_{ja}^i and C_{bc}^a transform like the components of a d-tensor field.

Now let W_{jb}^{ia} be a d-tensor field. Its h-covariant derivative is a new d-tensor field defined as follows (the rule is a general one)

(2.8) $$W^{ia}_{jb|k} = \frac{\delta W^{ia}_{jb}}{\delta x^k} + L^i_{hk}W^{ha}_{jb} + K^a_{dk}W^{id}_{jb} - L^h_{jk}W^{ia}_{hb} + K^d_{bk}W^{ia}_{jd},$$

while its v-covariant derivative is the d-tensor field (again the rule is a general one)

$$(2.9)\quad W^{ia}_{jb}\,|_d = \frac{\partial W^{ia}_{jb}}{\partial y^d} + M^{\ i}_{hd} W^{ha}_{jb} + C^{\ a}_{ed} W^{ie}_{jb} - M^{\ h}_{jd} W^{ia}_{hb} - C^{\ e}_{bd} W^{ia}_{je}\,.$$

Now let us suppose that W^{ia}_{jb} is a gauge d-tensor field. It is clear that in order to attain our purpose, that is, to find a gauge covariant derivative we must to preserve the form of $W^{ia}_{jb|k}$ and $W^{ia}_{jb}\,|_d$ and to postulate these are again gauge d-tensor fields.

Thus additional laws of transformation of the local coefficients $L^{\ i}_{jk}, K^{\ a}_{bi}, M^{\ i}_{jc}, C^{\ a}_{bc}$ will be implied. If we denote $\tilde{L}^{\ i}_{jk} = L^{\ i}_{jk}(\tilde{x},\tilde{y})$, $\tilde{K}^{\ a}_{bi} = K^{\ a}_{bi}(\tilde{x},\tilde{y})$ and proceed similarly for $M^{\ i}_{jc}$ and $C^{\ a}_{bc}$ these laws of transformation read

$$(2.10)\quad \begin{aligned} \tilde{L}^{\ i}_{jk} &= X^i_h \bar{X}^l_j \bar{X}^m_k L^{\ h}_{lm} - \bar{X}^m_j \bar{X}^l_k \frac{\partial X^i_l}{\partial x^m}\,, \\ \tilde{K}^{\ a}_{bk} &= Y^a_c \bar{Y}^d_b \bar{X}^j_k K^{\ c}_{dj} - \bar{Y}^d_b \bar{X}^j_k \frac{\delta Y^a_d}{\delta x^j}\,, \\ \tilde{M}^{\ i}_{ja} &= X^i_h \bar{X}^k_j \bar{Y}^b_a M^{\ j}_{kb}\,, \\ \tilde{C}^{\ a}_{eb} &= Y^a_c \bar{Y}^d_e \bar{Y}^f_b C^{\ c}_{df} - \bar{Y}^d_e \bar{Y}^c_b \frac{\partial Y^a_d}{\partial y^c}\,. \end{aligned}$$

Notice that in the second line appears $\dfrac{\delta}{\delta x^j}$ and that $C^{\ a}_{bc}$ has no tensorial nature under (1.1).

We remark that if the gauge transformation (1.9) is considered than (2.7) is very similar to (2.10).

Definition 2.2 *A set of real functions* $D\Gamma = (L^{\ i}_{jk}, K^{\ a}_{bk}, M^{\ i}_{ja}, C^{\ a}_{bc})$ *which satisfy (2.7) as well as (2.10) will be called a gauge d-connection.*

It is now clear that by using the local coefficients of a gauge d-connection the h- and v- covariant derivatives constructed as in the generic formulae (2.8) and (2.9) keep closed the algebra of gauge d-tensor fields.

In order to distinguish the gauge covariant derivatives from the usual ones we shall denote by $d_k W^{\ldots i \ldots a \ldots}_{\ldots j \ldots b \ldots}$ and $D_c W^{\ldots i \ldots a \ldots}_{\ldots j \ldots b \ldots}$ the gauge h- and v- covariant derivatives, respectively. Notice that D_a acts on gauge-scalars as $\dfrac{\partial}{\partial y^a}$ and d_k acts on gauge-scalars as $\dfrac{\delta}{\delta \tilde{x}^k}$ given by (2.5).

The torsions and curvatures of a gauge d-connection are all gauge d-tensor fields.

This follows directly from their forms given below as well as from the Ricci identities (3.23), (3.24) in Ch.III.

The torsions of a gauge d-connection are as follows (cf.(3.7) in Ch.III)

$$(2.11)\qquad \begin{aligned} & T^i{}_{jk} = L^i_{jk} - L^i_{kj}, \quad R^a{}_{jk} = \frac{\delta N^a_j}{\delta x^k} - \frac{\delta N^a_k}{\delta x^j}, \quad M^i_{jc}, \\ & P^a{}_{jb} = \frac{\partial N^a_j}{\partial y^k} - K^a_{bj}, \quad S^a{}_{bc} = C^a_{bc} - C^a_{cb}. \end{aligned}$$

The curvatures of a gauge d-connection are as in (3.13) from Ch.III with some obvious modifications of notations:

$$(2.12)\qquad \begin{aligned} R^{\ i}_{h\ jk} &= \frac{\delta L^i_{hj}}{\delta x^k} - \frac{\delta L^i_{hk}}{\delta x^j} + L^m_{hj} L^i_{mk} - L^m_{hk} L^i_{mj} + M^i_{ha} R^a{}_{jk}, \\ \tilde{R}^{\ a}_{b\ jk} &= \frac{\delta K^a_{bj}}{\delta x^k} - \frac{\delta K^a_{bk}}{\delta x^j} + K^c_{bj} K^a_{ck} - K^c_{bk} K^a_{cj} + C^a_{bc} R^c{}_{jk}, \\ \tilde{P}^{\ i}_{j\ ka} &= \frac{\partial L^i_{jk}}{\partial y^a} - d_k M^i_{ja} + M^i_{jb} P^b{}_{ka}, \\ P^{\ a}_{b\ kc} &= \frac{\partial K^a_{bk}}{\partial y^c} - \frac{\delta C^a_{bc}}{\delta x^k} - K^a_{ek} C^e_{bc} + K^e_{bk} C^a_{ec} + K^e_{ck} C^a_{be} + C^a_{bd} P^d{}_{kc}, \\ \tilde{S}^{\ i}_{j\ bc} &= \frac{\partial M^i_{jb}}{\partial y^c} - \frac{\partial M^i_{jc}}{\partial y^b} + M^h_{jb} M^i_{hc} - M^h_{jc} M^i_{hb}, \\ S^{\ a}_{b\ cd} &= \frac{\partial C^a_{bc}}{\partial y^d} - \frac{\partial C^a_{bd}}{\partial y^c} + C^e_{bc} C^a_{ed} - C^e_{bd} C^a_{ec}. \end{aligned}$$

Remark 2.1. If the gauge transformation (1.9) is considered it is easy to check that the Liouville vector field $y^a \frac{\partial}{\partial y^a}$ is a gauge d-vector field. It follows that the deflection tensors derived from it by using a gauge d-connection are gauge d-tensor fields.

§3. Metrical Gauge d-Connections

In order to obtain the concept of metrical gauge d-connection and to establish some of its properties we follow closely the ideas from §5 of Ch.III, especially those in which local coordinates are featured.

Let $g_{ij}(x,y)$ be a gauge d-tensor field symmetric, nondegenerate, that is, $\det(g_{ij}) \neq 0$ and such that the quadratic form $g_{ij}(x,y)\xi^i\xi^j$ has constant signature. We shall say $g_{ij}(x,y)$ is a *gauge h-metric* on E. A gauge d-tensor field $h_{ab}(x,y)$ having similar properties with $g_{ij}(x,y)$ will be called a *gauge v-metric* on E.

Let g^{hk} and h^{cd} be the reciprocals of the gauge d-tensor fields. We shall use g_{ij} and h_{ab} for raising and lowering indices of a gauge tensor field in accordance with the standard rules. It is clear that a tensor index can be raised or lowered only by its corresponding metric tensor: $T_{ijk} = g_{ih}T^{h}{}_{jk}$, $P_{ajb} = h_{ac}P^{c}{}_{jb}$, $S^{ab}_{cd} = h^{bc}S^{a}_{ecd}$, etc. Also it is obvious that the algebra of gauge d-tensor fields is closed under these operations.

Definition 3.1. *A pair* $(g_{ij}(x,y), h_{ab}(x,y))$, *where* g_{ij} *is a gauge h-metric and* h_{ab} *is a gauge v-metric is said to be a gauge (h,v)-metric on E.*

Now let us assume E is endowed with a gauge non-linear connection of local coefficients $(N_i^a(x,y))$.

Definition 3.2. *A gauge d-connection* $D\Gamma = (L^i_{jk}, K^a_{bk}, M^i_{jc}, C^a_{bc})$ *is said to be (h,v)-metrical or simple, metrical if*

(3.1) $d_k g_{ij} = 0$, $D_a g_{ij} = 0$, $d_k h_{ab} = 0$, $D_a h_{bc} = 0$.

For a reason to set up this definition see the Proposition 5.3 in Ch.III.

If $D\Gamma$ satisfies only the first two equations (3.1) we shall say it is h-metrical while it satisfies only the last two equations (3.1) we call it v-metrical.

It is not difficult to see that the Kawaguchi metrization process still work in the present framework. Thus if $D\dot{\Gamma} = (\dot{L}^i_{jk}, \dot{K}^a_{bj}, \dot{M}^i_{ja}, \dot{C}^a_{bc})$ is a gauge d-connection and $\dot{d}_k, \dot{D}_k$ denote the gauge h- and v-covariant derivatives with respect to it then the following gauge d-connection is a metrical one (see (5.12), Ch.III).

$$
\begin{aligned}
L^i_{jk} &= \dot{L}^i_{jk} + \frac{1}{2} g^{ih}\dot{d}_k g_{hj}, \quad K^a_{bk} = \dot{K}^a_{bk} + \frac{1}{2} h^{ac}\dot{d}_k h_{cb}, \\
M^i_{ja} &= \dot{M}^i_{ja} + \frac{1}{2} g^{ih}\dot{D}_a g_{hj}, \quad C^a_{bc} = \dot{C}^a_{bc} + \frac{1}{2} h^{ad}\dot{D}_c h_{db}.
\end{aligned} \tag{3.2}
$$

Now we proceed to construct a gauge d-connection $D\dot{\Gamma}$. Starting from it, by applying (3.2) we shall find a useful metrical gauge d-connection.

Let us set

$$\text{(3.3)}\quad \begin{aligned} \overset{c}{L}{}^{i}_{jk} &= \frac{1}{2} g^{ih}\left(\frac{\delta g_{hj}}{\delta x^{k}} + \frac{\delta g_{hk}}{\delta x^{j}} - \frac{\delta g_{jk}}{\delta x^{h}}\right), \\ \overset{c}{C}{}^{a}_{bc} &= \frac{1}{2} h^{ad}\left(\frac{\partial h_{db}}{\partial y^{c}} + \frac{\partial h_{dc}}{\partial y^{b}} - \frac{\partial h_{bc}}{\partial y^{d}}\right). \end{aligned}$$

Proposition 3.1. *The set of functions* $\overset{b}{D\Gamma} = (\overset{c}{L}{}^{i}_{jk}, \frac{\partial N^{a}_{k}}{\partial y^{b}}, 0, \overset{c}{C}{}^{a}_{bc})$ *defines a gauge d-connection with the properties*

$$\overset{b}{d}_{k} g_{ij} = 0, \quad \overset{b}{D}_{c} h_{ab} = 0, \quad \overset{c}{L}{}^{i}_{jk} = \overset{c}{L}{}^{i}_{kj}, \quad \overset{c}{C}{}^{a}_{bc} = \overset{c}{C}{}^{a}_{cb}.$$

Proof. As we know from §5 Ch.III, $\overset{b}{D\Gamma}$ is a d-connection of the Berwald type which satisfies $g_{ij|k} = 0$ and $h_{ab}|_{c} = 0$. It remains to check it is a gauge d-connection, that is, its coefficients satisfy (2.10). This is done by a tedious computation. We give an example of this checking for $\frac{\partial N^{a}_{k}}{\partial y^{b}}$. We have

$$\begin{aligned} \tilde{K}^{a}_{bj} &= \frac{\partial \tilde{N}^{a}_{j}}{\partial \tilde{y}^{b}} = \frac{\partial \tilde{N}^{a}_{j}}{\partial y^{c}} \bar{Y}^{c}_{b} = \left(\bar{X}^{k}_{j} \frac{\partial Y^{a}_{d}}{\partial y^{c}} N^{d}_{k} + \bar{X}^{k}_{j} Y^{a}_{d} \frac{\partial N^{d}_{k}}{\partial y^{c}} - \bar{X}^{k}_{j} \frac{\partial^{2} Y^{a}}{\partial x^{k} \partial y^{c}}\right) \bar{Y}^{c}_{b} = \\ &= Y^{a}_{d} \bar{Y}^{c}_{b} \bar{X}^{k}_{j} K^{d}_{ck} - \bar{Y}^{c}_{b} \bar{X}^{k}_{j} d_{k} Y^{a}_{c}. \end{aligned}$$

Theorem 3.1. *There exists on E a metrical d-connection depending on* N^{a}_{i}, g_{ij}, h_{ab} *only, whose h(hh)- and v(vv)-torsions vanish. Its local coefficients are* $\overset{c}{D\Gamma} = (\overset{c}{L}{}^{i}_{jk}, \overset{c}{K}{}^{a}_{bi}, \overset{c}{M}{}^{i}_{ja}, \overset{c}{C}{}^{a}_{bc})$, *where* $\overset{c}{L}{}^{i}_{jk}$ *and* $\overset{c}{C}{}^{a}_{bc}$ *are given by (3.3) and*

$$\text{(3.4)}\quad \begin{aligned} \overset{c}{K}{}^{a}_{bi} &= \frac{\partial N^{a}_{i}}{\partial y^{b}} + \frac{1}{2} h^{ac}\left(\frac{\delta h_{bc}}{\delta x^{i}} - \frac{\partial N^{d}_{i}}{\partial y^{b}} h_{dc} - \frac{\partial N^{d}_{i}}{\partial y^{c}} h_{db}\right), \\ \overset{c}{M}{}^{i}_{ja} &= \frac{1}{2} g^{ik} \frac{\partial g_{jk}}{\partial y^{a}}. \end{aligned}$$

Proof. The metrical gauge d-connection $\overset{c}{D\Gamma}$ is obtained from $\overset{b}{D\Gamma}$ by the Kawaguchi metrization process. The obvious symmetry of $\overset{c}{L}{}^{i}_{jk}$ *and* $\overset{c}{C}{}^{a}_{bc}$ with respect to the lower indices implies h(hh)- and v(vv)-torsions vanish.

It can be proved also the existence of a metrical gauge d-connection having h(hh)- and v(vv)-torsions prescribed. Their local coefficients are similar to those given in (5.22), Ch.III.

§4. Generalized Einstein-Yang Mills Equations

Let us assume the total space E of a vector bundle $\xi=(E,p,M)$ is endowed with a gauge non-linear connection, a gauge d-connection and a gauge (h,v)-metric (g_{ij}, h_{ab}). For moment there is not assumed any compatibility between the gauge (h,v)-metric and the gauge d-connection. Later we shall assume the gauge d-connection is metrical. Let us set

(4.1) $G=|\det(g_{ij})|^{1/2}|\det(h_{ab})|^{1/2}$.

Lowering and raising indices in an allowable and convenient manner and making contractions, one can obtain starting from gauge torsions and gauge curvatures of a gauge d-connection various gauge scalars:

(4.2) $L_1=T^i{}_{jk}T^{jk}_i,\ L_2=R^a{}_{jk}R^{jk}_a,\ L_3=M^i_{jc}M^{jc}_i,\ L_4=P^a{}_{jb}P^{jb}_a,\ L_5=S^a{}_{bc}S^{bc}_a,$

(4.3) $$L_{11}=R^{ij}_{ij},\ L_{12}=\tilde{R}_{abkl}\tilde{R}^{abkl},\ L_{13}=\tilde{P}_{ijkc}\tilde{P}^{ijkc},\ L_{14}=P_{abkc}P^{abkc},$$
$$L_{15}=\tilde{S}_{ijbc}\tilde{S}^{ijbc},\ L_{16}=S^{ab}_{ab},\ L_{21}=R_{ijkh}R^{ijkh},\ L_{22}=S_{abcd}S^{abcd}.$$

Using these gauge scalars the following generalized gauge Lagrangian may be considered (cf. G.S. Asanov [19])

(4.4) $$L=n_1L_1+n_2L_2+n_3L_3+n_4L_4+n_5L_5+n_{11}L_{11}+n_{12}L_{12}+$$
$$+n_{13}L_{13}+n_{14}L_{14}+n_{15}L_{15}+n_{16}L_{16}+n_{21}L_{21}+n_{22}L_{22},$$

where $n_1,\ldots,n_{12},\ldots n_{21},n_{22}$ are constants.

In order to formulate covariant differential equations for all generalized gauge field $\Phi\in\{N^A_i, L^i_{jk}, K^a_{bi}, M^i_{jc}, C^a_{bc}\}$ it is natural to set fourth the variational principle

(4.5) $\delta\int LG\,dx^1\ldots dx^n dy^1\ldots dy^n=0.$

The Euler-Lagrange equations derived from (4.5) read for any Φ as follows:

$$\frac{\delta LG}{\delta\phi} = \frac{\partial}{\partial x^j}\left(\frac{\partial LG}{\partial(\partial_j\phi)}\right) + \frac{\partial}{\partial y^a}\left(\frac{\partial LG}{\partial(\partial_a\phi)}\right) - \frac{\partial LG}{\partial\phi} = 0 \tag{4.6}$$

where ∂_i and ∂_a stand for $\frac{\partial}{\partial x^j}$ and $\frac{\partial}{\partial y^a}$, respectively.

For an explicit form of these equations, seven in number, as well as for their solutions we refer to G.S. Asanov [19], [20].

Chapter VII

Geometry of the Total Space of a Tangent Bundle

The geometry of the manifold TM, the total space of the tangent bundle over a smooth manifold M is very rich. This manifold carries a lot of interesting geometrical structures, which were intensively studied (see Yano, K. and Ishihara, S.[168] and the references therein). The development of the Finsler geometry brought in this field new ideas especially that of using systematically a non-linear connection in the tangent bundle(TM,τ,M). Also, a possibility to think the Finsler geometry as a subgeometry of the geometry of TM has appeared. This point of view on Finsler geometry, that allowed new approaches of its generalisations and applications, was clarified by R. Miron [103,104], M. Anastasiei [9], I. Popovici and M. Anastasiei [129].

In this chapter we shall present some facts from the geometry of the manifold TM which will be used later in order to develop the geometry of Finsler and Lagrange spaces as well as their applications to theoretical physics. Most of them will be obtained by particularising the results which we have proved for vector bundles. Of course, the specific features will be carefully treated.

First we shall introduce a notion of non-linear connection in (TM,τ,M) which will be intensively studied in §1-4. Then we shall treat d-connections in §5, metrical structures and their compatibility in §6. Some interesting particular cases will be pointed out.

1. Non-linear Connections in Tangent Bundle

Let (TM,τ,M) be the tangent bundle over M and (VTM,τ_V,TM) the vertical subbundle over TM i.e. the kernel of the mapping τ^T, the tangent mapping to τ. We recall that the local coordinates on TM are (x^i,y^i), $i=1,2,\ldots,n=\dim M$, where in fact $x^i=x^i\circ\tau$ and (y^i) are the components of a tangent vector to M in the point $x\in M$ in the natural frame $(\frac{\partial}{\partial x^i})_x$. A vector field on TM can be locally written as $X=X^i\frac{\partial}{\partial x^i}+Y^i\frac{\partial}{\partial y^i}$ and it results that the vertical distribution$u\to V_uTM=\mathrm{Ker}\tau^T_u$, $u\in TM$

is spanned by $(\frac{\partial}{\partial y^i})$. Hence it is an integrable distribution.

Let us consider the following exact sequence of vector bundles over TM:

$$(1.1) \quad 0 \longrightarrow VTM \xrightarrow{i} TTM \xrightarrow{\tau!} TTM/VTM \longrightarrow 0.$$

Definition 1.1. *A non-linear connection in the tangent bundle (TM, τ, M) is a splitting on the left of the exact sequence (1.1).*

Let C be the splitting morphism i.e. C:TTM →VTM such that $C \circ i$ is the identity on VTM. The kernel of this morphism is the horizontal subbundle over TM which will be denoted by $(\mathrm{HTM}, \tau_H, \mathrm{TM})$. We have

$$(1.2) \quad TTM = HTM \oplus VTM \qquad \text{(Whitney' sum).}$$

As in the case of any vector bundle we have

Theorem 1.1. *There exists a non-linear connection in (TM, τ, M) if and only if there exists a vector subbundle (HTM, τ_H, TM) of the tangent bundle over TM such that (1.2) holds good.*

Theorem 1.2. *If M is paracompact there exist non-linear connections in the tangent bundle (TM, τ, M).*

The local form of the morphism C introduces a set of n^2 real functions (N^i_j) defined on $\overset{-1}{\tau}(U)$, where U is a coordinate neighbourhood on M. These functions characterise a non-linear connection in the tangent bundle. That is, the following theorem is true (cf. §2, Ch.II):

Theorem 1.3. *To give a non-linear connection in the tangent bundle (TM, τ, M) is equivalent to giving a set of real functions (N^j_i), $i,j = 1,2,\dots,n$ on every coordinate neighbourhood on TM, which on the intersections of such coordinate neighbourhoods satisfies*

$$(1.3) \quad N^{j'}_{i'} \frac{\partial x^{i'}}{\partial x^i} = \frac{\partial x^{j'}}{\partial x^j} N^j_i - \frac{\partial^2 x^{j'}}{\partial x^i \partial x^j} y^j .$$

Homogenous and linear connections in the tangent bundle are defined as for any vector bundle. Of course, the homogeneous connections are defined in the bundle $(\tilde{TM}, \tau|_{\tilde{TM}}, M)$ where $\tilde{TM}$ is the manifold TM without the image of the null-section in the tangent bundle. A linear connection defined on this way is nothing but the usual concept of linear connection on M.

Of course, the characterisations of a non-linear connection by means of the vertical projector v, the horizontal projector h and the almost product structure P, respectively, proved in §3, Ch.II, remain valid for a non-linear connection in the tangent bundle.

Keeping the previous notations we set

$$(1.4)\quad X = hX + vX, \qquad X \in \chi(TM).$$

As for any vector bundle, the horizontal lift defined by a non-linear connection leads to the consideration of a local frame $(\frac{\delta}{\delta x^i}, \frac{\partial}{\partial y^i})$ adapted to the horizontal and vertical distributions, where

$$(1.5)\quad \frac{\delta}{\delta x^i} = \frac{\partial}{\partial x^i} - N_i^j(x,y)\frac{\partial}{\partial y^j}.$$

For the sake of brevity we set $\delta_i := \frac{\delta}{\delta x^i}$ and $\dot{\partial}_i := \frac{\partial}{\partial y^i}$.

The following equations are obvious

$$(1.6)\quad h(\delta_i) = \delta_i, \quad h(\dot{\partial}_i) = 0,$$

$$(1.7)\quad v(\delta_i) = 0, \quad v(\dot{\partial}_i) = \dot{\partial}_i,$$

$$(1.8)\quad P(\delta_i) = \delta_i, \quad P(\dot{\partial}_i) = -\dot{\partial}_i.$$

Let us consider $J : \chi(TM) \to \chi(TM)$ defined by

$$(1.9)\quad J(\frac{\partial}{\partial x^i}) = \frac{\partial}{\partial y^i}, \quad J(\frac{\partial}{\partial y^i}) = 0.$$

Notice the definition of J does not depend on the non-linear connection in the tangent bundle.

It is easy to check:

Proposition 1.1. *The endomorphism J has the properties*

(i) $J^2=0$, (ii) $\mathrm{Im}\, J=\mathrm{Ker} J=VTM$.

Thus J is a natural (or canonical) almost tangent structure on TM.

The Nijenhuis tensor field of the almost tangent structure J is

(1.10) $N_J(X,Y)=[JX,JY]-J[JX,Y]-J[X,JY], \quad X,Y\in\chi(TM)$.

A direct calculation shows that N_J identically vanishes. Thus we have

Proposition 1.2. *The natural almost tangent structure J on TM is integrable.*

Using the tangent structure J we get the following new characterization of a non-linear connection in the tangent bundle.

Theorem 1.4. *To give a non-linear connection in the tangent bundle(TM,τ,M) is equivalent to giving a morphism $\Gamma:\chi(TM)\to\chi(TM)$ satisfying the conditions*

(1.11) $J\Gamma=J, \quad \Gamma J=-J$.

Proof. If a non-linear connection is given, putting $\Gamma=P$ we see, by using (1.8) and (1.9) that(1.11) is verified.

Conversely, if Γ is of the form

$$(1.12)\quad \begin{aligned}\Gamma(\delta_i)&=A_i^j\delta_j+B_i^j\dot\partial_j,\\ \Gamma(\dot\partial_i)&=C_i^j\delta_j+D_i^j\dot\partial_j,\end{aligned}$$

by (1.11) we get $A_i^j=\delta_i^j$, $C_i^j=0$, $D_i^j=-\delta_i^j$. Consequently $\Gamma^2=I$ and $\Gamma(X)=-X$ if and only if X is vertical. Thus Γ is an almost product structure on TM whose eigenspaces corresponding to eigenvalue -1 are vertical subspaces. Therefore Γ defines a non-linear connection in the tangent bundle (TM,τ,M).

Let $\mathbb{C}=y^i\dfrac{\partial}{\partial y^i}$ be the Liouville vector field on TM. It is easy to check it globally defined. It is also clear that it is a vertical vector field.

Making explicit the integral curves of $\mathbb{C}$ we found that it generates an one-parameter group of homotheties on TM, i.e. of mappings

$$h_t:TM\to TM,\quad h_t(p,X_p)=(p,tX_p),\quad t\in R_+^*.$$

Using $\mathbb{C}$ we shall be able to express the homogeneity with respect to y^i of various geometrical objects on TM. A real function defined on TM is homogeneous of degree r if $h_t^* f = t^r f$. This equality is equivalent to

(1.13) $L_C f = rf$,

where L_C denotes the Lie derivative with respect to $\mathbb{C}$.

A p-form ω on TM is homogeneous of degree r with respect to y^i if $h_t^* \omega = t^r \omega$. This condition is equivalent to

(1.14) $L_C \omega = r\omega$.

We shall say that a tensor field T of type (1,1) on TM is homogeneous of degree r if

(1.15) $L_C T = (r-1)T$.

Proposition 1.3. *The endomorphism J is homogeneous of degree zero with respect to y^i, i.e. we have*

(1.16) $L_C J = -J$.

Proof. We have $(L_C J)(X) = [\mathbb{C}, JX] - J([\mathbb{C}, X])$. Thus we must show that $[\mathbb{C}, JX] - J([\mathbb{C}, X]) = -JX$ for every $X \in \chi(TM)$. This equality is easily verified in local coordinates by using (1.9).

The endomorphism $H = \frac{1}{2} L_C P$ is called the *tension* of the non-linear connection defined by the almost structure P. It is obvious that H is vanishing if and only if the automorphism P is homogeneous of degree 1 with respect to y^i. On this way we recast the condition of homogeneity introduced before. Indeed, by using a coordinate system we find

$$H = (N_i^j - y^k \frac{\partial N_i^j}{\partial y^k}) \dot{\partial}_k \otimes dx^j.$$

Thus H=0 if and only if $y^k \dot{\partial}_k (N_i^j) = N_i^j$ which means $N_i^j(x,y)$ is 1-homogeneous with respect to y^i.

The covariant derivative defined by a non-linear connection of local coefficients (N_j^i) is as follows:

(1.17) $$D_X Y = X^k\left(\frac{\partial Y^i}{\partial x^k} + N_k^i(x,Y)\right)\frac{\partial}{\partial x^i},$$

where $X = X^k\frac{\partial}{\partial x^k}$ and $Y = Y^i\frac{\partial}{\partial x^i}$ are vector fields on M. It is $\mathcal{F}(M)$ -linear with respect to X and nonlinear with respect to Y, except the case when the coefficients (N_j^i) are the local coefficients of a linear connection.

A vector field Y is parallel along of a smooth curve c in M if $D_{\dot c}Y = 0$. In a coordinate neighbourhood containing the curve c this condition becomes

(1.18) $$\frac{dY^i}{dt} + N_j^i(x,Y)\frac{dx^j}{dt} = 0.$$

Here $x^i = x^i(t)$, $t \in [a,b]$ are the equations of the curve c and $Y^i := Y^i(x(t))$ are the components of the vector field Y along the curve c.

A curve c in M is called autoparallel with respect to a non-linear connection of local coefficients (N_j^i) if $D_{\dot c}\dot c = 0$. We have

Proposition 1.4. *Let $c:[a,b]\to M$ be a curve in M such that c([a,b]) is contained in a coordinate neighbourhood U on M. Assuming c is given by the equations $x^i = x^i(t)$, $t \in [a,b]$ then c is autoparallel with respect to a non-linear connection $N(N_j^i)$ if and only if*

(1.19) $$\frac{d^2x^i}{dt^2} + N_j^i\left(x, \frac{dx}{dt}\right)\frac{dx^j}{dt} = 0.$$

§2. Semisprays, Sprays and Non-linear Connections

Let $\tau: TM \to M$ be the tangent bundle over M. Then TM is a 2n-dimensional manifold and we can consider the tangent bundle $\pi: T^2M \to TM$ over it. On the other hand we may consider the tangent mapping $\tau^T: T^2M \to TM$ This mapping provides a new vector bundle over TM whose total space is T^2M.

Let $(\overset{-1}{\tau}(U),(x,y))$ be a coordinate system on TM. A vector field on TM i.e. a cross-section of the tangent bundle $\pi: T^2M \to TM$ has the local form $X^i\partial_i + Y^i\dot\partial_i$. Thus a point of T^2M which belongs to the coordinate neighbourhood $\overset{-1}{\pi}(\overset{-1}{\tau}(U))$ has the local coordinates (x,y,X,Y). Then π maps (x,y,X,Y) to (x,y) and as is easy to see, τ^T maps (x,y,X,Y) to (x,Y). Thus we find the vector bundles (T^2M,π,TM) and (T^2M,τ^T,TM) are different. However these bundles are isomorphic. An isomorphism of them is given by the mapping $(x,y,X,Y)\to(x,X,y,Y)$ which, moreover, has the property that its square is the identity on T^2M.

Definition 2.1. *A semispray or a second order differential equation on M is a vector field on TM of the class C^∞ on $\tilde{TM}$ and only continuous on the null-section, which is also a cross-section of the vector bundle (T^2M,τ^T,TM).*

Let S be a semispray on M. Then in a coordinate system on TM we may set $S=s^i\partial_i+S^i\dot{\partial}_i$. Locally we have $S:(x,y)\to(x,y,s^i,S^i)$ and $\tau^T:(x,y,s^i,S^i)\to(x,s^i)$. In order that $\tau^T\circ S$ be identity on TM we must have $s^i=y^i$. Thus a semispray on TM has the local form $S=y^i\partial^i+S^i\dot{\partial}^i$. It is now very easy to prove

Proposition 2.1. *A vector field on TM is a semispray if and only if $JS=\mathbb{C}$.*

Definition 2.2. *a. Given a semispray S, the vector field $S^*=L_CS-S$ will be called the deviation of S.*

b. A semispray S is said to be a spray if it is of class C^1 on the null-section and *its deviation S^* vanishes. The latter condition means S^i is homogeneous of degree 2 with respect to (y^i).*

c. A spray S is called a quadratic spray if it is of class C^2 on the null-section.

Now we proceed to relate semisprays to non-linear connections.

Theorem 2.1. *a) Given a non-linear connection of tension H, there exists a semispray S on M such that $S^*=H(S)$.*

b) If the given non-linear connection is homogeneous and the almost product structure P associated to it is of class C^1 on the null-section then S becomes a spray.

c) If the given non-linear connection becomes linear and P is of class C^2 on null-section then S becomes a quadratic spray.

Proof. Let be S' any semispray on M and $h=\frac{1}{2}(I+P)$ the horizontal projector associated to the given non-linear connection. Let us consider the vector field $S=h(S')$. It follows S' is a semispray since $JS=Jh(S')=JS'=\mathbb{C}$. Moreover, S does not depend on S' because if S'' is another semispray, we have $J(S'-S'')=0$ which means $S'-S''$ is vertical.

Hence $hS' = hS''$. It remains to prove that $S^* = H(S)$ that is, $[C,P(S)] - P[C,S] = 2[C,S] - 2S$. This last equality is equivalent to $h([C,hS'] - S') = 0$. In order to prove it we shall prove that $[C,hS'] - S'$ is a vertical vector field by showing that $J([C,hS'] - S') = 0$. We need the following identity

$$(2.1)\quad J[JX,S] = JX, \quad X \in \chi(TM),$$

for any semispray S on M. This follows by combining the identities $N_J(S,X) = 0$ and $L_C J = -J$. Applying (2.1) we get

$$J([C,hS'] - hS') = J([JhS',hS']) - JhS' = JhS' - JhS' = 0.$$

Thus the proof of a) is complete.

b) If the given non-linear connection is homogeneous with respect to (y^i) we have $H=0$, hence $S^* = 0$. If P is of class C^1 on the null-section, then h is so. Thus $S = h(S')$, where S' is of class C^∞ on TM, is of class C^1 on the null-section. Therefore S is a spray on M.

c) We have again $S^* = 0$ and as in the case b) we find that S is of class C^2 on the null-section. Thus S is a quadratic spray.

Now we shall prove a converse of the Theorem 2.1.

Theorem 2.2. *a) If S is a semispray on M, the automorphism $\Gamma = -L_S J$ gives a non-linear connection on M, whose associated semispray is $S + \frac{1}{2}S^*$.*

b) If S is a spray, then the connection Γ defined by S is homogeneous with respect to y^i and its associated spray is just S.

c) If S is a quadratic spray, then the connection given by Γ is linear and symmetric.

Proof. We shall begin by proving that $\Gamma = -L_S J$ satisfies (1.11). We have $J\Gamma(X) = -J[S,JX] + J^2[S,X] = JX$ by (2.1) and $\Gamma J(X) = -[S,J^2X] + J[S,JX] = -JX$, again by (2.1). The semispray associated to the non-linear connection Γ is $hS' = \frac{1}{2}(S' + \Gamma(S')) = \frac{1}{2}(S' - [S,JS'] + J[S,S']) = \frac{1}{2}(S' - [S,C] + S - S') =$ $= S + \frac{1}{2}([C,S] - S) = S + \frac{1}{2}S^*$ because of $J([S,S']) = S - S'$. The last equality can be obtained taking $X = S - S'$ in the identity

$$(2.2)\quad X = J[X,S] - [JX,S] - J[JX,S], \quad X \in \chi(TM),$$

which was deduced from (2.1) by replacing JX by X-[JX,S].

b) If S is a spray, then $S^*=0$, hence $[\mathbb{C},S]=S$. As to the tension of Γ we get $2H(X)=[\mathbb{C},\Gamma X]-\Gamma[\mathbb{C},X]=[\mathbb{C},J[S,X]]-[\mathbb{C},[S,JX]]-J[S,[\mathbb{C},X]]+[S,J[\mathbb{C},X]]=0$ by virtue of the Jacobi identity and of $[\mathbb{C},S]=S$. Thus the connection Γ is homogeneous of degree 1.

c) We shall do the analysis in a coordinate system. As we have seen, a semispray S is of the form $S=y^i\partial_i+S^i(x,y)\dot\partial_i$. Let now $X=X^i\partial^i+Y^i\dot\partial_i$ be a vector field on TM. A direct calculation gives that $\Gamma=-L_SJ$ is of the form

$$\Gamma(X)=X^i\partial_i+((\dot\partial_jS^i)X^j-Y^i)\dot\partial_i.$$

On the other hand the almost product structure given by (1.8), takes by using two times (1.5) the following form $P(X)=X^i\partial_i-(Y^i+2N^i_jX^j)\dot\partial_j$. Comparing these forms of P and Γ we see necessarily must have

$$N^i_j(x,y)=-\frac{1}{2}\dot\partial_jS^i(x,y). \tag{2.3}$$

If S is a spray then the functions S^i are 2-homogeneous with respect to (y^i) and of class C^1 on the null-section. By (2.3) we get $N^i_j(x,y)$ are homogeneous of degree one with respect to (y^i) and only continuous on the null-section. It follows the connection induced by S is homogeneous. Thus a new proof of b) is obtained. Now, if S is a quadratic spray the functions S^i are 2-homogeneous and of class C^2 on the null-section. By (2.3) we see N^i_j are homogeneous of degree one and of class C^1 on the null-section. It follows these functions are linear with respect to y^i. Hence we may write

$$N^i_j(x,y)=\Gamma^i_{kj}(x)y^k,\ \text{where}\ \Gamma^i_{kj}=-\frac{1}{2}\dot\partial_k\dot\partial_jS^i. \tag{2.4}$$

It is obvious that $\Gamma^i_{kj}=\Gamma^i_{jk}$. Thus S determines in this case a symmetric linear connection.

A smooth curve $c:I\subseteq\mathbf{R}\to M$ is said to be a path of a semispray S on M if the curve $\dot c:I\to TM$ is an integral curve of S , that is:

$$\ddot c=S(\dot c) \tag{2.5}$$

holds.

If we assume that c(I) is contained in a coordinate neighbourhood on M and we take $x^i = x^i(t)$, $t \in I$ as equations for the curve c, then we have

$$(2.6)\quad \ddot{c}(t) = (x^i(t), \frac{dx^i}{dt}, \frac{dx^i}{dt}, \frac{d^2x^i}{dt^2}).$$

The local form of $\ddot{c}$ could have suggested the notion of semispray (spray). The equation (2.5) becomes

$$(2.7)\quad \frac{d^2x^i}{dt^2} = S^i(x, \frac{dx}{dt}).$$

If we set $S^i = -2G^i$, then (2.7) takes the form

$$(2.8)\quad \frac{d^2x^i}{dt^2} + 2G^i(x, \frac{dx}{dt}) = 0$$

and (2.3) becomes

$$(2.9)\quad N_j^i(x,y) = \dot{\partial}_j G^i.$$

If S is a spray, by using the property of homogeneity of degree 2 of G^i we get the following form of (2.8):

$$(2.10)\quad \frac{d^2x^i}{dt^2} + N_j^i(x, \frac{dx}{dt})\frac{dx^j}{dt} = 0.$$

Comparing (2.10) to (1.19) we get

Theorem 2.3. *The autoparallel curves of a strict homogeneous connection in* (TM, τ, M) *coincide with the paths of the spray associated to it.*

§3. Torsions and Curvature of a Non-linear Connection

As in the general case of a vector bundle we may consider the Berwald connection associated with a non-linear connection in the tangent bundle (TM, τ, M).

If (N_j^i) are the local coefficients of a non-linear connection N, then the operator of covariant derivation $\hat{D}$ of the Berwald connection associated with it is given by

$$(3.1)\quad \begin{aligned} &\hat{D}_{\partial_i}\dot{\partial}_j = \partial_j(N_i^k)\dot{\partial}_k,\\ &\hat{D}_{\dot{\partial}_i}\dot{\partial}_j = 0. \end{aligned}$$

Recall that the Berwald connection is in fact a linear connection in the vertical bundle (VE,τ_{VE},E). Let ∇ be the absolute differentiation operator associated to $\hat{D}$ (defined as in (5.6), Ch.II). Recall it acts on q-forms which are vertical valued. J appears as a remarkable vertical valued 1-form. Thus is quite natural to introduce

Definition 3.1. *The vertical valued 2-form $t=-\nabla J$ is said to be the weak torsion of the non-linear connection N.*

Expliciting the formula for t we find

$$(3.2)\quad t(X,Y) = -\hat{D}_{hX}JY + \hat{D}_{hY}JX + J[hX,hY],\ \ X,Y \in \chi(TM).$$

In a coordinate system we get

$$(3.3)\quad \begin{aligned} &t(\partial_i,\partial_j) = [\dot{\partial}_i(N_j^k) - \dot{\partial}_j(N_i^k)]\dot{\partial}_k,\\ &t(\partial_i,\dot{\partial}_j) = 0,\quad t(\dot{\partial}_i,\dot{\partial}_j) = 0. \end{aligned}$$

Let S be the canonical semispray associated to the non-linear connection N.

Definition 3.2. *The strong torsion of a non-linear connection N is the 1-form $T = i_S t - H$, where i_S is the contraction with S.*

Recall that $H = \frac{1}{2}L_C P$ is the tension of the non-linear connection N.

In a coordinate system we have

$$(3.4)\quad H(\partial_i) = [N_i^k - y^j\dot{\partial}_j(N_i^k)]\dot{\partial}_k,\ \ H(\dot{\partial}_i) = 0.$$

Using (3.3) and (3.4) one easily deduces

$$(3.5)\quad T(\partial_j) = [y^i\dot{\partial}_j(N_i^k) - N_j^k]\dot{\partial}_k,\ \ T(\dot{\partial}_j) = 0.$$

Proposition 3.1. *The strong torsion T of a non-linear connection N has the property*

(3.6) $i_S T + S^* = 0.$

Proof. We have $i_S T = -i_S H$ and by the Theorem 2.1 we get $i_S H = H(S) = S^*$.

Proposition 3.2. *The strong torsion T vanishes if and only if the weak torsion and tension of the non-linear connection N vanish.*

Proof. If $t=0$, then $\dot\partial_j N_i^k = \dot\partial_i N_j^k$ and if $H=0$ then $N_j^k = y^i \dot\partial_i N_j^k$. Taking into account (3.5) we get $T=0$.

Conversely, if $T=0$, by deriving the both sides of the equality $N_j^k = y^s \dot\partial_j (N_s^k)$ with respect to y^i we get $\dot\partial_i N_j^k - \dot\partial_j N_i^k = y^s \dot\partial_i \dot\partial_j N_s^k$. The left hand of the latter equality is sqewsymmetric while the right hand is symmetric. Hence the both must be zero, that is, $t=0$ and $H=0$.

Of course, the curvature of a non-linear connection in the tangent bundle is as in the general case $\Omega = -\nabla C$.

A structure equation similar to (5.8), Ch.II, as well as the Bianchi identity (5.11), Ch.II hold.

The formula (5.9) Ch. II giving the local coefficients of the curvature takes here the form

(3.7) $$\Omega^i_{kj} = \partial_k N_j^i - \partial_j N_k^i + N_j^s \dot\partial_s N_k^i - N_k^s \dot\partial_s N_j^i .$$

Notice the following brackets

(3.8) $$[\delta_j, \delta_k] = R^i{}_{jk} \dot\partial_i \,, \quad [\delta_j, \dot\partial_k] = (\dot\partial_k N_j^i) \dot\partial_i \,,$$

where $R^i{}_{jk} = -\Omega^i_{jk}$.

Let us define a tensor field F of type (1,1) on TM by

(3.9) $$F(\delta_i) = -\dot\partial_i \,, \quad F(\dot\partial_i) = \delta_i .$$

It is obvious that F is well-defined on TM and $F^2 = -I$ (Kronecker's tensor field). Thus F is an almost complex structure associated with the non-linear connection N.

We shall end this section with

Proposition 3.3. *The almost complex structure F associated to a non-linear connection N is integrable if and only if $t=0$, $\Omega = 0$.*

Proof. The almost complex structure F is integrable if the Nijenhuis tensor field N_F

associated to it vanishes. Recall that this tensor field is given by

(3.10) $N_F(X,Y)=[FX,FY]-[X,Y]-F[FX,Y]-F[X,FY],\ X,Y\in\chi(TM).$

By (3.7) and (3.8) we find

$$
(3.11)\quad
\begin{aligned}
N_F(\delta_j,\delta_k) &= -t^i_{jk}\delta_i - \Omega^i_{jk}\dot\partial_i,\\
N_F(\delta_j,\dot\partial_k) &= -\Omega^i_{jk}\delta_i + t^i_{jk}\dot\partial_i,\\
N_F(\dot\partial_j,\dot\partial_k) &= t^i_{jk}\delta_i + \Omega^i_{jk}\dot\partial_i,
\end{aligned}
$$

where

(3.12) $t^i_{jk}=\dot\partial_j(N^i_k)-\dot\partial_k(N^i_j).$

Notice that t=o if and only if $t^i_{jk}=0$.

From (3.11) one reads immediately that $N_F=0$ if and only if $t=0$ and $\Omega=0$.

§4. Transformations of Non-linear Connections

As we have seen, if M is a paracompact manifold there exist non-linear connections in the tangent bundle. From the definition of a non-linear connection, as well as from the proof of the Theorem 1.4 it comes out that it is not unique. The set of all non-linear connections is characterised by (1.11). We have

Proposition 4.1. *Let be* Γ,Γ' *two morphisms of* $\chi(TM)$ *satisfying (1.11). Then the morphism* $A:\chi(TM)\to\chi(TM)$ *given by* $A=\Gamma'-\Gamma$ *verifies*

(4.1) $JA=0,\ AJ=0.$

Conversely, given a non-linear connection Γ *and a morphism* A *of* $\chi(TM)$ *satisfying (4.1), the morphism* $\Gamma'=\Gamma+A$ *defines a non-linear non-linear in* (TM,τ,M).

Proof. The equations (4.1) follow immediately taking into account that Γ and Γ' satisfy (1.11). Next we have

$$J\Gamma'=J\Gamma+JA=J \quad\text{and}\quad \Gamma'J=\Gamma J-AJ=-J.$$

that is, Γ' satisfies (1.11).

Remark 4.1. As we know from the proof of the Theorem 1.4, $\Gamma'=\Gamma+A$ is an almost product structure on TM. Hence necessarily we have

(4.2) $A^2+A\Gamma+\Gamma A=0.$

Let **A** be the set of all morphisms of $\chi(TM)$ which satisfy (4.1). This may be also thought of as vectorial valued 1-forms. Let us denote by **N** the set of all non-linear connections in (TM,τ,M).

Definition 4.1. *A mapping* $t_A:N\to N$, $\Gamma\to\Gamma+A$, *where* $A\in \mathbf{A}$ *will be called a transformation of non-linear connections.*

It is natural to consider the composition of two such transformations $t_{A'}\circ t_A$. It follows easily $t_{A'}\circ t_A=t_{A+A'}$. Thus we have

Proposition 4.2. *The set* $\{t_A|A\in \mathbf{A}\}$ *of all transformations of non-linear connections and the composition of mappings is an abelian group isomorphic to the additive group* **A**.

Let us fix a non-linear connection of local coefficients $(N_j^i(x,y))$. Then the almost product structure P associated to it satisfies (1.11) as is easily to check using (1.8). In a coordinate system, taking into account (1.5) and (1.8) we find

(4.3) $P(\partial_i)=\partial_i-2N_i^j(x,y)\dot{\partial}_j,\quad P(\dot{\partial}_i)=-\dot{\partial}_i.$

Let now Γ be another non-linear connection, that is, a morphism of $\chi(TM)$ which satisfies (1.11). Then there exists a morphism A of $\chi(TM)$ such that $\Gamma=P+A$. Since A satisfy (4.1) it takes the following local form

(4.4) $A(\partial_i)=2A_i^j\dot{\partial}_j,\quad A(\dot{\partial}_j)=0.$

The factor 2 was taken for the sake of convenience.

By (4.3) easily deduces

(4.5) $\Gamma(\partial_i)=\partial_i-2(N_i^j-A_i^j)\dot{\partial}_j,\quad \Gamma(\dot{\partial}_i)=-\dot{\partial}_i.$

Comparing (4.3) to (4.5) we read that the local coefficients $(\tilde{N}_i^j)$ of the non-linear connection Γ are related to (N_i^j) by

$$\tilde{N}_i^j = N_i^j + A_i^j(x,y). \tag{4.6}$$

The equations (4.6) give the local form of a transformation of non-linear connections.

Let S be a semispray and let us consider

$$G_S = \{t_A \mid A \in \mathcal{A}, \quad A(S) + S^* = 0\}.$$

Note G_S is nonvoid since the strong torsion T of any non-linear connection belongs to it. We have:

Proposition 4.3. *Let be Γ the non-linear connection given by $\Gamma = -L_S J$. All non-linear connections obtained from Γ by using the transformations from G_S have the same associated semispray S.*

Proof. Let $\Gamma'(X) = -(L_S J)(X) + A(X)$ a non-linear connection obtained from Γ by the transformation t_A, $A \in G_S$. The semispray associated to Γ' (S' is any semispray on TM) is: $2h(S') = S' + J[S,S'] - [S,JS'] + A(S')$. But $J[S,S' = S - S'$ (see the proof of the Theorem 2.2). Then $AJ = 0$ implies $A(S) = A(S')$ and $h(S') = S + \frac{1}{2}(S^* + A(S)) = S$.

Notice that the semispray associated to $\Gamma = -L_S J$ is not S but adding $A \in G_S$ to Γ the associated semispray of this non-linear connection becomes S.

J. Grifone [47] proved the following decomposition theorem for non-linear connections.

Theorem 4.1. *Given a semispray S and a morphism T of $\chi(TM)$ such that $t_T \in G_S$, then there exists an unique non-linear connection whose associated semispray is S and whose strong torsion is T. It is given by $\Gamma = -L_S J + T$.*

Remark 4.2. By the Theorem 4.1 an element $A \in \mathcal{A}$ such that $t_A \in G_S$ is the strong torsion of the non-linear connection obtained from $-L_S J$ by using t_A. Such we have a geometrical meaning for a subset of **A**. If $t_0 \in G_S$ then S must be a spray. But $T = 0$ implies the weak torsion and tension are vanishing. Thus we obtain again that in this case $-L_S J$ gives an homogeneous connection.

§5. Normal d-Connections on TM

Let the tangent bundle (TM,τ,M) be again endowed with a non-linear connection N. As in the case of any vector bundle we may speak about d-tensor (or M- tensor) fields on TM. These are nothing but the elements of the tensorial algebra of the Whitney sum vector bundle $HTM\oplus VTM\to TM$. This algebra is spanned by $(1,\delta_i,\dot{\partial}_i)$ over the ring $\mathcal{F}(TM)$.

A change of coordinates $(x^i,y^i)\to(x'^i,y'^i)$ on TM is of the form

$$(5.1)\qquad \begin{aligned} &x'^i = x'^i(x^1,\dots,x^n), \quad \operatorname{rank}\left(\frac{\partial x'^i}{\partial x^k}\right) = n,\\ &y'^i = \frac{\partial x'^i}{\partial x^k} y^k. \end{aligned}$$

We have

$$(5.2)\qquad \begin{aligned} &\delta_j = \frac{\partial x'^i}{\partial x^j}\delta'_i, \quad \dot{\partial}_j = \frac{\partial x'^i}{\partial x^j}\dot{\partial}'_i,\\ &dx'^i = \frac{\partial x'^i}{\partial x^k}dx^k, \quad \delta y'^i = \frac{\partial x'^i}{\partial x^k}\delta y^i, \end{aligned}$$

where

$$(5.3)\qquad \delta y^i = dy^i + N^i_k(x,y)dx^k.$$

Definition 5.1. *A linear connection on TM is said to be a distinguished connection (shortly a d-connection or M-connection) if it preserves by parallelism the vertical and horizontal distributions on TM.*

We have given a study of this concept in the case of the total space of any vector bundle in Ch.III. All results from there remain unchanged excepting that in all local representations indices a,b,c and i,j,k,... have the same range: 1,2,...,n.

Notice that in using the local coordinates it is convenient to preserve two kinds of indices in order to mark elements related to vertical and horizontal bundles, respectively.

Now we shall consider a particular class of d-connections.

Definition 5.2. *A d-connection D on TM is said to be normal (or natural) if the almost complex structure F associated with the non-linear connection N is absolute parallel with respect to it, that is,* $D_X F = 0$, $X\in \chi(TM)$.

Theorem 5.1. *A normal d-connection D is characterised by a pair of local coefficients* $D\Gamma = (L^i_{jk}(x,y), C^i_{jk}(x,y))$ *given by*

$$(5.4)\quad \begin{array}{ll} D_{\delta_k}\delta_j = L^i_{jk}\delta_i, & D_{\dot\partial_k}\delta_j = C^i_{jk}\delta_i, \\ D_{\delta_k}\dot\partial_j = L^i_{jk}\dot\partial_i, & D_{\dot\partial_k}\dot\partial_j = C^i_{jk}\dot\partial_i \end{array}$$

such that L^i_{jk} *and* C^i_{jk} *change under (5.1) as follows*

$$(5.5)\quad \begin{aligned} L'^i_{jk}(x',y') &= \frac{\partial x'^i}{\partial x^h}\frac{\partial x^s}{\partial x'^j}\frac{\partial x^r}{\partial x'^k}L^h_{sr} + \frac{\partial x'^i}{\partial x^r}\frac{\partial^2 x^r}{\partial x'^j \partial x'^k}, \\ C'^i_{jk}(x',y') &= \frac{\partial x'^i}{\partial x^h}\frac{\partial x^s}{\partial x'^j}\frac{\partial x^r}{\partial x'^k}C^h_{sr}. \end{aligned}$$

Proof. As we know, a d-connection is characterised by a set of local coefficients $D\Gamma = (L^i_{jk}, K^a_{bk}, M^i_{jc}, C^a_{bc})$ which transform under (5.1) by the laws (2.9)-(2.12), Ch.III, respectively.

The condition $(D_X F)(Y) = D_X F(Y) - F(D_X Y) = 0$ for any $X, Y \in \chi(TM)$ is equivalent to $K^a_{bk} = \delta^a_i \delta^j_b L^i_{jk}$ and $M^i_{jc} = \delta^i_a \delta^b_j C^a_{bc}$. Thus (5.4) holds. Then (5.5) follows from (2.9)-(2.12), Ch.III. Conversely, if $D\Gamma = (L^i_{jk}, C^i_{jk})$ is given such that (5.5) be satisfied, then we define a covariant derivative D by (5.4). It will be well-defined because of (5.5) and it is easy to check that $D_X F = 0$, $X \in \chi(TM)$.

The components of the torsion of a normal d-connection in the adapted frame $(\delta_i, \dot\partial_i)$ are as follows:

$$(5.6)\quad \begin{array}{lll} T^i{}_{jk} = L^i_{jk} - L^i_{kj}, & R^i{}_{jk} = \delta_k N^i_j - \delta_j N^i_k, & C^i_{jk}, \\ P^i{}_{jk} = \dot\partial_k (N^i_j) - L^i_{kj}, & S^i{}_{jk} = C^i_{jk} - C^i_{kj}. & \end{array}$$

The curvature tensor field of a normal d-connection has three essential components in the frame $(\delta_i, \dot\partial_i)$: $R_j{}^i{}_{kh}, P_j{}^i{}_{kh}, S_j{}^i{}_{kh}$.

Definition 5.3. *A normal d-connection* $D\Gamma = (L^i_{jk}, C^i_{jk})$ *is said to be of the Cartan type if* $y^i_{|k} = 0$, $y^i|_k = \delta^i_k$, *where short and long bar denote the h- and v- covariant derivatives with respect to* $D\Gamma$.

Proposition 5.1. *A normal d-connection on TM is of the Cartan type if and only if*

(5.7) $N^i_k = L^{\ i}_{jk} y^j, \quad C^{\ i}_{jk} y^j = 0.$

The Bianchi and Ricci identities for a normal d-connection simplify correspondingly.

The Ricci identities applied to the Liouville vector field $y^i \dot{\partial}_i$ lead to the following

Proposition 5.2. *If a normal d-connection on TM is of the Cartan type, then*

(5.8) $R^i_{\ jk} = y^s R_{s\ jk}^{\ i}, \quad P^i_{\ jk} = y^s P_{s\ jk}^{\ i}, \quad S^i_{\ jk} = y^s S_{s\ jk}^{\ i},$

hold.

§6. Metrical Structures on TM

Recall that a given metrical structure G on the total space E of a vector bundle determines a non-linear connection and with respect to it G decomposes into a sum of two d-tensor fields which may be viewed as defining metrical structures in horizontal and vertical bundle, respectively. Conversely, if a non-linear connection, as well as some metrical structures in horizontal and vertical bundles are given, a metrical structure on the total space E may be obtained. We follow this latter way in the case of the total space TM of the tangent bundle (TM,τ,M). In order to indicate elements related to vertical and horizontal bundle, we shall use two kinds of indices a,b,c,.. and i,j,k,.. respectively. Recall they all range from 1 to n.

From now on we fix a non-linear connection N in the tangent bundle (TM,τ,M).

Definition 6.1. *a) An h-metric on TM is a d-tensor field* $hG = g_{ij}(x,y) dx^i \otimes dx^j$, *where* $g_{ij}(x,y) = g_{ji}(x,y)$, $\det(g_{ij}(x,y)) \neq 0$ *and the quadratic form* $g_{ij}\xi^i\xi^j$ *has constant signature.*

b) A v-metric on TM is a d-tensor field $vG = h_{ab}(x,y)\delta^a \otimes \delta y^b$, *where* h_{ab} *has the same properties as* g_{ij}.

c) An (h,v)-metric on TM is the d-tensor field $G = hG + vG$ *i.e.*

(6.1) $G = g_{ij}(x,y)\, dx^i \otimes dx^j + h_{ab}(x,y)\, \delta y^a \otimes \delta y^b.$

Let D be a d-connection on TM whose local coefficients are $D\Gamma = (L^{\ i}_{jk}(x,y), K^{\ a}_{bi}(x,y), M^{\ i}_{ja}(x,y), C^{\ a}_{bc}(x,y))$. As usual it is said D is a metrical connection with respect to the metric G if $D_X G = 0$ for every $X \in \chi(TM)$.

An easy calculation in local coordinates gives

Proposition 6.1. *The d-connection $D\Gamma$ is metrical with respect to the metric G given by (6.1) if and only if*

(6.2) $g_{ij|k}=0,\ g_{ij}|_a=0,\ h_{ab|k}=0,\ h_{ab}|_c=0,$

where short and long bar denote as usual the h- and v-covariant derivatives, respectively.

As we know, starting from any d-connection, by using a method proposed by A. Kawaguchi, [76], a metrical d-connection can be obtained (cf. §5, Ch.III).

Let us put

$$\overset{c}{L}{}^i_{jk}=\frac{1}{2}g^{ih}(\delta_k g_{hj}+\delta_j g_{hk}-\delta_h g_{jk}),$$
(6.3)
$$\overset{c}{C}{}^a_{bc}=\frac{1}{2}h^{ad}(\dot\partial_b h_{dc}+\dot\partial_c h_{bd}-\dot\partial_d h_{bc})$$

and let us consider the following d-connection: $\overset{o}{D\Gamma}=(\overset{c}{L}{}^i_{jk},\dot\partial_b N^a_k,0,\overset{c}{C}{}^a_{bc})$.

We have

Theorem 6.1. *The d-connection $C\Gamma=(\overset{c}{L}{}^i_{jk},K^a_{bi},M^i_{ja},\overset{c}{C}{}^d_{bc})$, where $\overset{c}{L}{}^i_{jk}$ and $\overset{c}{C}{}^a_{bc}$ are given by (6.3) and the other two coefficients are given by*

$$K^a_{bk}=\dot\partial_b N^a_k+\frac{1}{2}h^{ad}[\delta_k h_{bd}-(\dot\partial_b N^c_k)h_{cd}-\dot\partial_d(N^c_k)h_{bc}],$$
(6.4)
$$M^i_{ja}=\frac{1}{2}g^{ih}\dot\partial_a g_{jh},$$

is a metrical d-connection on TM.

Proof. If $\overset{\circ}{|}$ and $\overset{\circ}{\big|}$ denote the h- and v-covariant derivatives, associated to $\overset{o}{D\Gamma}$ respectively, by (6.3) we get $g_{ij\overset{\circ}{|}k}=0,\ h_{ab}\overset{\circ}{\big|}_c=0$. Now using (5.12), Ch.III we obtain the metrical d-connection $C\Gamma$ with the coefficients

$$\overset{c}{L}{}^i_{jk},\ \overset{c}{C}{}^a_{bc}\ \text{and}\ K^a_{bk}=\dot\partial_b N^a_k+\frac{1}{2}h^{ad}h_{db\overset{\circ}{|}k},\ M^i_{ja}=\frac{1}{2}g^{ih}g_{jh}\overset{\circ}{\big|}_a.$$

The last two coefficients take the form (6.4) as it is easy to be seen.

The d-connection $C\Gamma$ depends on N^i_j, g_{ij}, h_{ab} only. It will be called the canonical metrical d-connection on TM. Notice that its torsions $\overset{c}{T}{}^i_{jk}$ and $\overset{c}{S}{}^a_{bc}$ vanish. This fact allows us to determine from the equations $g_{ij|k}=0$ and $h_{ab}|_c=0$ the coefficients $\overset{c}{L}{}^i_{jk}$ and $\overset{c}{C}{}^a_{bc}$, respectively, in the form (6.3). The procedure which sometimes is called the Christoffel process is as follows: we permute cyclicly i,j,k in the equation $g_{ij|k}=0$ then adding two equations and subtracting the third one gets, after contravecting by g^{jk}, the coefficients $\overset{c}{L}{}^i_{jk}$. Thus if we keep fixed K^a_{bk} and M^i_{ja} as in (6.4) then we may speak about the uniqueness of $C\Gamma$, that is, we have

Theorem 6.2. *There exists an unique metrical d-connection* $C\Gamma = (\overset{c}{L}{}^i_{jk}, K^a_{bi}, M^i_{ja}, \overset{c}{C}{}^a_{bc})$ *with the properties:*

i) K^a_{bi} *and* M^i_{ja} *are given by (6.4),*

ii) $\overset{c}{T}{}^i{}_{jk}=0$, $\overset{c}{S}{}^i{}_{jk}=0$.

Then $\overset{c}{L}{}^i_{jk}$ and $\overset{c}{C}{}^i_{jk}$ *are necessarily of the form (6.3).*

Proof. By contradiction.

Also, we have the following more general

Theorem 6.3. *There exists an unique metrical d-connection* $D\Gamma = (L^i_{jk}, K^a_{bi}, M^i_{ja}, C^a_{bc})$ *with the properties*

i) K^a_{bi} *and* M^i_{ja} *are given (6.4),*

ii) Its torsions $T^i{}_{jk}$ *and* $S^a{}_{bc}$ *are apriori given.*

Then the other two coefficients of $D\Gamma$ *are given by*

$$
\begin{aligned}
L^i_{jk} &= \overset{c}{L}{}^i_{jk} + \frac{1}{2} g^{ir}(g_{rs} T^s{}_{jk} - g_{js} T^s{}_{rk} + g_{ks} T^s{}_{jr}), \\
C^a_{bc} &= \overset{c}{C}{}^a_{bc} + \frac{1}{2} h^{ad}(h_{de} S^e{}_{bc} - h_{be} S^e{}_{dc} + h_{ce} S^e{}_{bd}).
\end{aligned}
\tag{6.5}
$$

Proof. By (6.4) one gets $g_{ij}|_a=0$ and $h_{ab|k}=0$. Then $g_{ij|k}=0$ and $h_{ab}|_c=0$ yield, by a Cristoffel process taking into account ii), the coefficients L^i_{jk} and C^a_{bc}, respectively, as they appear in (6.5). The uniqueness follows easily by contradiction.

The curvatures of the metrical d-connection satisfy some identities apart from Bianchi identities. These involve the curvatures with four indices of covariance obtained by lowering the upper indices with the help of g_{ij} and h_{ab}, respectively. Indeed, the Ricci identities for g_{ij} and h_{ab} lead to

Theorem 6.4. *The curvatures of any metrical d-connection $D\Gamma$ satisfy the following identities*

$$\begin{aligned} &R_{ijkh}+R_{jikh}=0, \quad P_{ijkd}+P_{jikd}=0, \quad \tilde{S}_{ijbc}+\tilde{S}_{jibc}=0, \\ &\tilde{R}_{abjk}+\tilde{R}_{bajk}=0, \quad \tilde{P}_{abkd}+\tilde{P}_{bakd}=0, \quad S_{abcd}+S_{bacd}=0. \end{aligned} \tag{6.6}$$

§7. Some Remarkable Metrics on TM

1. We shall say that the (h,v)-metric G given by (6.1) is a generalised Lagrange metric on TM if $g_{ij}=\delta^a_i\delta^b_j h_{ab}$.

These metrics will be extensively studied in a next chapter.

2. The (h,v)-metric G given by (6.1) is said to be h-Riemanian if (g_{ij}) do not depend on y^i and it is said to be v-Riemannian if (h_{ab}) do not depend on y^i. It is now clear what means G is (h,v)-Riemannian. We have

a. G is an h-Riemannian metric if and only if M^i_{ja} from (6.4) vanishes.

b. G is a v-Riemannian metric if and only if $\overset{c}{C}{}^a_{bc}$ from (6.3) vanishes.

c. G is (h,v)-Riemannian metric if and only if

$$M^i_{ja}=0, \quad \overset{c}{C}{}^a_{bc}=0.$$

3. The (h,v)-metric G given by (6.1) is said to be Riemannian-locally Minkowski if it is h-Riemannian and v-locally Minkowsky.

An example of Riemannian-locally Minkowsky metric was given in §4, Ch.V for any vector bundle subject to some additional assumptions. That example can be worked out in this context, too.

Until now the non-linear connection on TM was fixed. We come back to the viewpoint followed in §5 of Ch.III and we give a version of it as follows.

Let us assume TM is endowed with an d-tensor field $h_{ab}(x,y)\delta y^a \otimes \delta y^b$ such that $h_{ab}(x,y) = h_{ba}(x,y)$, $\det(h_{ab}) \neq 0$ and the quadratic form $h_{ab}\xi^a\xi^b$ has constant signature. Using it we construct the following function

$$(6.7) \quad \varepsilon = h_{ab}(x,y)\, y^a y^b .$$

It will be called the *absolute energy* associated to (h_{ab}).

Definition 6.2. *The d-tensor field* $h_{ab}\delta y^a \otimes \delta y^b$ *is said to be weakly regular if the d-tensor field whose components are*

$$(6.8) \quad \tilde{h}_{ab} = \frac{1}{2}\frac{\partial^2 \varepsilon}{\partial y^a \partial y^b}$$

is nondegenerate, i.e. $\det(\tilde{h}_{ab}) \neq 0$.

We have

Theorem 6.5. *If* $h_{ab}\delta y^a \otimes \delta y^b$ *is a weakly regular v-metric on TM then the functions*

$$(6.9) \quad N^a_i(x,y) = \dot{\partial}_b G^a(x,y)\delta^b_i, \; G^a = \frac{1}{2}\tilde{h}^{ab}[(\dot{\partial}_b \partial_k \varepsilon)\delta^k_c y^c - (\partial_k \varepsilon)\delta^k_b]$$

are the coefficients of a non-linear connection completely determined by (h_{ab}).

Proof. A tedious computation shows that under a change of coordinates on TM, (N^a_i) from (6.9) transforms by the desired law.

Coming back to the v-locally Minkowski metric on TM we have

Proposition 6.2. *If the (h,v)-metric G given by (6.1) is v-locally Minkowski and* (h_{ab}) *is a weakly regular v-metric, then in a coordinate neighbourhood in which* (h_{ab}) *does not depend on* x^i *we have also*

i) $N^a_j = 0,\; \overset{c}{L}{}^i_{jk} = \left\{ {i \atop j\,k} \right\},\; K^a_{bj} = 0,$

ii) $\overset{c}{T}{}^i{}_{jk} = 0,\; \overset{c}{S}{}^a{}_{bc} = 0,\; \overset{c}{R}{}^a{}_{ij} = 0,\; \overset{c}{P}{}^a{}_{jb} = 0,$

iii) $\overset{c}{R}_j{}^i{}_{hk} = r_j{}^i{}_{hk}$, $\overset{c}{\tilde{R}}_b{}^a{}_{jk} = 0$, $\overset{c}{\tilde{P}}_b{}^a{}_{kc} = 0$,

where $\left\{ {i \atop j\ k} \right\}$ *are the Christoffel symbols and* $r_j{}^i{}_{hk}$ *the curvature tensor constructed with*(g_{ij}). *The superscript c refers to the metrical canonical connection* $C\Gamma$.

Proof. Since (h_{ab}) does not depend on x^i then G^a in (6.9) vanishes, hence N_i^a vanishes, too. It follows $\delta_i = \partial_i$ and by (6.3) one gets $\overset{c}{L}{}^i_{jk} = \left\{ {i \atop j\ k} \right\}$. Then a glance to (6.4) says us $K^a_{bj} = 0$. The other equations are now clear.

Proposition 6.3. *If the (h,v)-metric G of the form (6.1) is a Riemannian-locally Minkowski metric such that* (h_{ab}) *is a weakly regular v-metric, then*

i) The metrical canonical connection $C\Gamma$ *has the coefficients:*

$$\overset{c}{L}{}^i_{jk} = \left\{ {i \atop j\ k} \right\},\ K^a_{bj} = 0,\ \overset{c}{C}{}^a_{bc} = \frac{1}{2} h^{ad}(\dot{\partial}_c h_{db} + \dot{\partial}_b h_{dc} - \dot{\partial}_d h_{bc}).$$

ii) The metrical connection $C\Gamma$ *coincide to the Levi-Civita connection of G, that is, all its torsions vanish.*

iii) The curvature of $C\Gamma$ *are as follows*

$$R_j{}^i{}_{hk} = r_j{}^i{}_{hk},\quad \tilde{R}_b{}^a{}_{jk} = 0,\quad P_j{}^i{}_{kc} = 0,\quad \tilde{P}_b{}^a{}_{kc} = 0$$

$\tilde{S}_j{}^i{}_{bc} = 0$ and $S_b{}^a{}_{cd}$ is given by $(3.13)_6$ in Ch.III.

Proof. As $N_i^a = 0$ we get $\delta_i = \partial_i$ and then by (6.3) and (6.4) we obtain i). Then an easy computation shows that the torsions of $C\Gamma$ vanish, that is $C\Gamma$ coincide to the Levi-Civita connection of the Riemannian metric G on TM. By i) and ii) one obtain iii) on account of (3.13) in Ch.III.

Chapter VIII

Finsler Spaces

Discovered in 1938 by Paul Finsler, the spaces bearing his name came into science due mainly to L. Synge, L. Berwald and E. Cartan. Two outstanding books written by H. Rund and M. Matsumoto respectively, definitely established their place in contemporary geometry.

The concept of Finsler space was introduced as a very natural generalization of that of Riemann space, but the efforts in studying it by using the methods from Riemannian geometry have met a lot of complications. Some adequate geometrical models have finally enabled one to understand the intricate nature of these spaces. One of them, a creation of Romanian geometers ([109]), will be extensively described in this chapter.

A permanent impulse for researching the geometry of Finsler spaces has been their applications to mechanics, variational calculus, theoretical physics including gauge theories, optimal control etc.

The idea to present, even in a concise form, the geometry of Finsler spaces is compelling because of the arguments above as well as the fact that the methods which will be used are of great importance in the researching of the geometry of Lagrange spaces, as we will see in the next chapter.

Thus we shall present the notion of Finsler space together with the fundamental geometrical objects associated with it: the Cartan non-linear connection, the metrical Cartan connection, the Berwald, Rund and Hashiguchi connections and the commutative diagram these satisfy. The structure equations of the metrical Cartan connection are established. The almost Kählerian model of a Finsler space will be gradually constructed. Finally the theory of subspaces in Finsler spaces, including the Gauss-Codazzi equations, is outlined.

§1. The Notion of Finsler Space

Let M be a smooth n-dimensional real manifold, (TM,τ,M) its tangent bundle and $\tilde{TM} = TM\setminus\{0\}$, where 0 means the image of the null cross-section of the projection $\tau : TM \to M$. The restriction to $\tilde{TM}$ of the vertical distribution will be denoted by the same letter V. If (x^i,y^i) is a coordinate system on TM, then (x^i,y^i) with not all y^i vanishing is a coordinate system on $\tilde{TM}$. As usual, the indices i,j,k,... run from 1 to n. The local vector fields $\dot{\partial}_1 = \frac{\partial}{\partial y^1}, \ldots, \dot{\partial}_n = \frac{\partial}{\partial y^n}$ give a local basis of the vertical distribution V.

The vector field

$$(1.1)\quad C = y^i \dot{\partial}_i$$

is globally defined and nowhere zero on $\tilde{TM}$. It will be called the *Liouville vector field on* $\tilde{TM}$.

We notice that if $f \in \mathcal{F}(\tilde{TM})$ then $\dot{\partial}_i \dot{\partial}_j f$ gives a symmetric d-tensor field of type (0,2).

Now we are prepared to formulate

Definition 1.1. *A Finsler metric on M is a function $F: TM \to \mathbf{R}$ having the properties:*

1. The restriction of F to $\tilde{TM}$ is of the class C^∞ and F is only continuous on the image of the null cross-section in the tangent bundle to M.

2. The restriction of F to $\tilde{TM}$ is positively homogeneous of degree 1 with respect to (y^i).

3. The restriction of F to $\tilde{TM}$ is a positive function.

4. The quadratic form on $\mathbf{R}^n$ with the coefficients

$$(1.2)\quad g_{ij}(x,y) = \frac{1}{2} \frac{\partial^2 F^2}{\partial y^i \partial y^j},$$

defined on $\tilde{TM}$, is positive definite.

The conditions listed in the above are very strong.

Let us notice that 2. reads on $\tilde{TM}$ as follows

$$(1.3)\quad F(x, \lambda y) = \lambda F(x,y),\ \lambda \in \mathbf{R}_+^*.$$

Thus, $g_{ij}(x,y)$ are the components of a symmetric d-tensor field of type (0,2), 0-homogeneous with respect to (y^i).

The condition 4. tell us that on $\tilde{TM}$:

$$(1.4)\quad \operatorname{rank}\left(g_{ij}(x,y)\right) = n,$$

holds.

This is a reason to say the d-tensor field $g_{ij}(x,y)$ is nondegenerate or nonsingular.

Definition 1.2. *A pair $(M,F)=F^n$ which consists of a smooth n-dimensional real manifold M and a Finsler metric F on M is said to be a Finsler space.*

The following terms are classical ones. The function F, written often as F(x,y) is called the fundamental function of F^n and $(g_{ij}(x,y))$ from (1.2) is called the metric tensor, or the fundamental tensor of the Finsler space F^n.

Examples. 1. The *Riemannian spaces* $\mathfrak{R}^n=(M,\gamma_{ij}(x))$ are Finsler spaces with the fundamental function $F(x,y)=\sqrt{\gamma_{ij}(x)y^iy^j}$.

2. The *Randers spaces* are Finsler spaces with the fundamental function $F(x,y)=\sqrt{\gamma_{ij}(x)y^iy^j}+b_i(x)y^i$, where $\gamma_{ij}(x)$ is the metric tensor of a Riemannian space and $b_i(x)y^i\neq 0$ on $\tilde{TM}$.

3. *The Kropina spaces* are Finsler spaces with the fundamental function $F(x,y)=\gamma_{ij}(x)y^iy^j/\beta_i(x)y^i$ where again $\gamma_{ij}(x)$ is the metric tensor of a Riemannian space and $b_i(x)y^i\neq 0$ on $\tilde{TM}$.

The properties 1. - 4. from the Definition 1.1 are easily verified for all these three functions F.

Now we shall characterise the Riemannian space among the Finsler spaces. To this aim let be the following d-tensor field

$$(1.5)\quad C_{ijk}=\frac{1}{4}\dot{\partial}_i\dot{\partial}_j\dot{\partial}_kF^2 .$$

It is obvious this is totally symmetric on $\tilde{TM}$ and we have

$$(1.6)\quad C_{ijk}=\frac{1}{2}\dot{\partial}_kg_{ij},\ C_{0jk}:=y^iC_{ijk}=0.$$

Thus it easily follows

Proposition 1.1. *A Finsler space $F^n=(M,F(x,y))$ is a Riemannian space if and only if the d-tensor field C_{ijk} vanishes on $\tilde{TM}$.*

From now on we set

$$(1.7)\quad y_i=\frac{1}{2}\dot{\partial}_iF^2.$$

Then y_i is a d-tensor field on $\tilde{TM}$ and the positive homogeneity of degree 2 with respect to (y^i) of F^2 implies

(1.8) $y_i = g_{ij} y^j$.

If we denote by $g^{ij}(x,y)$ the reciprocal of the d-tensor field $g_{ij}(x,y)$ given by

$$g_{ij}(x,y) g^{jk}(x,y) = \delta_i^k$$

we have also

(1.8)′ $y^i = g^{ij} y_j$.

Consequently,

(1.9) $\omega = y_i dx^i = g_{ij} y^j dx^i = \frac{1}{2}\dot{\partial}_i F^2 dx^i$

is an 1-form on $\tilde{TM}$. It is called the *Liouville form* of the space F^n.

We finish this section with the remark that when the applications will be involved, we shall drop the conditions 3. and 4. requiring only $F(x,y) > 0$ on a domain of $\tilde{TM}$ and on that domain (1.4) holds and the signature of the quadratic form with the coefficients $g_{ij}(x,y)$ from (1.2) be constant. Even in these weaker assumptions we shall preserve the terminology in the above. In fact, the considerations which follow are valid, with some cautions, under these new assumptions.

§2. Non-linear Cartan Connection

The fundamental tensor $g_{ij}(x,y)$ of the Finsler space F^n allows us to define the so-called *absolute energy* of F^n:

(2.1) $\varepsilon(x,y) = g_{ij}(x,y) y^i y^j$.

Using (1.7) and (1.8) one easily obtains

Proposition 2.1. *The absolute energy of the Finsler space F^n is equal to the square of the fundamental function i.e.*

(2.2) $\varepsilon(x,y) = F^2(x,y)$.

If we denote by $\|X\|^2 = g_{ij}(x,y)X^iX^j$, the square of the length of a d-vector field, then for the Liouville vector field we have

(2.2)′ $\|C\|^2 = F^2(x,y) = y_i y^i$.

We shall determine the non-linear Cartan connection by the help of a variational problem formulated for the absolute energy ε.

A parametrized curve of class C^k, $k \geq 2$, in M is a pair (Γ,γ), where $\gamma: J \to M$, $t \to x(t)$ is a mapping of class C^k, J is an interval in $\mathbb{R}$ and $\Gamma = \mathrm{Im}\,\gamma$ is the image of γ. We shall take $J=[0,1]$ and we shall consider

(2.3) $$I(\Gamma) = \int_0^1 \varepsilon(x,\frac{dx}{dt})dt.$$

$I(\Gamma)$ will be called the action integral of the absolute energy. Here $x^i = x^i(t)$ is the local representation of γ in a coordinate system and $\frac{dx^i}{dt}$ are the components of its tangent vector.

Let A and B be the ends $\gamma(0)$ and $\gamma(1)$ of Γ, respectively, and $v^i(x(t))$ a vector along Γ such that $v^i(A)=v^i(B)=0$. Let us consider the set of curves Γ_ε given by all mappings $t \to x^i(t) + \varepsilon v^i(x(t))$, with $|\varepsilon|$ small. The curves Γ_ε have as end the points A and B and in a clear sense these are neighbours of Γ. The action integral $I(\Gamma_\varepsilon)$ is as follows

$$I(\Gamma_\varepsilon) = \int_0^1 \varepsilon(x(t)+\varepsilon v(t),\frac{dx}{dt}+\varepsilon\frac{dv}{dt})dt.$$

A necessary condition for Γ be an extremal of the action integral is $\frac{dI(\Gamma_\varepsilon)}{d\varepsilon}\Big|_{\varepsilon=0} = 0$ This equality is equivalent to

$$\int_0^1 (\frac{\partial\varepsilon}{\partial x^i}v^i + \frac{\partial\varepsilon}{\partial v^i}\frac{dv^i}{dt})dt = 0.$$

Integrating by parts the second term one gets

$$\int_0^1 v^i[\frac{\partial\varepsilon}{\partial x^i} - \frac{d}{dt}(\frac{\partial\varepsilon}{\partial y^i})]dt = 0, \quad (y^i = \frac{dx^i}{dt}).$$

Since v^i is an arbitrary d-vector field we get:

A necessary condition that the parametrized curve (Γ,γ) be an extremal of the action integral is that the mapping $t \to x(t)$, $t \in [0,1]$ satisfy the system of Euler-Lagrange equations

$$(2.4) \quad \frac{d}{dt}(\frac{\partial\varepsilon}{\partial y^i}) - \frac{\partial\varepsilon}{\partial x^i} = 0, \quad y^i = \frac{dx^i}{dt}.$$

The consideration of the parametrized curves is required by certain applications to theoretical physics where the parameter t is a specific one e.g. proper time.

Lemma 2.1. *The system of Euler-Lagrange equations(2.4) is equivalent to*

$$(2.5) \quad \frac{d^2x^i}{dt^2} + 2G^i(x, \frac{dx}{dt}) = 0,$$

where

$$(2.6) \quad G^i(x,y) = \frac{1}{4}g^{ij}(\frac{\partial^2\varepsilon}{\partial y^j \partial x^k}y^k - \frac{\partial\varepsilon}{\partial x^j}), \quad y^i = \frac{dx^i}{dt}.$$

Proof. Expanding the derivative with respect to t and taking into account (2.2), the system of equations (2.4) take the form

$$2g_{ik}\frac{d^2x^k}{dt^2} + (\frac{\partial^2\varepsilon}{\partial y^i \partial x^k}y^k - \frac{\partial\varepsilon}{\partial x^i}) = 0, \quad y^i = \frac{dx^i}{dt},$$

which is transformed into (2.5) with the notation (2.6) by expliciting $\frac{d^2x^i}{dt^2}$.

Now let $\gamma^i_{jk}(x,y)$ be the Christoffel symbols constructed by using the fundamental tensor $g_{ij}(x,y)$, i.e.

$$(2.7) \quad \gamma^i_{jk}(x,y) = \frac{1}{2}g^{ih}(\partial_j g_{hk} + \partial_k g_{jh} - \partial_h g_{jk}),$$

where ∂_i stands for $\frac{\partial}{\partial x^i}$. We consider also

(2.7)′ $\gamma^i_{00} = \gamma^i_{jk}(x,y)y^j y^k.$

Then we have

Theorem 2.1. *The extremal curves of the action integral in a Finsler space are solutions of the following system of differential equations*

$$(2.8)\quad \frac{d^2x^i}{dt^2} + \gamma^i_{jk}(x, \frac{dx}{dt})\frac{dx^j}{dt}\frac{dx^k}{dt} = 0.$$

Proof. Inserting $\varepsilon(x,y)$ given by (2.1) in (2.6) and using the notations introduced in the above, G^i becomes

$$G^i = \frac{1}{4}g^{ij}[2(\partial_k C_{jrs})y^k y^r y^s + (\partial_s g_{jr} + \partial_r g_{js} - \partial_j g_{rs})y^r y^s].$$

On the other hand we know that $C_{0ij} = 0$ and thus

$$(2.9)\quad G^i = \frac{1}{2}\gamma^i_{00}.$$

Consequently, the system of equations (2.5) takes the form(2.8).

Now we are ready to prove

Theorem 2.2. *The real functions*

$$(2.10)\quad N^i_j(x,y) = \frac{\partial G^i}{\partial y^j} = \frac{1}{2}\frac{\partial \gamma^i_{00}}{\partial y^j}$$

are the local coefficients of a non-linear connection on $\tilde{TM}$, which depends only on the fundamental function $F(x,y)$ of the Finsler space F^n.

Proof. When a change of coordinates $(x^i, y^i) \rightarrow (\bar{x}^i, \bar{y}^i)$ on $\tilde{TM}$ is performed we see by using (2.1) that G^i transforms as follows

(2.11) $\bar{G}^i = \frac{\partial \bar{x}^i}{\partial x^k} G^k + \frac{1}{2} \frac{\partial \bar{x}^i}{\partial x^r} \frac{\partial^2 x^r}{\partial \bar{x}^k \partial \bar{x}^h} \bar{y}^k \bar{y}^h .$

On account of $\frac{\partial}{\partial \bar{y}^j} = \frac{\partial x^i}{\partial \bar{x}^j} \frac{\partial}{\partial y^i}$ it comes out that $N^i_j(x,y)$ from (2.10) transforms as the coefficients of a non-linear connection (see (1.3), ch. VII).

The non-linear connection N whose local coefficients are given by (2.10) is an important tool in the study of the Finsler spaces. It is called the *non-linear Cartan connection* of the space F^n. Notice that it is a strict homogeneous connection.

As we have seen in §2 of ch. VII, the autoparallel curves of the non-linear connection $N^i_j(x,y)$ are solutions of the following system of differential equations

(2.12) $\frac{d^2x^i}{dt^2} + N^i_j(x, \frac{dx}{dt}) \frac{dx^j}{dt} = 0 .$

Thus we have

Theorem 2.3. *The autoparallel curves of the non-linear Cartan connection of a Finsler space F^n coincide to the extremal curves of the integral action of the absolute energy of F^n.*

Proof. Inspecting (2.7), (2.7)′ and (2.9) we find that the functions $G^i(x,y)$ are positively homogeneous of degree 2 with respect to (y^i). Thus by the Euler theorem we get

$$N^i_j y^j = \frac{\partial G^i}{\partial y^j} y^j = 2G^i = \gamma^i_{00} .$$

Hence the systems of differential equations (2.8) and (2.10) coincide.

§3. Geodesics

The fundamental function F(x,y) of a Finsler space F^n generalizes the metric function $\sqrt{\gamma_{ij}(x) y^i y^j}$ of a Riemannian space whose metric tensor is $\gamma_{ij}(x)$. This fact suggests to us to define the length $l(\Gamma)$ of a parametrized curve (Γ,γ), $\gamma:[0,1]\rightarrow M$ as

$$(3.1)\quad l(\Gamma) = \int_0^1 F(x, \frac{dx}{dt})dt.$$

The hypothesis 1.-4. in the Definition 1.1 assure $l(\Gamma)$ does not depend on the parametrization γ of Γ, as it can be directly checked. Let us consider the function $s(t) = \int_{t_0}^{t} F(x, \frac{dx}{d\sigma})d\sigma$, with $t_0, t \in [0,1]$. We have $\frac{ds}{dt} = F(x, \frac{dx}{dt}) > 0$. It follows the function $t \to s(t)$ is invertible on $[0,1]$ and its inverse h is of the class C^k. The parametrization $\gamma \circ h$ of Γ is called a natural one. Also it is said that Γ is parametrized by the arc-length s, a parameter called natural and sometimes affine parameter. When the natural parameter, for Γ, is used we have

$$(3.2)\quad F(x, \frac{dx}{ds}) = 1.$$

Starting from the integral (3.1) an usual variational problem (with fixed end points) may be formulated using, of course, parametrized curves. Proceeding as for the integral action we find now that the extremals $t \to x(t)$, $t \in [0,1]$ must necessarily satisfy the following system of Euler-Lagrange equations

$$(3.3)\quad \frac{d}{dt}(\frac{\partial F}{\partial y^i}) - \frac{\partial F}{\partial x^i} = 0, \quad y^i = \frac{dx^i}{dt}.$$

These extremals are called very often the *geodesics* of the Finsler space F^n.

An easy calculation on account of $F(x,y) = \sqrt{\varepsilon(x,y)} = \sqrt{g_{ij}(x,y)y^i y^j}$ shows that (3.3) is equivalent to

$$(3.3)'\quad \frac{d}{dt}(\frac{\partial \varepsilon}{\partial y^i}) - \frac{\partial \varepsilon}{\partial x^i} = \frac{1}{F(x,y)}\frac{\partial \varepsilon}{\partial y^i}\frac{dF}{dt}, \quad y^i = \frac{dx^i}{dt}.$$

Passing now to the natural parameter s, we have $\frac{d}{ds}F(x, \frac{dx}{ds}) = 0$ and the equations (3.3)′ read

$$(3.4)\quad \frac{d^2x^i}{ds^2} + \gamma^i_{jk}\frac{dx^j}{ds}\frac{dx^k}{ds} = 0,$$

by virtute of certain formulae found in the above. Thus we proved

Theorem 3.1. *In a Finsler space F^n, the geodesics, the autoparallel curves of the non-linear Cartan connection and the extremals of the action integral are solutions of the system of differential equations (3.4) with s as natural parameter given by the arc-length.*

§4. Metrical Cartan Connection

We shall prove that in a Finsler space F^n there exists a metrical d-connection with few torsions. It is basic in Finsler geometry. We shall use throughout in what follows the non-linear Cartan connection given by (2.10).

Recall that

(4.1) $\delta_i = \partial_i - N_i^k(x,y)\dot{\partial}_k$

give a local basis for the horizontal distribution N on $\tilde{TM}$ and $(\delta_i, \dot{\partial}_i)$ is a local basis of the $\mathcal{F}$(TM)-module $\chi(\tilde{TM})$ adapted to the horizontal and vertical distributions. The basis dual to it is $(dx^i, \delta y^i)$, where

(4.2) $\delta y^i = dy^i + N_k^i(x,y)dx^k$.

When a normal d-connection on $\tilde{TM}$ is given, using the methods given in §5 ch. VII an algorithm of h- and v- covariant derivation of the d-tensor fields on $\tilde{TM}$ can be found. As usual the h-covariant derivative will be denoted by a small vertical bar and the v-covariant derivative by a long vertical bar e.g. $t^i_{j|k}$, $t^i_j|_k$, respectively. Recall also that a d-connection on TM or $\tilde{TM}$ can be defined only by using a non-linear connection.

Now we are prepared to prove

Theorem 4.1. *i) There exists an unique d-connection $C\Gamma = (N^i_j, F^i_{jk}, C^i_{jk})$ with the properties:*

1. *N^i_j is the Cartan non-linear connection (2.10);*
2. *$C\Gamma$ is compatible with the fundamental tensor $g_{ij}(x,y)$, that is*

(4.3) $g_{ij|k} = 0, \quad g_{ij}|_k = 0.$

3. *Its h(hh)- and v(vv)- torsions, $T^i_{\ jk}$ and $S^i_{\ jk}$, respectively, identically vanish.*

ii) The coefficients of $C\Gamma$ are given by (2.10) and by the "generalized Christoffel symbols":

(4.4) $F^i_{jk} = \frac{1}{2} g^{ir}(\delta_j g_{rk} + \delta_k g_{jr} - \delta_r g_{jk}) \quad C^i_{jk} = \frac{1}{2} g^{ir}(\dot{\partial}_j g_{rk} + \dot{\partial}_k g_{jr} - \dot{\partial}_r g_{jk})$.

Proof. In order to prove the existence of the desired d-connection it is enough to check that F^i_{jk} and C^i_{jk} from (4.4) verify 2. and 3. when, of course, N^j_i is given by (2.10). First, it is obvious that $F^i_{jk} = F^i_{kj}$ and $C^i_{jk} = C^i_{kj}$, hence $T^i{}_{jk} = 0$ and $S^i{}_{jk} = 0$. Noticing that (4.3) is equivalent to

(*) $\delta_k g_{ij} - F^h_{ik} g_{hj} - F^h_{jk} g_{ih} = 0$,

(**) $\dot{\partial}_k g_{ij} - C^h_{ik} g_{hj} - C^h_{jk} g_{ih} = 0$

it is a matter of a simple calculation to check it and thus to confirm 2. Now, let us assume there exists another d-connection $C\tilde{\Gamma} = (N^i_j, \tilde{F}^i_{jk}, \tilde{C}^i_{jk})$ verifying 1.-3. in i). If in (*), written for $\tilde{F}^i_{jk}$ we cyclicly permute the indices i,j,k we obtain two new equations. Adding them and subtracting that initial one, on account of $\tilde{F}^i_{jk} = \tilde{F}^i_{kj}$ one gets $\tilde{F}^i_{jk} = F^i_{jk}$ given in (4.4). Starting from (**) written for $\tilde{C}^i_{jk}$, on a similar way we find $\tilde{C}^i_{jk} = C^i_{jk}$. Thus the uniqueness of $C\Gamma$ is proved.

Let us notice that $C\Gamma$ is completely determined by the fundamental function F(x,y) of the Finsler space F^n. It is called the *metrical Cartan connection* of F^n, or canonical metrical d-connection of F^n. We call it, for brevity, the Cartan connection of F^n.

Now we proceed to establish some properties of $C\Gamma$.

Proposition 4.1. *The h- and v- deflection tensor fields of the Cartan connection* $C\Gamma$ *are given by*

(4.5) $D^i_j = 0, \quad d^i_j = \delta^i_j$.

Proof. We have $D^i_j = y^i{}_{|j} = y^r F^i_{rj} - N^i_j =$

$$= \frac{1}{2} g^{is}[(\partial_j - N^h_j \dot{\partial}_h) g_{rs} + (\partial_r - N^h_r \dot{\partial}_h) g_{sj} - (\partial_s - N^h_s \dot{\partial}_h) g_{rj}] y^r - \frac{1}{2} \dot{\partial}_j \gamma^i_{00} = \gamma^i_{0j} - \gamma^i_{0j} = 0$$

because of $y^r \dot{\partial}_r \gamma^h_{00} = 0$. The equality $d^i_j = \delta^i_j$ is immediate as a consequence of the equations $C_{0ij} = C_{i0j} = 0$.

Let us recall that $C\Gamma$ has a torsion $P^i{}_{jk} = \dot{\partial}_k N^i_j - F^i_{kj}$ and let us set $P_{ijk} = g_{ih} P^h{}_{jk}$.

A direct calculation leads us to

Theorem 4.2. *The following properties of the Cartan connection* $C\Gamma$ *hold good:*

1. $F_{|k}=0,\ F|_k=\frac{1}{F}y_k,$
2. $F^2_{|k}=0,\ F^2|_k=2y_k,$
3. $y_{i|k}=0,\ y_i|_k=g_{ik},$
4. $P_{ijk}=C_{ijk|0},$

where $C_{ijk}=g_{jk}C_i{}^h{}_k$ *and* $C_{ijk|0}=C_{ijk|h}y^h$.

Throughout the indice 0 means the contraction by y^i.

As we have seen in §5, Ch.VII, another torsion of $C\Gamma$ is

$$(4.6)\quad R^i{}_{jk}=\delta_k N^i_j-\delta_j N^i_k.$$

This is also the integrability tensor of the horizontal distribution of the non-linear Cartan connection in a sense that it vanishes if and only if that horizontal distribution is integrable (cf. §5, Ch. II).

Let us recall the following Ricci identities satisfied by any d-connection, particularly by $C\Gamma$.

$$\begin{aligned} &X^i{}_{|k|h}-X^i{}_{|h|k}=X^rR_r{}^i{}_{kh}-X^i|_rR^r{}_{kh},\\ (4.7)\quad &X^i{}_{|k}|_h-X^i|_{h|k}=X^rP_r{}^i{}_{kh}-X^i{}_{|r}C^r_{kh}-X^i|_rP^r{}_{kh},\\ &X^i|_k|_h-X^i|_h|_k=X^rS_r{}^i{}_{kh}.\end{aligned}$$

In these identities $R_j{}^i{}_{kh}$, $P_j{}^i{}_{kh}$ and $S_j{}^i{}_{kh}$ are the curvatures of $C\Gamma$ given by the formulae (3.13), Ch. III, in which some few obvious changes must be operated.

Applying the Ricci identities to the Liouville vector field y^i and taking into account $y^i_{|k}=0$ and $y^i|_k=\delta^i_k$ one obtains

Proposition 4.2. *The following identities hold good*

$$(4.8)\quad R_0{}^i{}_{jk}=R^i{}_{jk},\ P_0{}^i{}_{jk}=P^i{}_{jk},\ S_0{}^i{}_{jk}=0.$$

For a d-tensor field $t_{ij}(x,y)$ the Ricci identities, i.e. the commutation formulae produced by h- and v- covariant derivative associated to any d-connection read as follows

$$
\begin{aligned}
& t_{ij|k|h} - t_{ij|h|k} = -t_{rj} R_i{}^r{}_{kh} - t_{ir} R_j{}^r{}_{kh} - t_{ij}|_r R^r{}_{kh}, \\
(4.9)\quad & t_{ij|k}|_h - t_{ij}|_{h|k} = -t_{rj} P_i{}^r{}_{kh} - t_{ir} P_j{}^r{}_{kh} - t_{ij|r} C^r_{kh} - t_{ij}|_r P^r{}_{kh}, \\
& t_{ij}|_k|_h - t_{ij}|_h|_k = -t_{rj} S_i{}^r{}_{kh} - t_{ir} S_j{}^r{}_{kh}.
\end{aligned}
$$

Applying these identities to g_{ij} and $C\Gamma$ we get

Proposition 4.3. *The curvatures of $C\Gamma$ satisfy the following identities*

$$(4.10)\quad R_{ijkl} + R_{jikh} = 0,\ P_{ijkh} + P_{jikh} = 0,\ S_{ijkh} + S_{jikh} = 0.$$

The lowering of indices was done by using g_{ij} e.g. $R_{ijkh} = g_{jr} R_i{}^r{}_{kh}$.

Some tedious computations but without any conceptual difficulties give

Proposition 4.4. *The curvature of $C\Gamma$ satisfy also the following identities*

$$
\begin{aligned}
& S_{jikh} = C_{iks} C^s_{jh} - C_{ihs} C^s_{jk}, \\
(4.11)\quad & P_{ihkj} = C_{ijk|h} - C_{hjk|i} + C^r_{hj} C_{rik|0} - C^r_{ij} C_{rkh|0}, \\
& P_{ijkh} - P_{ijhk} = -S_{ijkh|0}.
\end{aligned}
$$

We shall associate to $C\Gamma$ the following geometrical object

$$(4.12)\quad \theta = g_{ij}(x,y)\delta y^i \wedge dx^j.$$

We have

Theorem 4.3. *i) θ is a 2-form on $\tilde{TM}$.*

ii) The exterior differential of the Liouville 1-form ω coincide with θ, that is:

$$(4.13)\quad d\omega = \theta.$$

iii) θ is a symplectic structure on $\tilde{TM}$.

Proof. i) is an immediate consequence of the fact that $g_{ij}(x,y)$ are the local components of a d-tensor field of type (0,2).

ii) It is easy to see that for any function $f \in \mathcal{F}(\tilde{TM})$ we have

$$df = \delta_k f\, dx^k + \dot{\partial}_k f\, \delta y^k.$$

Hence the exterior differential of $\omega = y_i\, dx^i$ is

$$d\omega = dy_i \wedge dx^i = (\delta_k y_i dx^k + \dot{\partial}_k y^i \delta y^k) \wedge dx^i.$$

Taking into account 3. in Theorem 4.2. and $\dot{\partial}_k y_i = g_{ki}$ one gets

$$d\omega = \frac{1}{2}(\delta_k y_i - \delta_i y_k) dx^k \wedge dx^i + g_{ki} \delta y^k \wedge dx^i = \theta.$$

iii) The matrix of the 2-form θ in the adapted basis $(dx^i, \delta y^i)$ is $\frac{1}{2}\begin{bmatrix} 0 & -g_{ij} \\ g_{ij} & 0 \end{bmatrix}$ whose determinant is clear not null. Thus θ is an almost symplectic structure. But ii) tells us θ is closed, hence it is a symplectic structure.

A consequence of $d\theta = 0$ reads as follows

Proposition 4.5. *The Cartan connection $C\Gamma$ has the following properties*
(4.14) $\sum_{ijk} R_{ijk} = 0$, P_{ijk} *is totally symmetric.*

Here $\sum_{ijk}$ means cyclic sum with respect to the indices i, j, k.

§5. Structure Equations. Bianchi Identities

If $X^i(x,y)$ is a d-vector field and $C\Gamma$ is the Cartan connection of the Finsler space F^n, then the covariant differential of X^i, with respect to $C\Gamma$ is given by

(5.1) $DX^i = X^i_{|k} dx^k + X^i|_k \delta y^k = dX^i + X^j \omega^i_j$,

where ω^i_j are the 1-forms of connection:

(5.2) $\omega^i_j = F^i_{jk} dx^k + C^i_{jk} \delta y^k$.

These 1-forms, together with the exterior differentials of the 1-forms δy^i:

(5.3) $d(\delta y^i) = -\frac{1}{2}R^i{}_{jk}dx^j \wedge dx^k - \dot{\partial}_k N_j^i dx^j \wedge \delta y^k$

allow us to prove:

Theorem 5.1. *The 1-forms of the Cartan connection $C\Gamma$ satisfy the structure equations*

$$(5.4)\quad \begin{aligned} d(dx^i) - dx^h \wedge \omega_h^i &= -\Omega^i \\ d(\delta y^i) - \delta y^h \wedge \omega_h^i &= -\tilde{\Omega}^i \end{aligned}$$

$$(5.4)'\quad d\omega_j^i - \omega_j^h \wedge \omega_h^i = -\Omega_j^i,$$

where the 2-forms of torsion Ω^i and $\tilde{\Omega}^i$ are:

$$(5.5)\quad \Omega^i = C_{jk}^i dx^j \wedge \delta y^k \ , \ \tilde{\Omega}^i = \frac{1}{2}R^i{}_{jk}dx^j \wedge dx^k + P^i{}_{jk}dx^j \wedge \delta y^k$$

while the 2-forms of curvature Ω_j^i are as follows

$$(5.5)'\quad \Omega_j^i = \frac{1}{2}R_j{}^i{}_{kh}dx^k \wedge dx^h + P_j{}^i{}_{kh}dx^k \wedge \delta y^h + \frac{1}{2}S_j{}^i{}_{kh}\delta y^k \wedge \delta y^h .$$

Proof. It is clear that

$$d(dx^i) - dx^h \wedge \omega_h^i = -dx^h \wedge \omega_h^i$$

and by (5.2) it follows $(5.4)'$ with Ω^i given by (5.5). Recall that v(vv)-torsion of $C\Gamma$ vanishes. Using (5.3) in the calculation of $d(\delta y^i) - \delta y^h \wedge \omega_h^i$ we find it is $-\tilde{\Omega}^i$. Exterior differentiating (5.2) and using again (5.2) we arrive at $(5.4)'$ with Ω_j^i given by $(5.5)'$.

The structure equations (5.4) and $(5.4)'$ give certain geometrical meanings for the torsions and the curvatures of $C\Gamma$. We shall use these equations in deducing the Bianchi identities satisfied by the torsions and curvatures of $C\Gamma$. These will be obtained as differential consequences of the system of exterior equations (5.4) and $(5.4)'$ in which these equations are used again. Writing all these consequences and taking into account that

$$dx^i \wedge dx^j \wedge dx^k,\ i<j<k,\ dx^i \wedge dx^j \wedge \delta y^k,\ i<j,\ dx^i \wedge \delta y^j \wedge \delta y^k,$$
$$j<k\ \ \delta y^i \wedge \delta y^j \wedge \delta y^k\ \ i<j<k\}$$

is a local basis of the $\mathscr{F}(\tilde{TM})$-module of 3-forms on TM, one finds eleven Bianchi identities:

$$\begin{aligned}
&\sum_{(jkl)} \{R_{j\ kl}^{\ i} - R^{h}_{\ jk}C^{i}_{hl}\} = 0\\
(5.6)_1\quad &\sum_{(jkl)} \{R_{j\ k|l}^{\ i} + R^{h}_{\ jk}P^{i}_{\ hl}\} = 0\\
&\sum_{(jkl)} \{R_{s\ jk|l}^{\ i} + R^{h}_{\ jk}P_{s\ lh}^{\ i}\} = 0
\end{aligned}$$

$$(5.6)_2\quad \sum_{(jkl)} S_{j\ kl}^{\ i} = 0,\ \sum_{(jkl)} S_{s\ jk}^{\ i}|_{l} = 0$$

Next $\underset{(jk)}{a}$ denotes the interchange of the indices j,k and substraction.

$$\begin{aligned}
&\underset{(jk)}{a}\{C^{i}_{kl|j} + C^{i}_{jh}P^{h}_{\ kl} - P_{j\ kl}^{\ i}\} = 0,\\
(5.6)_3\quad &\underset{(jk)}{a}\{R^{i}_{\ jk}C^{h}_{kl} + P^{i}_{\ jh}P^{h}_{\ kl} + P^{i}_{kl|j}\} = R_{l\ jk}^{\ i} - R^{i}_{jk|l},\\
&\underset{(jk)}{a}\{R_{s\ jh}^{\ i}C^{h}_{kl} + P_{s\ jh}^{\ i}P^{h}_{\ kl} + P_{s\ kl|j}^{\ i}\} = -S_{s\ lh}^{\ i}R^{h}_{\ jk} - R_{s\ jk}^{\ i}|_{l},
\end{aligned}$$

$$\begin{aligned}
&\underset{(jk)}{a}\{\dot{\partial}_{k}C^{i}_{sj} + C^{h}_{sj}C^{i}_{hk}\} = S_{s\ jk}^{\ i},\\
(5.6)_4\quad &\underset{(jk)}{a}\{P_{j\ lk}^{\ i} + P^{i}_{hj}C^{h}_{lk} - P^{i}_{lk}|_{j}\} = 0,\\
&\underset{(jk)}{a}\{P_{s\ hj}^{\ i}C^{h}_{lk} - S_{s\ jk}^{\ i}P^{h}_{lk} - P_{s\ lk}^{\ i}|_{j}\} = -S_{s\ jk|l}^{\ i}.
\end{aligned}$$

The Ricci tensors of $C\Gamma$ are as follows:

$$R_{ij} = R_{i\ jh}^{\ i},\quad S_{ij} = S_{i\ jh}^{\ h},\quad {}'P_{ij} = P_{i\ jh}^{\ h},\quad {}''P_{ij} = P_{i\ hj}^{\ h}$$

The Bianchi identities lead to

Proposition 5.1. *The Ricci tensors of the Cartan connection satisfy the following identities*

$$\begin{aligned}
&R_{ij} - R_{ji} = R^{h}_{\ ij}C^{s}_{hs} + R^{h}_{\ si}C^{s}_{hj} - R^{h}_{\ sj}C^{s}_{hi},\\
(5.7)\quad &S_{ij} - S_{ji} = 0,\\
&{}'P_{ij} - {}''P_{ij} = S_{ij|0}.
\end{aligned}$$

The last equation in (5.7) is a consequence of the Proposition 4.5.

On account of the symmetry of the v-Ricci tensor S_{ij} one proves a very important result in the applications to theoretical physics.

Theorem 5.2. *(M. Matsumoto [98]) If a four-dimensional Finsler space F^4 has the property*

$$S_{ij} = \frac{S}{3} h_{ij} \ , \quad h_{ij} = F\dot{\partial}_i \dot{\partial}_j F \ ,$$

then the v-curvature tensor S_{ijkh} may be written in the form

$$S_{ijkh} = \frac{S}{6}(h_{ik}h_{jh} - h_{ih}h_{jk}).$$

Therefore we have

Corollary 5.1. *If the v-Ricci tensor S_{ij} of a four-dimensional Finsler space vanishes, then the v-curvature tensor S_{ijkh} of this space vanishes.*

Another important result is concerning to the h- and v- Einstein tensors:

$$(5.8) \quad \overset{H}{E}_{ij} = R_{ij} - \frac{1}{2} R g_{ij} \ , \quad \overset{V}{E}_{ij} = S_{ij} - \frac{1}{2} S g_{ij} .$$

Theorem 5.3. *The following identities hold*

$$(5.9) \quad \begin{aligned} &\overset{H}{E}{}^i_{j|i} = -\frac{1}{2}(P^{ih}{}_{js} R^s{}_{hi} + 2R^s{}_{ij} {'P}^i_s) \ , \\ &\overset{V}{E}{}^i_j|_i = 0 \ . \end{aligned}$$

Proof. A simple calculation based on the third Bianchi identity in $(5.6)_1$ and the second in $(5.6)_2$.

Remarks. 1. The second equation in (5.9) has suggested that a Finslerian theory of relativity could be founded on the equation

$$S_{ij} - \frac{1}{2} S g_{ij} = \kappa \overset{V}{T}_{ij} \ ,$$

where the energy-momentum tensor $\overset{V}{T}_{ij}$ verifies the conservation law $\overset{V}{T}{}^i_j|_i = 0$ and κ is a constant.

However, these equations are not enough since the "horizontal part" it was not considered in any form. We shall see this fact in a more general theory of relativity developed in the framework of the Lagrange geometry.

2. Since the deflection tensor D_{ij} vanishes it is not possible to introduce the electromagnetic tensor as we shall do in the case of Lagrange spaces.

An important class of Finsler spaces is introduced by

Definition 5.1. *A Finsler space $F^n = (M, F(x,y))$ is called locally-Minkowski if around of any point in $\tilde{TM}$ there exists a local chart in which the fundamental function F does not depend on (x^i).*

As we have seen, the local charts on TM are induced by the local charts on M and the bundle structure of $\tau: TM \to M$. It is easy to see that the changes of coordinates on M which preserve the property in the Definition 5.1 are affine.

We have

Theorem 5.4. *A Finsler space F^n is locally-Minkowski if and only if*

(5.10) $R_j{}^i{}_{kh} = 0, \quad C^i_{jk|h} = 0.$

Proof. If in a coordinate system on $\tilde{TM}$ we have $F(x,y) = F(y)$, then by (2.10) and (4.4) it follows $N^i_j = 0$, $F^i_{jk} = 0$, hence (5.10) holds.

Conversely, by (5.10), the Theorem 4.2 and the Proposition 4.4 one gets $P^{\ i}_{jk} = 0$, $P_j{}^i{}_{kh} = 0$ and consequently, $\dot{\partial}_k F^i_{jh} = 0$. Thus F^i_{jk} depends on x^i only. Since $R_j{}^i{}_{kh} = 0$ implies by Proposition 4.3 that $R^i_{jk} = 0$, one obtains

$$\partial_h F^i_{jk} - \partial_k F^i_{jh} + F^r_{jk} F^i_{rh} - F^r_{jh} F^i_{rk} = 0.$$

Using a method very known in the theory of Riemannian spaces we find it is possible to choose a coordinate system on $\tilde{TM}$ in which $F^i_{jk} = 0$. In this coordinate system we find $N^i_j = 0$ and from $g_{ij|k} = 0$ we infer $\dot{\partial}_k g_{ij} = 0$ hence g_{ij} do not depend on x^i. Then $F^2(x,y) = g_{ij}(y) y^i y^j$ does not depend on (x^i).

Remark 5.1. For Finsler spaces $F^n = (M, F(x,y))$ the requirement that in a special coordinate system F does not depend on (x^i) is equivalent to the requirement that the fundamental tensor field $g_{ij}(x,y)$has the same property.

§6. Remarkable Finslerian Connections

The metrical Cartan connection of a Finsler space F^n is one of the most important for the geometry of these spaces, as well as for applications. It was characterized in various ways. One of them is as follows.

Theorem 6.1. *(M. Matsumoto [97]). There exists an unique d-connection $C\Gamma=(N^i_j, F^i_{jk}, C^i_{jk})$ which satisfies the following axioms*

$$1.\ D^i_j=N^i_j-F^i_{0j}=0\ ,\ 2.\ g_{ij|k}=0\ ,\ 3.\ g_{ij}|_k=0\ ,\ 4.\ T^i{}_{jk}=0\ ,\ 5.\ S^i{}_{jk}=0.$$

This theorem will appear as a consequence of two much deeper facts on metrical d-connections.

We have

Theorem 6.2. *A transformation from the Cartan non-linear connection (N^i_j) to any non-linear connection $\bar{N}^i_j=N^i_j-A^i_j$ carries the Cartan connection $C\Gamma$ into a metrical d-connection*

$$C\bar{\Gamma}=(N^i_j-A^i_j, F^i_{jk}+C^i_{jr}A^r_k, C^i_{jk}).$$

Proof. A transformation of non-linear connections $N^i_j \to N^i_j - A^i_j$, where A^i_j is any d-tensor field of type $(1,1)$ implies a transformation of the Cartan connection $C\Gamma$ into the connection $C\bar{\Gamma}=(\bar{N}^i_j, \bar{F}^i_{jk}, \bar{C}^i_{jk})$, where the local coefficients are given, cf. (4.6), (5.4) Ch.VII, by

$$\begin{aligned} &\bar{N}^i_j = N^i_j - A^i_j,\\ (6.1)\quad &\bar{F}^i_{jk} = F^i_{jk}+C^i_{jr}A^r_k,\\ &\bar{C}^i_{jk} = C^i_{jk}. \end{aligned}$$

If we denote by $\bar{|}$ and $\bar{|}$ the h- and v- covariant derivatives with respect to $C\bar{\Gamma}$ we obtain

$$g_{ij\bar{|}k}=g_{ij|k}+g_{ij}|_rA^r_k\ ,\ g_{ij}\bar{|}_k=g_{ij}|_k\ .$$

But $g_{ij|k}=0$ and $g_{ij}|_k=0$ imply $g_{ij\bar{|}k}=0$, $g_{ij}\bar{|}_k=0$.

Let now be the Obata operators

$$(6.2)\quad \Omega^{ij}_{hk}=\frac{1}{2}(\delta^i_h\delta^j_k-g_{hk}g^{ij})\ ,\ \Omega^{*ij}_{hk}=\frac{1}{2}(\delta^i_h\delta^j_k+g_{hk}g^{ij}).$$

These are supplementary projectors on the $\mathcal{F}(\tilde{T}M)$-module of d-tensor fields of type (p,q), $p\geq 1$, $q\geq 1$.

We have

Theorem 6.3. *The set of all d-connections* $F\bar{\bar{\Gamma}}=(\bar{\bar{N}}^i_j,\bar{\bar{F}}^i_{jk},\bar{\bar{C}}^i_{jk})$ *with the properties*

1. $\bar{\bar{N}}^i_j$ *is a non-linear connection apriori given* $(=\bar{N}^i_j)$,

2. $F\bar{\bar{\Gamma}}$ *is a metrical d-connection,*

is given by

$$(6.3)\quad \bar{\bar{F}}^i_{jk}=\bar{F}^i_{jk}+\Omega^{ir}_{sj}X^s_{rk}\ ,\ \bar{\bar{C}}^i_{jk}=\bar{C}^i_{jk}+\Omega^{ir}_{sj}Y^s_{rk},$$

where $F\bar{\Gamma}=(\bar{N}^i_j,\bar{F}^i_{jk},\bar{C}^i_{jk})$ *is a fixed metrical d-connection and* X^i_{jk} , Y^i_{jk} *are arbitrary d-tensor fields.*

Proof. Straightforward calculation.

We notice that the transformations (6.1) have structure of group with respect to the composition of mappings. The same fact is true for the transformations (6.3).

By the Theorems 6.2 and 6.3 we obtain

Theorem 6.4. *(Miron-Hashiguchi, [110]). The set of all d-connections* $F\bar{\Gamma}=(\bar{N}^i_j,\bar{F}^i_{jk},\bar{C}^i_{jk})$ *compatible with the fundamental tensor* (g_{ij}) *of a Finsler space* F^n *i.e. which are metrical, is given by*

$$\bar{N}^i_j=N^i_j-A^i_j\ ,\ \bar{F}^i_{jk}=F^i_{jk}+C^i_{jr}A^r_k+\Omega^{ir}_{sj}X^s_{rk}\ ,\ \bar{C}^i_{jk}=C^i_{jk}+\Omega^{ir}_{sj}Y^s_{rk},$$

where $(N_j^i, F_{jk}^i, C_{jk}^i)$ is the Cartan connection of the space F^n and A_j^i , X_{jk}^i , Y_{jk}^i are arbitrary d-tensor fields.

This theorem leads to

Theorem 6.5. *i) If $C\Gamma = (N_j^i, F_{jk}^i, C_{jk}^i)$ is the Cartan connection of a Finsler space, then there exists an unique d-connection $C\bar{\Gamma} = (\bar{N}_j^i, \bar{F}_{jk}^i, \bar{C}_{jk}^i)$ with the following properties*

1. $\bar{N}_j^i = N_j^i$ (the non-linear connection of $C\bar{\Gamma}$ is the Cartan non-linear connection),

2. $g_{ij\bar{|}k} = 0$, $g_{ij}\bar{|}_k = 0$ ($C\bar{\Gamma}$ is a metrical d-connection),

3. $\bar{T}^i{}_{jk} = T^i{}_{jk}$, $\bar{S}^i{}_{jk} = S^i{}_{jk}$ where $T^i{}_{jk} = -T^i{}_{kj}$ and $S^i{}_{jk} = -S^i{}_{kj}$ are d-tensor fields apriori given.

ii) The d-connection $C\bar{\Gamma}$ has the coefficients

$$\bar{F}_{jk}^i = F_{jk}^i + \tau_{jk}^i , \quad \bar{C}_{jk}^i = C_{jk}^i + \sigma_{jk}^i \tag{6.4}$$

where the deformation tensors are

$$\begin{aligned} \tau_{jk}^i &= -\frac{1}{2} g^{ih}(g_{jr}T^r{}_{hk} + g_{kr}T^r{}_{hj} - g_{hr}T^r{}_{jk}), \\ \sigma_{jk}^i &= -\frac{1}{2} g^{ih}(g_{jr}S^r{}_{hk} + g_{kr}S^r{}_{hj} - g_{hr}T^r{}_{jk}). \end{aligned} \tag{6.5}$$

The d-tensor τ_{jk}^i and σ_{jk}^i will be called h- and v-contorsions, respectively.

Remarks. 1. In a Finslerian theory of relativity the metrical d-connection$C\bar{\Gamma}$ with h- and v- torsions apriori given was used (cf. R.Miron [108]).

2. R.G. Beil [28,a] has showed that the Finslerian gauge theory based on the Cartan non-linear connection is equivalent to a Kaluza-Klein gauge theory.

3. We note that it is possible to find the transformations carrying the Cartan connection $C\Gamma = (N_j^i, F_{jk}^i, C_{jk}^i)$ to the Berwald connection $B\Gamma = (N_j^i, \dot{\partial}_k N_j^i, 0)$, to the Rund connection $R\Gamma = (N_j^i, F_{jk}^i, 0)$ or to the Hashiguchi connection $H\Gamma = (N_j^i, \dot{\partial}_k N_j^i, C_{jk}^i)$.

§7. Almost Kählerian Model of a Finsler Space

We shall see that the Finsler space can be thought of as an almost Kähler space on the smooth manifold $\tilde{TM} = TM \setminus \{0\}$. We say such an almost Kähler space is a model for the Finsler space F^n. Thus the theory of Finsler spaces gets a much more geometrical consistence. Moreover, a Finslerian theory of relativity (see. R. Miron, [108]), can be understand much better on such a model since the almost symplectic structure of the space is in fact a symplectic structure. Such a model puts together the symplectic structure θ (see the Theorem 4.3) and the metrical Finsler structure given by $g_{ij}(x,y)$. Therefore the following prediction of R.S. Ingarden [70] is fulfilled.

"Symplectic geometry is a geometry of mechanical phase space showing the sense of the canonical transformations of positions and momenta. But in mechanics except of momenta there exist also potentials. So except the symplectic geometry we need also geometry of potentials and that is namely a metric geometry".

We consider a Finsler space $F^n = (M, F(x,y))$, its fundamental tensor $g_{ij}(x,y)$ and Cartan non-linear connection $N^i_j(x,y)$. The Cartan non-linear connection allows us to define an almost complex structure **F** on TM as follows

$$(7.1)\quad F(\delta_i) = -\dot{\partial}_i,\ F(\dot{\partial}_i) = \delta_i.$$

It is obvious that **F** is well-defined and $\mathbf{F}^2 = -I$. Since the weak torsion of the Cartan non-linear connection vanishes, it follows:

Theorem 7.1. *The almost complex structure **F** is integrable if and only if the torsion $R^i{}_{jk}$ of $C\Gamma$ vanishes.*

Recall that the vanishing of $R^i{}_{jk}$ is a necessary and sufficient condition for the integrability of the horizontal distribution N determined by the Cartan non-linear connection.

Let $(dx^i, \delta y^i)$ be the dual basis of $(\delta_i, \dot{\partial}_i)$. Then

$$(7.2)\quad G = g_{ij}(x,y)dx^i \otimes dx^j + g_{ij}(x,y)\delta y^i \otimes \delta y^j$$

is a tensor field of type (0,2) on $\tilde{TM}$, symmetric and not degenerate. Moreover, G is a Riemannian metric on $\tilde{TM}$ and the horizontal and vertical distributions are orthogonal with respect to it.

We have

Theorem 7.2. *i) The pair (G,F) is an almost Hermitian structure on* $\tilde{TM}$.

ii) The almost symplectic 2-form associated with the almost Hermitian structure (G,F) is

$$(7.3)\quad \theta = g_{ij}(x,y)\,\delta y^i \wedge dx^j .$$

Proof. i) A direct calculation using (7.1) and (7.2) shows that the equality $G(FX,FY) = G(X,Y)$ holds for any $X, Y \in \chi(\tilde{TM})$.

ii) In adapted basis $(\delta_i, \dot{\partial}_i)$ the almost symplectic structure $\theta(X,Y) = G(FX,Y)$, $\forall\, X,Y \in \chi(\tilde{TM})$ reduces to that given by (7.3), because of $G(\delta_i,\delta_j) = g_{ij}$, $G(\delta_i,\dot{\partial}_j) = 0$, $G(\dot{\partial}_i,\dot{\partial}_j) = g_{ij}$.

We notice that (G,F) completely determines the considered Finsler space. Indeed, it is enough to consider the term $g_{ij}(x,y)\,dx^i \otimes dx^j$ in (7.2). This well determines a Finsler function F by

$$(*)\quad F^2(x,y) = g_{ij}(x,y)\,y^i y^j .$$

But not every d-tensor field $g_{ij}(x,y)$ symmetric and non -degenerate has this property. In order for this happen the system of partial differential equations

$$(**)\quad g_{ij}(x,y) = \frac{1}{2}\dot{\partial}_i\dot{\partial}_j F^2$$

must have solutions F^2 which to fulfil the conditions in the Definition 1.1. One of the necessary conditions for the existence of the solution of (**) is that the d-tensor field $C_{ijk} = \frac{1}{2}\dot{\partial}_k g_{ij}$ be totally symmetric. If this condition holds and g_{ij} is 0-homogeneous with respect to (y^i) then F^2 from (*) is a solution.

Let H^{2n} be the almost Hermitian space obtained in the Theorem 7.2. i). Furthermore we have

Theorem 7.3. *The almost Hermitian space* H^{2n} *is almost Kählerian.*

Proof. It is sufficient to prove that θ is closed. By (7.3) and the Theorem 4.3 we get θ is an exact 2-form, hence it is closed.

The almost Kählerian space $K^{2n} = (\tilde{TM}, G, F)$ will be called the *almost Kählerian model* of the Finsler space F^n. This model particularises for locally-Minkowski Finsler spaces.

Theorem 7.4. *If the Finsler space F^n is a locally-Minkowski Finsler space then its model K^{2n} is a Kähler space.*

Proof. We recall that for a locally-Minkowski Finsler space we have $R_j{}^i{}_{kh} = 0$, which implies $R^i{}_{jk} = 0$. By the Theorem 7.1., the almost complex structure $\mathbf{F}$ is integrable, hence K^{2n} is a Kähler space.

We shall see now that important geometrical objects in F^n are characterized by geometrical objects in its model K^{2n} .

Definition 7.1. *An almost Kähler connection of Finsler type is a linear connection ∇ on $\tilde{TM}$ with the properties*

1. *∇ preserves by parallelism the vertical distribution.*
2. *∇ is compatible with the almost Kähler structure (G,F), i.e.*

(7.4) $\nabla_X G = 0, \ \nabla_X F = 0, \ \forall X \in \chi(\tilde{TM}).$

Note that an almost Kähler connection of Finsler type, shortly a FAK-connection, preserves by parallelism the horizontal distribution, too. In the adapted basis $(\delta_i, \dot{\partial}_i)$ a FAK-connection ∇ is given by its coefficients $(F^i_{jk}, \bar{F}^i_{jk}, C^i_{jk}, \bar{C}^i_{jk})$ from

$$(7.5)\quad \begin{array}{l} \nabla_{\delta_k}\delta_j = F^i_{jk}\delta_i, \ \nabla_{\delta_k}\dot{\partial}_j = \bar{F}^i_{jk}\dot{\partial}_i, \\ \nabla_{\dot{\partial}_k}\delta_j = \bar{C}^i_{jk}\delta_i, \ \nabla_{\dot{\partial}_k}\dot{\partial}_j = C^i_{jk}\dot{\partial}_i. \end{array}$$

But $\nabla_X F = 0$ gives $(\nabla_X F)\delta_j = -\nabla_X \dot{\partial}_j - F(\nabla_X \delta_j)$ and taking $X = \delta_k$ and then $X = \dot{\partial}_k$ we find that $\nabla_X F = 0$ implies

$$(7.5)' \quad F^i_{jk} = \bar{F}^i_{jk}, \ C^i_{jk} = \bar{C}^i_{jk}.$$

Thus ∇ is determined by the coefficients F^i_{jk}, C^i_{jk} only, which must satisfy the conditions implied by $\nabla_X G = 0$.

We have

Theorem 7.5. *There exists an unique almost Kählerian connection of Finsler type whose h(hh)- and v(vv)- torsions vanish. The local coefficients (F^i_{jk}, C^i_{jk}) of this connection are the coefficients of the Cartan connection $C\Gamma$ of the Finsler space F^n. Conversely, the*

Cartan connection $C\Gamma$ *determines via (7.5) and* $(7.5)'$ *an unique almost Kähler connection of Finsler type on* $\tilde{TM}$ *whose h(hh)- and v(vv)- torsions vanish.*

Proof. The torsion **T** of ∇ is given by

$$(*)\ \mathbf{T}(X,Y) = \nabla_X Y - \nabla_Y X - [X,Y]\ ,\ X,Y \in \chi(\tilde{TM}).$$

But every vector field X on TM decomposes as $X = hX + vX$. The h(hh)-torsion is the horizontal component of $\mathbf{T}(hX, hY)$ and v(vv)-torsion is the vertical component of $\mathbf{T}(vX, vY)$. In the basis $(\delta_i, \dot{\partial}_i)$ these torsions are $T^i{}_{jk} = F^i_{jk} - F^i_{kj}$ and $S^i{}_{jk} = C^i_{jk} - C^i_{kj}$, respectively. On account of $G(\delta_i, \delta_j) = g_{ij}$, $G(\delta_i, \dot{\partial}_j) = 0$, $G(\dot{\partial}_i, \dot{\partial}_j) = g_{ij}$ it follows that $\nabla_X G = 0$ is equivalent to $g_{ij|k} = 0$, $g_{ij}|_h = 0$. The latter equations admit, under the conditions $T^i{}_{jk} = S^i{}_{jk} = 0$ an unique solution

$$(7.6)\ F^i_{jk} = \frac{1}{2} g^{ih}(\delta_j g_{hk} + \delta_k g_{hj} - \delta_h g_{jk})\ ,\ C^i_{jk} = \frac{1}{2} g^{ih}(\dot{\partial}_j g_{hk} + \dot{\partial}_k g_{hj} - \dot{\partial}_h g_{jk}).$$

Thus $C\Gamma = (N^i_j, F^i_{jk}, C^i_{jk})$ is the Cartan connection of the Finsler space F^n.

Conversely, given $C\Gamma$ and defining the linear connection ∇ by (7.5), $(7.5)'$ one obtains a linear connection ∇ on $\tilde{TM}$ which clearly satisfies 1. and 2. in Definition 7.1. Furthermore h(hh)- and v(vv)- torsions of ∇ vanish. The uniqueness follows easily by contradiction.

Concluding we may say that the geometry of the Finsler space F^n based on the Cartan connection coincides with the geometry of the almost Kählerian model K^{2n} based on the almost Kählerian connection of Finsler type whose h(hh)- and v(vv)- torsions vanish.

Remark 7.1. Interpreting the components of the fundamental tensor $g_{ij}(x,y)$ as gravitational potentials (y^i are considered as internal variables by many physicists) we shall take the equations of the gravitational field as the Einstein equations associated with the almost Kähler connection of Finsler type found in the previous theorem. This idea is explored in several papers of the first author [107, 108, 114].

§8. Subspaces in a Finsler Space

We shall present the theory of subspaces in Finsler spaces closely following the theory of embeddings of vector bundles developed in Ch.IV. Certainly the almost

Kählerian model K^{2n} of a Finsler space has a special role in a sense the theory becomes much more a geometrical one and some ambiguous concepts from the previous treatments are eliminated. The results which follow are to a great extent a particularisation of those of Ch.IV. Consequently, the proofs will be omitted. Main ideas regarding the moving frame, its moving equations and differential consequences of them will be underlined.

Let $F^n = (M, F)$ be a Finsler space, $C\Gamma = (N^i_j, F^i_{jk}, C^i_{jk})$ its Cartan connection and $K^{2n} = (\tilde{T}M, G, \mathbb{F})$ its almost Kählerian model.

As we have seen in Ch.IV, an embedding $f': M' \to M$ (dim $M' = n'$, $n =$ dim M) determines an embedding of the corresponding tangent bundles

(8.1) $(f'^T, f') : (TM', \tau', M') \to (TM, \tau, M)$,

where $f'^T : TM' \to TM$ is the differential of the mapping f'. In a coordinate system, (8.1) reads

$$(8.1)' \quad \begin{aligned} & x^i = x^i(u^1, \dots, u^{n'}) \ , \ \operatorname{rank} \left(\frac{\partial x^i}{\partial u^\alpha} \right) = n' , \\ & y^i = \frac{\partial x^i}{\partial u^\alpha} v^\alpha . \end{aligned}$$

Reviewing the notations in Ch.IV it becomes clear that now the indices a, b, c,.... run from 1 to n and A, B, C,.... run from 1 to n'. Thus we have, in conformity with §1, Ch.IV

(8.2) $y^i = \delta^i_a y^a \ , \ v^\alpha = \delta^\alpha_A v^A$.

Let be $\tilde{T}M' = TM' \setminus \{0'\}$, where $0'$ is the image of the null cross-section of the projection $\tau' : TM' \to M'$.

By (8.1)$'$ we have the embedding

(8.3) $f'^T_{|\tilde{T}M'} : \tilde{T}M' \to \tilde{T}M$.

The Riemannian metric G in K^{2n} appears as a particular case of the metric G from Ch.IV, given by

(8.4) $g_{ij} = \delta^a_i \delta^b_j h_{ab}$.

Then the natural moving frame in $\tilde{T}M$ along of $\tilde{T}M'$, by (1.3) Ch.IV and (8.1) has the property

$$(8.5)\quad B^i_\alpha(u)=\frac{\partial x^i}{\partial u^\alpha}=\delta^i_a\delta^A_\alpha B^a_A(u).$$

Thus only B^i_α remain in this frame. Completing this set of vector fields with the vector fields $\{B^i_{\bar\alpha}(u,v)\}$, $\bar\alpha,\bar\beta,\ldots=1,\ldots,n-n'$ given by

$$(8.6)\quad g_{ij}B^i_\alpha B^j_{\bar\beta}=0\ ,\ g_{ij}B^i_{\bar\alpha}B^j_{\bar\beta}=\delta_{\bar\alpha\bar\beta},$$

one obtains that the natural moving frame in TM along TM' is

$$(8.7)\quad \mathfrak{R}=\{(u,v);B^i_\alpha(u),B^i_{\bar\alpha}(u,v)\}.$$

By (8.4) and (2.2)′ we have

$$(8.5)'\quad B^i_{\bar\alpha}=\delta^i_a\delta^{\bar A}_{\bar\alpha}B^a_{\bar A},$$

where the indices $\bar A,\bar B,\ldots$ run from 1 to n-n′.

We denote $\mathfrak{R}^*=\{(u,v);B^\alpha_i(u,v),B^{\bar\alpha}_i(u,v)\}$ the frame dual to $\mathfrak{R}$.

The equations (2.5) in Ch.IV with the modifications implied by (8.5) and (8.5)′ are verified.

The embedding (8.1) allows us to define the function

$$(8.8)\quad \bar F(u,v)=F(x(u),B_\alpha(u)v^\alpha),$$

which obviously has the properties of a fundamental function on M'. Thus we may say that (8.1) is an embedding of the Finsler space $\bar F^{n'}=(M',\bar F(u,v))$ into the Finsler space $F^n=(M,F(x,y))$.

By (8.8) one deduces

$$\frac{\partial\bar F^2}{\partial v^\alpha}=\frac{\partial F^2}{\partial y^i}B^i_\alpha(u)\ ,\ \frac{\partial^2\bar F^2}{\partial v^\alpha\partial v^\beta}=\frac{\partial^2F^2}{\partial y^i\partial y^j}B^i_\alpha(u)B^j_\alpha(u).$$

Thus the fundamental tensor field g_{ij} from F^n induces via (8.1) the fundamental tensor field $g_{\alpha\beta}(u,v)$ in $\bar F^{n'}$ and we have

(8.9) $g_{\alpha\beta}(u,v)=B^i_\alpha(u)B^j_\beta(u)g_{ij}(x(u),B^i_\alpha(u)v^\alpha)$.

It follows that in the frame $\mathfrak{R}$ the tensor field $g_{ij}(x,y)$ can be represented as

(8.10) $g_{ij}=g_{\alpha\beta}B^\alpha_i B^\beta_j+\delta_{\bar\alpha\bar\beta}B^{\bar\alpha}_i B^{\bar\beta}_j$.

From now on the whole theory from Ch.IV applies with clear changes keeping in mind the particularizations given in the above.

Thus the Gauss-Weingarten formulae of $F^{n'}$ embedded in F^n are

(8.11) $\nabla B^i_\alpha=B^i_{\bar\beta}\Pi^{\bar\beta}_\alpha\ ,\ \nabla B^i_{\bar\alpha}=-B^i_\beta\Pi^\beta_{\bar\alpha}$,

where

$$\Pi^{\bar\beta}_\alpha=H_\alpha{}^{\bar\beta}{}_\gamma du^\gamma+K_\alpha{}^{\bar\beta}{}_\gamma\delta v^\gamma\ ,\ \Pi^\beta_{\bar\alpha}=g^{\beta\gamma}\delta_{\bar\alpha\bar\beta}\Pi^{\bar\beta}_\gamma$$

while $H_\alpha{}^{\bar\beta}{}_\gamma$, $K_\alpha{}^{\bar\beta}{}_\gamma$ are the fundamental tensors of the second type of the embedding.

The Gauss-Codazzi equations are as follows

$$\begin{aligned}&B^i_\alpha B^j_\beta\Omega_{ij}-\Omega_{\alpha\beta}=\Pi_{\beta\bar\varphi}\wedge\Pi^{\bar\varphi}_\alpha\ ,\\ (8.12)\quad &B^i_{\bar\alpha}B^j_{\bar\beta}\Omega_{ij}-\Omega_{\bar\alpha\bar\beta}=\Pi_{\varphi\bar\beta}\wedge\Pi^\varphi_{\bar\alpha}\ ,\\ &-B^i_\alpha B^j_\beta\Omega_{ij}=\delta_{\bar\beta\bar\varphi}(d\Pi^{\bar\varphi}_\alpha+\Pi^{\bar\varphi}_\gamma\wedge\omega^\gamma_\alpha-\Pi^{\bar\gamma}_\alpha\wedge\omega^{\bar\varphi}_{\bar\gamma})\ .\end{aligned}$$

We left to the reader the explicit writing of the connection 1-forms and of the structure equations induced on $F^{n'}$ by their counterparts in the Finsler space F^n.

Chapter IX

Lagrange Spaces

The Lagrange spaces were introduced by J. Kern [82] in order to geometrize a fundamental concept in mechanics, that of Lagrangian. A Lagrange space $L^n = (M, L(x,y))$ is defined as a pair which consists of a real, smooth n-dimensional manifold M and a regular Lagrangian $L: TM \to R$. It comes out that a Finsler space is a Lagrange space, but not conversely since the Lagrangian L may be not homogeneous with the respect to the variables $(y^i), i = 1, 2, \ldots, n$.

The fact that the Finsler spaces are particular Lagrange spaces suggested the developing of the geometry of the Lagrange spaces by extending the methods which have been used in the study of the geometry of Finsler spaces. In this way one can study sufficiently general regular Lagrangians which appear in mechanics, electrodynamics, optimal control etc.

The geometry of Lagrange spaces gives a model for both the gravitational and the electromagnetic field in a very natural blending of the geometrical structure of the space with the characteristic properties of these physical fields. This is possible due to of the utilisation of some specific Lagrangians together with some fundamental concepts from the geometry of the total space TM as is for instance, the Liouville vector field.

As is expected, the variational problem formulated for the action integral of the regular Lagrangian $L(x,y)$ of a space L^n leads to the Euler-Lagrange equations which are very useful in the geometry of L^n. First, these are used in introducing a canonical non-linear connection and then a canonical metrical d-connection. These two connections are basic in the geometry of L^n. Let us notice the considered notions capture both the symplectic structure induced by L^n on the manifold TM and the vertical structure on TM. These give together an almost Kählerian structure. The almost Kählerian space K^{2n} determined on TM as in the above is a geometrical model of the Lagrange space L^n. It gives a geometrical legitimacy to the whole study of the Lagrange space L^n. Otherwise the study remains on the analytical level.

The problems enumerated in the above will be discussed in this chapter. We mention that in solving them an important contribution has been made by the Romanian school of geometry in the framework of the National Seminar on Finsler, Lagrange and Hamilton spaces which held once at every two years since 1980, in University of Braşov. It is organized by the Universities from Iaşi and Braşov. In this Seminar the geometry of Lagrange spaces was extensively studied and notions as generalized Lagrange spaces and Hamilton spaces were introduced for the first time. In parallel a substantial geometrical theory of vector bundles and their duals was created.

§1. The Notion of Lagrange Space

As usual M will denote a real, smooth n-dimensional manifold and (TM,τ,M) its tangent bundle. Also we shall use frequently the manifold $\tilde{TM} = TM\setminus\{0\}$. We shall keep the notation (x^i, y^i) for the local coordinates on TM.

Definition 1.1. *A regular Lagrangian is a function $L: TM \to R$ which fulfils the following conditions:*

a) L is of class C^∞ on $\tilde{TM}$ and continuous on the image of the null section in the tangent bundle.

b) The matrix with the entries

$$(1.1) \quad g_{ij}(x,y) = \frac{1}{2}\frac{\partial^2 L}{\partial y^i \partial y^j}$$

has rank n on $\tilde{TM}$.

If we pass to another coordinate system on TM it comes out that $g_{ij}(x,y)$ from (1.1) may be thought of as components of a symmetric d-tensor field of type (0,2) on $\tilde{TM}$. Furthermore, in what follows we shall assume that the quadratic form

$$(1.1)' \quad \Phi(\xi) = g_{ij}\xi^i\xi^j, \quad \xi \in \mathbb{R}^n$$

has constant signature on $\tilde{TM}$.

Remark 1.1. The condition a) in the Definition 1.1 is merely a caution in view of the applications. In fact the general results which follow remain the same if we assume L is C^∞ on TM. Then $g_{ij}(x,y)$ define a d-tensor field on TM.

Definition 1.2. *A Lagrange space is a pair* $L^n = (M, L(x,y))$ *which consists of a smooth manifold M and a regular Lagrangian L which fulfils* (1.1)$'$.

The Lagrangian L with the proprieties a) and b) from the Definition (1.1) will be called the fundamental (or metric) function of the Lagrange space L^n and the d-tensor field whose components are given by (1.1) will be called the fundamental (or metric) tensor field of L^n.

Examples.

1. Every Finsler space $F^n=(M,F(x,y))$ is a Lagrange space with the fundamental function $L(x,y)=F^2(x,y)$. It is clear this L is a regular Lagrangian and $(1.1)'$ is satisfied. Conversely, if a Lagrange space L^n has its fundamental function L(x,y) positive and 2-homogeneous with respect to (y^i) on $\tilde{TM}$ then it becomes a Finsler space $F^n=(M,\sqrt{L(x,y)})$. Thus we have

Theorem 1.1. *A Lagrange space $L^n=(M,L(x,y))$ is a Finsler space if and only if its fundamental function L(x,y) is positive and 2-homogeneous with respect to (y^i) on $\tilde{TM}$.*

2. A Lagrange space $L^n=(M,L(x,y))$ whose fundamental function is given by

$$(1.2)\quad L(x,y)=F^2(x,y)+A_i(x)y^i+U(x),$$

where F is the fundamental function of a Finsler space (M,F(x,y)) will be called an almost Finslerian Lagrange space, shortly AFL-space. In (1.2) $A_i(x)$ are the components of a covector field (sometimes it is called electromagnetic vector potential) and U(x) is a smooth function on M (sometimes called potential function).

The fundamental metric tensor field $g_{ij}(x,y)$ of an AFL-space coincides with the fundamental metric tensor of the Finsler space (M,F(x,y)).

A classical example of AFL-space is provided by the well-known Lagrangian from electrodynamics

$$(1.3)\quad L(x,y)=mc\gamma_{ij}(x)y^iy^j+\frac{2e}{m}A_i(x)y^i+U(x),$$

where γ_{ij} is a Lorentz metric tensor on M (gravific potentials), $A_i(x)$ a field of covectors (electromagnetic potentials), U(x) a function on M (potential function) and $m\neq 0$, c,e are the usual constants from Physics.

§2. Euler-Lagrange Equations. Canonical Non-linear Connection

Let $L^n=(M,L(x,y))$ be a Lagrange space and $(\Gamma,\gamma), \gamma:[0,1]\to M$ a smooth parametrized curve on the manifold $M, \Gamma=\operatorname{Im}\gamma$.

The action integral associated with L along (Γ,γ) is defined as follows

$$(2.1)\quad I(\Gamma)=\int_0^1 L(x,\dot{x})dt,$$

where $\dot{x}$ denotes the tangent vector field to (Γ,γ).

Exactly as in §2, Ch.VIII we can prove:

A necessary condition which (Γ,γ) must satisfy in order that it affords an extreme value for the action integral $I(\Gamma)$ is that the mapping $\gamma:t\to x(t)$, $t\in[0,1]$ be solution of the Euler-Lagrange equations

$$(2.2)\quad \frac{d}{dt}\left(\frac{\partial L}{\partial y^i}\right)-\frac{\partial L}{\partial x^i}=0,\quad y^i=\frac{dx^i}{dt}.$$

We warn that we have considered here parametrized curves. In other words $I(\Gamma)$ depends on the parametrization γ of the curve (Γ,γ).

Some obvious transformations of (2.2) give

Theorem 2.1. *The Euler-Lagrange equations (2.2) are equivalent to the differential equations of order two*

$$(2.3)\quad \frac{d^2x^i}{dt^2}+2G^i\left(x,\frac{dx}{dt}\right)=0,$$

where

$$(2.4)\quad G^i(x,y)=\frac{1}{4}g^{ij}\left(\frac{\partial^2 L}{\partial y^j\partial x^k}y^k-\frac{\partial L}{\partial x^j}\right),\quad y^i=\frac{dx^i}{dt}.$$

Performing a change of coordinates $(x,y)\to(x',y')$ on the manifold $\tilde{TM}$ we find G^i transform as in (2.11), Ch.VIII. Thus we get

Theorem 2.2. *The functions*

$$(2.5)\quad N^i_j(x,y) = \frac{\partial G^i(x,y)}{\partial y^j}$$

are the local coefficients of a non-linear connection on $\tilde{TM}$.

The non-linear connection given by (2.5) is completely determined by the fundamental function L(x,y) of the Lagrange space L^n. Notice that generally may exist many regular Lagrangians on M which to determine the same non-linear connection.

It is obvious that the two regular Lagrangians L(x,y) and L′(x,y) will determine the same non-linear connection if they are equivalent from the viewpoint of the variational calculus, that is,

$$(2.2)'\quad \frac{d}{dt}\left(\frac{\partial L}{\partial y^i}\right) - \frac{\partial L}{\partial x^i} = \frac{d}{dt}\left(\frac{\partial L'}{\partial y^i}\right) - \frac{\partial L'}{\partial x^i},\quad y^i = \frac{dx^i}{dt}$$

holds for any parametrized curve (Γ, γ). The equality (2.2)′ provides an equivalence relation (see M. Kirkovits [86]) on the set of regular Lagrangians. We have

Theorem 2.3. *(Caratheodory [32]). The regular Lagrangians L and* L′ *are equivalent if and only if*

$$(2.6)\quad L'(x,y) = L(x,y) + \frac{\partial \varphi}{\partial x^i} y^i + a\,,$$

where φ *is an arbitrary function on M and a any constant.*

Proof. If (2.6) holds, then $\dfrac{\partial L'}{\partial y^i} = \dfrac{\partial L}{\partial y^i} + \dfrac{\partial \varphi}{\partial x^i}$ and $\dfrac{\partial L'}{\partial x^i} = \dfrac{\partial L}{\partial x^i} + \dfrac{\partial^2 \varphi}{\partial x^i \partial x^j}\dfrac{dx^j}{dt}$. Inserting these in (2.2)′ it comes out it is verified.

Conversely, if the regular Lagrangian L and L′ are equivalent, then (2.2)′ show us that the extremals of the action integral associated to L′ coincide to the extremals of the action integral associated to L′. Therefore, the system of equations (2.3) is equivalent to a corresponding one for L′(x,y). It follows that $G^i(x,y) = G'^i(x,y)$ and as an easy to prove consequence we find $g_{ij}(x,y) = g'_{ij}(x,y)$ or equivalently, $\dot{\partial}_i \dot{\partial}_j (L - L') = 0$. Resolving this system of partial differential equations one gets $L'(x,y) = L(x,y) + A_i(x) y^i + U(x)$ where $A_i(x)$ is an arbitrary covector and U a smooth function. Inserting L′ of this form in (2.2)′ it comes out A_i must be a gradient and U a constant, q.e.d.

The correspondence L→L′ given by (2.6) gives a (gauge) transformation on the set of regular Lagrangians which sometimes is called the Caratheodory transformation.

The previous considerations may be formulated as follows:

Theorem 2.4. *The canonical non-linear connection of a regular Lagrangian L(x,y) is preserved by a Caratheodory transformation (2.6) of it.*

Of course, the canonical non-linear connection determined by L(x,y) defines an horizontal distribution on TM which is locally spanned by the vector fields

$$(2.7) \quad \delta_i = \partial_i - N_i^j \dot{\partial}_j$$

in such a way that $(\delta_i, \dot{\partial}_i)$ is an adapted basis to the horizontal and vertical distributions on $\tilde{TM}$.

Since the bracket of two local vector fields δ_i and δ_j is as follows

$$(2.8) \quad [\delta_i, \delta_j] = R^h{}_{ij} \dot{\partial}_h ,$$

where

$$(2.8)' \quad R^h{}_{ij} = \delta_j N_i^h - \delta_i N_j^h$$

one deduces

Theorem 2.5. *The horizontal distribution of the non-linear connection N is integrable if and only if $R^h{}_{ij} = 0$ on $\tilde{TM}$.*

Let $(dx^i, \delta y^i)$ be the dual basis to $(\delta_i, \dot{\partial}_i)$. Thus we have

$$(2.9) \quad \delta y^i = dy^i + N_j^i(x,y)\, dx^j .$$

We notice also

Theorem 2.6. *The autoparallel curves of the canonical non-linear connection N are the solutions of the following system of differential equations*

$$(2.10) \quad \frac{d^2x^i}{dt^2} + N_j^i(x,\dot{x}) \frac{dx^j}{dt} = 0, \quad \dot{x}^i = \frac{dx^i}{dt}$$

which is equivalent to

$$(2.10)' \quad \frac{\delta y^i}{dt} = \frac{dy^i}{dt} + N^i_j(x,y)\frac{dx^j}{dt} = 0, \quad y^i = \frac{dx^i}{dt}.$$

In general, the autoparallel curves of the non-linear connection N do not coincide to the extremals of the action integral associated to the Lagrangian L.

As in the case of Finsler spaces, in a Lagrange space L^n we may consider the following differential forms

$$(2.11) \quad \omega = \frac{1}{2}\frac{\partial L}{\partial y^i}dx^i, \quad \theta = g_{ij}(x,y)\delta y^i \wedge dx^j.$$

These differential forms have the properties of their counterparts met in the Finsler spaces (cf. Theorem 4.3, Ch.VIII). We recall these properties for the sake of completeness.

Theorem 2.7. *(i) θ is a field of 2-forms on TM,*
(ii) $d\omega = \theta$,
(iii) θ is a symplectic structure on $\tilde{TM}$.

An immediate consequence of (ii), that is $d\theta = 0$, leads to

Theorem 2.8. *The canonical non-linear connection N has the following properties*

$$(2.12) \quad \sum_{(ijk)} R_{ijk} = 0, \quad g_{ij|k} - g_{ik|j} = 0, \quad g_{ij}\|_k - g_{ik}\|_j = 0,$$

where $|$ *and* $\|$ *denote h- and v- covariant derivatives with respect to the Berwald connection* $B\Gamma = (N^i_j, \dot\partial_k N^i_j, 0)$.

§3. Canonical Metrical d-Connection

As we have seen in Ch.VII a normal d-connection $D\Gamma = (N^i_j, L^i_{jk}, C^i_{jk})$ allows us to consider the algorithms of h- and v- covariant derivation in the algebra of d-tensor fields on TM. Recall that $D\Gamma$ have five torsions $T^i_{\ jk}$, $S^i_{\ jk}$, $R^i_{\ jk}$, $P^i_{\ jk}$ and $C^i_{\ jk}$ and three curvatures $R_j{}^i{}_{kh}$, $P_j{}^i{}_{kh}$ and $S_j{}^i{}_{kh}$ whose forms may be seen in Ch.VII.

The Ricci formulae hold, too.

Now we shall point out a d-connection completely determined by the fundamental function L(x,y) of a Lagrange space L^n.

Theorem 3.1. *There exists an unique normal d-connection* $C\Gamma=(N^i_j, L^i_{jk}, C^i_{jk})$ *satisfying the following axioms:*

1° $N^i_j(x,y)$ *are the coefficients of the canonical non-linear connection (2.5),*

2° $C\Gamma$ *is a metrical d-connection i.e.*

$$(3.1) \qquad g_{ij|k}=0, \quad g_{ij}|_k=0,$$

3° The torsions $T^i{}_{jk}$ and $S^i{}_{jk}$ of $C\Gamma$ vanish.

Proof. The canonical non-linear connection being given by (2.5) we shall consider the "generalized Christoffel symbols"

$$(3.2) \qquad \begin{aligned} L^i_{jk} &= \frac{1}{2}g^{ih}(\delta_k g_{jh}+\delta_j g_{hk}-\delta_h g_{jk}), \\ C^i_{jk} &= \frac{1}{2}g^{ih}(\dot{\partial}_k g_{jh}+\dot{\partial}_j g_{hk}-\dot{\partial}_h g_{jk}). \end{aligned}$$

It is a matter of calculation to see that the d-connection $C\Gamma$ whose coefficients are given by (2.5) and (3.2) satisfies 1°-3°. Thus the existence of $C\Gamma$ is proved. The uniqueness results by contradiction noticing that L^i_{jk} and C^i_{jk} from (3.2) may be deduced from (3.1) as in the proof of the Theorem 4.1, Ch.VIII.

The d-connection satisfying the axioms 1°-3°. in the Theorem 3.1 is clearly determined by the Lagrangian L only. This is a reason to call it the *canonical metrical d-connection* of the Lagrange space $L^n=(M,L(x,y))$.

It is interesting to note

Corollary 3.1. *If the Lagrange space* L^n *is a Finsler space* F^n *then the canonical metrical d-connection* $C\Gamma$ *coincide with the metrical Cartan connection of* F^n.

This is the reason why the structure equations and the Bianchi identities satisfied by $C\Gamma$ have the same form as those given in Ch. VIII for a Finsler space.

In the present case, the connection 1-forms are

$$(3.3) \qquad \omega^i_j = L^i_{jk}dx^k + C^i_{jk}\delta y^k.$$

We have

Theorem 3.2. *The connection 1-forms* ω^i_j *of the canonical metrical d-connection* $C\Gamma$

satisfy the following structure equations

$$(3.4)\qquad \begin{aligned} d(dx^i) - dx^h \wedge \omega^i_h &= -\Omega^i, \\ d(\delta y^i) - dy^h \wedge \omega^i_h &= -\tilde{\Omega}^i, \end{aligned}$$

$$(3.4)'\qquad d\omega^i_j - \omega^h_j \wedge \omega^i_h = -\Omega^i_j,$$

in which the torsion 2-forms Ω^i *and* $\tilde{\Omega}^i$ *are*

$$\Omega^i = C^i{}_{jk} dx^j \wedge \delta y^k, \quad \tilde{\Omega}^i = \frac{1}{2} R^i{}_{jk} dx^j \wedge dx^k + P^i{}_{jk} dx^j \wedge \delta y^k$$

while the curvature 2-forms Ω^i_j *are*

$$\Omega^i_j = \frac{1}{2} R_j{}^i{}_{kh} dx^k \wedge dx^h + P_j{}^i{}_{kh} dx^k \wedge \delta y^h + \frac{1}{2} S_j{}^i{}_{kh} \delta y^k \wedge \delta y^h .$$

Exterior differentiating (3.4) and (3.4)′ and using them again one obtains the Bianchi identities in a form identical to those from the Theorem 5.2, Ch.VIII.

We notice that starting from CΓ some remarkable d-connections may be derived by using convenient transformations of connections. These are the Berwald connection $B\Gamma(N) = (\dot{\partial}_j N^i_k, 0)$ the Rund connection $R\Gamma(N) = (L^i_{jk}, 0)$ and the Hashiguchi connection $H\Gamma(N) = (\dot{\partial}_j N^i_k, C^i_{jk})$. The following commutative diagram holds

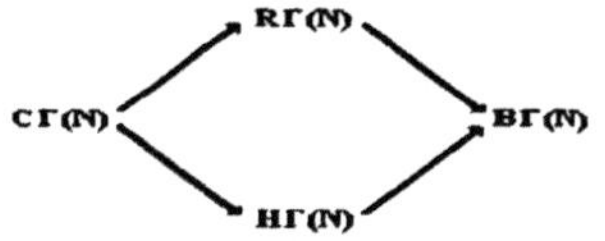

The corresponding transformations of connections may be easily deduced.

Here are some properties of CΓ.

Proposition 3.1. 1° $\sum_{(ijk)} R_{ijk} = 0$ *and* $P_{ijk} = g_{ih} P^h{}_{jk}$ *is totally symmetric d-tensor field.*

2° $C_{ijk} = \frac{1}{4} \dot{\partial}_i \dot{\partial}_j \dot{\partial}_k L, \quad C_{ijk} = g_{jh} C^h_{ik}.$

3° *The covariant curvature tensors* $R_{ijkh}, P_{ijkh}, S_{ijkh}$ *are skewsymmetric in the first two indices.*

4° $S_{jikh} = C_{iks}C^h_{jh} - C_{ihs}C^s_{jk}$.

We shall end with the following theorem whose proof is similar to that of Theorem 6.5, Ch.VIII.

Theorem 3.3. *i) The canonical non-linear connection (2.5) being fixed, there exists an unique normal d-connection $\overline{L\Gamma}(N) = (\overline{L}^i_{jk}, \overline{C}^i_{jk})$ compatible with the metric tensor $g_{ij}(x,y)$ whose torsions $\overline{T}^i{}_{jk}$ and $\overline{S}^i{}_{jk}$ are given apriori.*

ii) The coefficients of this normal d-connection are

$$\text{(3.5)}\quad \overline{L}^i_{jk} = L^i_{jk} + \tau^i_{jk},\quad \overline{C}^i_{jk} = C^i_{jk} + \sigma^i_{jk},$$

where (L^i_{jk}, C^i_{jk}) are the coefficients of the canonical metrical d-connection $C\Gamma(N)$ and

$$\text{(3.5)}'\quad \begin{aligned} \tau^i_{jk} &= -\frac{1}{2}g^{ih}(g_{ir}\overline{T}^r{}_{hk} + g_{kr}\overline{T}^r{}_{hj} - g_{hr}\overline{T}^r{}_{jk}),\\ \sigma^i_{jk} &= -\frac{1}{2}g^{ih}(g_{ir}\overline{S}^r{}_{hk} + g_{kr}\overline{S}^r{}_{hj} - g_{hr}\overline{S}^r{}_{jk}). \end{aligned}$$

It is clear that τ^i_{jk} and σ^i_{jk} are the contorsions of $\overline{L\Gamma}(N)$.

§4. Gravitational and Electromagnetic Fields

By an obvious particularisation, the Einstein equations written and discussed in Ch.V may be attached to the one of the metrical d-connections $C\Gamma(N)$ or $\overline{L\Gamma}(N)$. In what follows we shall discuss the Einstein equations attached to the canonical metrical d-connection $C\Gamma(N)$, where N is given by (2.5) and L^i_{jk}, C^i_{jk} are given by (3.2).

Thus for $n > 2$ we shall admit that the Einstein equations of the Lagrange space L^n corresponding to $C\Gamma(N)$ are as follows:

$$\text{(4.1)}\quad \begin{aligned} R_{ij} - \frac{1}{2}R g_{ij} &= \kappa \overset{H}{T}_{ij}, & {}'P_{ij} &= \kappa \overset{1}{T}_{ij},\\ S_{ij} - \frac{1}{2}S g_{ij} &= \kappa \overset{V}{T}_{ij}, & {}''P_{ij} &= -\kappa \overset{2}{T}_{ij}, \end{aligned}$$

where $R_{ij} = R_i{}^h{}_{jh}$, $S_{ij} = S_i{}^h{}_{jh}$, ${}'P_{ij} = P_i{}^h{}_{jh}$, ${}''P_{ij} = P_i{}^h{}_{hj}$ are the Ricci tensors of

$C\Gamma(N)$, $R = g^{ij}R_{ij}$, $S = g^{ij}S_{ij}$ are the scalar curvatures, $\overset{H}{T}_{ij}$, $\overset{V}{T}_{ij}$, $\overset{1}{T}_{ij}$, $\overset{2}{T}_{ij}$ are the components of the energy-momentum tensor and κ is a constant (equal to 0 for vacuum).

This assumption is completely supported by the almost Kählerian model of the Lagrange space L^n, which will be described later. It requires also the condition $n > 2$.

First, let us notice that not all Ricci tensors of $C\Gamma$ are symmetric. Indeed, on account of the Bianchi identities we get

Theorem 4.1. *The Ricci tensors of the canonical metrical d-connection* $C\Gamma$ *satisfy the following identities*

$$1^\circ \quad R_{ij} - R_{ji} = C^r_{rs}R^s{}_{ij} + C^r_{is}R^s{}_{jr} - C^r_{js}R^s{}_{ir},$$
$$2^\circ \quad S_{ji} = S_{ji},$$
$$3^\circ \quad {}'P_{ij} - {}'P_{ji} = C^r_{ri|j} - C^r_{rj|i} + C^s_{jr}P^r{}_{is} - C^s_{ir}P^r{}_{js},$$
$$4^\circ \quad {}''P_{ij} - {}''P_{ji} = -(C^r_{ij}|_r - C^r_{rj}|_i + C^s_{sr}P^r{}_{ij} - C^s_{ir}P^r{}_{sj}).$$

Using again the Bianchi identities one obtains

Theorem 4.2. *The energy-momentum tensors* $\overset{H}{T}_{ij}$ *and* $\overset{V}{T}_{ij}$ *satisfy the following equalities*

$$(4.2) \quad \kappa \overset{H}{T}{}^i_{j|i} = -\frac{1}{2}(P^{ih}{}_{js}R^s{}_{hi} + 2R^s{}_{ij}P^i_s), \quad \kappa \overset{V}{T}{}^i_{j|i} = 0.$$

Next we have

Corollary 4.1. *If the curvature tensor* $P_j{}^i{}_{kh}$ *of the canonical metrical d-connection* $C\Gamma$ *vanishes, then*

i) For $n > 2$ *the Einstein equations of the Lagrange space* L^n *take the form*

$$(4.1)' \quad \begin{aligned} R_{ij} - \frac{1}{2}Rg_{ij} &= \kappa \overset{H}{T}_{ij}, \\ S_{ij} - \frac{1}{2}SG_{ij} &= \kappa \overset{V}{T}_{ij}. \end{aligned}$$

ii) The following equalities are verified

$$(4.3) \quad \overset{H}{T}{}^i_{j|i} = 0, \quad \overset{V}{T}{}^i_j|_i = 0.$$

Corollary 4.2. *If the horizontal distribution of the canonical non-linear connection of the Lagrange space L^n is integrable then*

i) The Ricci tensor R_{ij} is symmetric,

ii) $\overset{H}{T}_{ij}$ and $\overset{V}{T}_{ij}$ satisfy (4.3).

Now let us consider the h- and v- deflection tensors of the connection $C\Gamma(N)$:

$$D^i_j = y^i_{|j} = L^i_{oj} - N^i_j, \quad d^i_j = y^i|_j = \delta^i_j + C^i_{oj}, \tag{4.4}$$

where the index "o" means a contraction by y^i.

These tensors satisfy some identities provided by the Ricci identities when these are applied to the Liouville vector field y^i. We shall write such identities for the covariant deflection tensors

$$D_{ij} = g_{ih} D^h_j, \quad d_{ij} = g_{ih} d^h_j. \tag{4.5}$$

Theorem 4.3. *The covariant deflection tensors D_{ij} and d_{ij} of the canonical metrical d-connection $C\Gamma(N)$ satisfy the identities*

$$\begin{aligned} &D_{ij|k} - D_{ik|j} = R_{0ijk} - d_{ir} R^r{}_{jk}, \\ &D_{ij}|_k - d_{ik|j} = P_{0ijk} - D_{ir} C^r_{jk} - d_{ir} P^r{}_{jk}, \\ &d_{ij}|_k - d_{ik}|_j = S_{0ijk}. \end{aligned} \tag{4.6}$$

Some reasons which will be seen later lead us to introduce

Definition 4.1. *The tensor fields*

$$F_{ij} = \frac{1}{2}(D_{ij} - D_{ji}), \quad f_{ij} = \frac{1}{2}(d_{ij} - d_{ji}) \tag{4.7}$$

are said to be h- and v-electromagnetic tensor fields of the Lagrange space L^n endowed with the canonical metrical d-connection $C\Gamma(N)$.

By (4.4) and (4.5) immediately follows, on account of the symmetry of the tensor C_{ijk}:

(4.7)′ $f_{ij}=0.$

The Bianchi identities together with (4.6), (4.7) and (4.7)′ give the following important result

Theorem 4.4. *The h- electromagnetic tensor field* F_{ij} *of the Lagrange space* L^n *corresponding to the canonical metrical d-connection* $C\Gamma$ *verifies the following Maxwell equations*

$$(4.8)\qquad \begin{aligned} &F_{ij|k}+F_{jk|i}+F_{ki|j}=-\sum_{(ijk)}C_{i0r}R^r{}_{jk},\\ &F_{ij}|_k+F_{jk}|_i+F_{ik}|_j=0. \end{aligned}$$

Corollary 4.3. *If* L^n *is a Finsler space* F^n *then h- and v-electromagnetic tensor fields,* F_{ij} *and* f_{ij} *vanish identically.*

Corollary 4.4. *If the fundamental tensor* $g_{ij}(x,y)$ *of a Lagrange space* L^n *is 0-homogeneous with respect to* y^i *then the Maxwell equations satisfied by the h-electromagnetic tensor field* F_{ij} *take the form*

$$(4.8)'\qquad \begin{aligned} &F_{ij|k}+F_{jk|i}+F_{ki|j}=0,\\ &F_{ij}|_k+F_{jk}|_i+F_{ki}|_j=0. \end{aligned}$$

Proof. If $g_{ij}(x,y)$ is 0-homogeneous with respect to y^i then $C_{i0k}=0$ and (4.8) take the form (4.8)′.

Notice that the AFL-spaces with the Lagrangians of type (1.2) have $g_{ij}(x,y)$ 0-homogeneous with respect to y^i.

Now we define h- and v- charge current densities hJ^i and vJ^i, respectively, by

$$(4.9)\qquad hJ^i=F^{ij}{}_{|j},\quad vJ^i=F^{ij}|_j,$$

where $F^{ij}=g^{is}g^{jr}F_{sr}$.

We have

Theorem 4.5. *The vector fields* hJ^i *and* vJ^i *have the properties*

(4.10) $hJ^i{}_{|i} = \frac{1}{2}\{F^{ij}(R_{ij} - R_{ji}) + F^{ij}|_r R'^r{}_{ij}\}, \quad vJ^i|_i = 0.$

Proof. It is clear that $hJ^i{}_{|i} = F^{ij}{}_{|j|i} = \frac{1}{2}(F^{ij}{}_{|j|i} - F^{ij}{}_{|i|j})$.

Now the Ricci identities apply and the first formula (4.10) is obtained. The second results on a similar way.

By the Corollary 4.2 one gets

Corollary 4.5. *If the horizontal distribution of the canonical non-linear connection of the Lagrange space L^n is integrable then hJ^i and vJ^i verify the equations*

$$hJ^i{}_{|i} = 0, \quad vJ^i|_i = 0.$$

An important particularisation of the theory will be discussed in the next section.

The theory presented in this section was developed by R.Miron, R.Roşca, K.Buchner, M.Anastasiei [114] and R.Miron, M.Radivoiovici [112]. Its physical grounds were discussed in depth by S.Ikeda in the paper [67].

§5. Lagrange Space of Electrodynamics

We shall apply the previous theory of the gravitational and electromagnetic fields to the following Lagrangian from electrodynamics

(5.1) $L_0(x,y) = mc\gamma_{ij}(x)y^iy^j + \frac{2e}{m}A_i(x)y^i,$

which is nothing but the Lagrangian (3.1) with $U(x) \equiv 0$. Recall that $\gamma_{ij}(x)$ is a Lorentz metric tensor, $A_i(x)$ is a covector field and m,c,e are the physical constants: mass of particle, light speed and electric charge, respectively.

First, we notice that $L_0(x,y)$ is a fundamental function for a Lagrange space $L_0^n = (M, L_0(x,y))$. Indeed L_0 is a regular Lagrangian whose fundamental tensor is given by

(5.2) $g_{ij}(x,y) = mc\gamma_{ij}(x)$

and the signature of the quadratic form with the coefficients $g_{ij}(x,y)$ is the same as that of the quadratic form $\Phi_0 = \gamma_{ij}(x)\xi^i\xi^j$. We call L_0^n the Lagrange space of electrodynamics.

By the Theorem 2.1. we have

Theorem 5.1. *The extremals of the action integral associated to the Lagrangian* $L_0(x,y)$ *are the solutions of the following system of differential equations*

$$(5.3)\qquad \frac{d^2x^i}{dt^2} + 2G^i(x,\dot{x}) = 0\,, \quad (\dot{x}^i = \frac{dx^i}{dt})\,,$$

where

$$(5.3)'\qquad G^i = \frac{1}{2}\gamma^i_{00} + \frac{e}{2m^2c}\gamma^{ij}(A_{j;k} - A_{k;j})y^k\,.$$

Here ";" denotes the covariant derivative defined with the help of the Christoffel symbols $\gamma^i_{jk}(x)$ of the Lorentz metric tensor $\gamma_{ij}(x)$.

Indeed, from (2.4) one easily deduces (5.3)′ on account of

$$g^{ij} = \frac{1}{mc}\gamma^{ij}.$$

Introducing the electromagnetic tensor defined by A_i, that is,

$$(5.4)\qquad \dot{F}_{ij} = \frac{e}{2m}(A_{j;i} - A_{i;j}),\quad \dot{F}^i_j = g^{ik}\dot{F}_{kj}$$

one gets that the canonical non-linear connection of the Lagrange space L_0^n has the coefficients

$$(5.5)\qquad N^i_j = \gamma^i_{jk}(x)y^k - \dot{F}^i_j.$$

Thus we have obtained

Theorem 5.2. *The autoparallel curves of the canonical non-linear connection given by (5.5) are the solutions of the Lorentz equations*

$$(5.6)\qquad \frac{d^2x^i}{dt^2} + \gamma^i_{jk}(x)\frac{dx^j}{dt}\frac{dx^k}{dt} = \dot{F}^i_j(x)\frac{dx^j}{dt}.$$

The integrability tensor $R^h_{\ ij}$ of canonical non-linear connection is

(5.7) $R^{i}{}_{jk} = r_{0}{}^{i}{}_{jk} - (\dot{F}^{i}{}_{j;k} - \dot{F}^{i}{}_{k;j})$,

where the $r_{0}{}^{i}{}_{jk}$ is the curvature tensor built by using γ^{i}_{jk}. Thus we have

Theorem 5.3. *The horizontal distribution of the canonical non-linear connection (5.5) is integrable if and only if the Riemannian space* $R^{n} = (M, \gamma_{ij}(x))$ *is flat and the electromagnetic tensor* $\dot{F}^{i}_{j}$ *is absolute parallel with respect to Levi-Civita connection of* R^{n}.

Now we shall determine the canonical metrical d-connection of the space L^{n}_{0}.

Theorem 5.4. *The canonical metrical d-connection* $C\Gamma(N)$ *of the Lagrange space* L^{n}_{0} *has the coefficients*

(5.8) $L^{i}{}_{jk}(x,y) = \gamma^{i}{}_{jk}(x), \quad C^{i}{}_{jk} = 0.$

Proof. By (5.2), $\delta_{k} g_{ij} = mc\dfrac{\partial \gamma_{ij}}{\partial x^{k}}$ and $\dot{\partial}_{k} g_{ij} = 0.$
Inserting these derivatives in (3.2) we obtain (5.8).

The torsions of $C\Gamma(N)$ are given by $T^{i}{}_{jk} = S^{i}{}_{jk} = C^{i}_{jk} = 0$ and $R^{i}{}_{jk}$ as in (5.7).
The curvatures of $C\Gamma(N)$ are the following ones

$$R_{j}{}^{i}{}_{kh} = r_{j}{}^{i}{}_{kh}, \quad P_{j}{}^{i}{}_{kh} = 0, \quad S_{j}{}^{i}{}_{kh} = 0.$$

Therefore, by (4.1) one deduces

Theorem 5.5. *The Einstein equations in* L^{n}_{0} *reduce to the classical Einstein equations associated to the Levi-Civita connection* γ^{i}_{jk} *of the Lorentz space* $(M, \gamma_{ij}(x))$ *and the conservation law of the energy-momentum tensor is identically verified.*

The covariant deflection tensors D_{ij} and d_{ij} are

$$D_{ij} = \dot{F}_{ij}(x), \quad d_{ij} = g_{ij}.$$

Noticing that $D_{ij} - D_{ji} = 2\dot{F}_{ij}$ and $F_{ij}|_{k} = 0$ one immediately gets

Theorem 5.6. *The h-electromagnetic field of the canonical metrical d-connection* $C\Gamma(N)$ *in* L^{n}_{0} *coincides to the electromagnetic tensor field* $\dot{F}_{ij}(x)$ *and the Maxwell equations (4.8)*

reduce to the classical ones.

These considerations might go further for hJ^i and vJ^i.

§6. Almost Finslerian Lagrange Spaces

The almost Finslerian Lagrange spaces, shortly AFL-spaces, were introduced in the section 1 as Lagrange spaces whose fundamental function L are of the form

$$(6.1)\quad L(x,y) = F^2(x,y) + \frac{2e}{m}A_i(x)y^i + U(x),$$

where $F(x,y)$ is the fundamental function of a Finsler space $F^n = (M, F(x,y))$, $A_i(x)$ is a covector field (electromagnetic potentials), $U(x)$, $x \in M$, is a potential function and e, $m \neq 0$ are constants.

From a geometrical point of view the AFL-spaces are very close to Finsler spaces with respect to their properties. From the point of view of the applications these spaces are very suitable for a substantial theory of gravitation and electromagnetism (R.Miron and M.Radivoiovici [112], S.Ikeda [67]). Notice that if A_i is a gradient and U is a constant then (6.1) is nothing but a Caratheodory transformation of Lagrangians. Therefore in this case AFL-spaces will be called Caratheodory AFL-spaces.

An example of AFL-space is obtained if in (6.1) one takes

$$F^2(x,y) = \varepsilon(x,y) = \gamma_{ij}(x)y^iy^j,$$

that is, F^2 is the absolute energy of a Riemannian space $(M, \gamma_{ij}(x))$. Such a Lagrange space will be called an almost Riemannian Lagrange space, shortly ARL-space. Going further and taking $U \equiv 0$ one obtains the spaces studied in the previous section. In what follows we shall treat the Caratheodory AFL-spaces or the ARL-spaces as particular cases of AFL-spaces.

Now we shall associate with the Lagrange space $L^n = (M, L(x,y))$, where the Lagrangian L is given by (6.1), the Finsler space $\dot{F}^n = (M, F(x,y))$ and we shall indicate by an index "0" placed over them all geometrical objects related to it. For example, the fundamental tensor field of $\dot{F}^n$ is

$$(6.2)\quad \dot{g}_{ij}(x,y) = \frac{1}{2}\dot{\partial}_i\dot{\partial}_j F^2,$$

while the Cartan connection is $\dot{C}\Gamma(\dot{N}) = (\dot{F}^i_{jk}, \dot{C}^i_{jk})$. Then h- and v-covariant derivatives with respect to $\dot{C}\Gamma(\dot{N})$ will be denoted by $\overset{\circ}{|}$ and $\overset{\circ}{|}$, respectively.

Thus we have

$$(6.3)\quad \dot{C}_{ijk} = \frac{1}{4}\dot{\partial}_i\dot{\partial}_j\dot{\partial}_k F^2, \quad \dot{N}^i_j = \frac{1}{2}\dot{\partial}_j(\dot{\gamma}^i_{rs}y^r y^s),$$

$$(6.4)\quad \dot{g}_{ij\overset{\circ}{|}k} = 0, \quad \dot{g}_{ij}\overset{\circ}{|}_k = 0, \quad F^2_{\overset{\circ}{|}k} = 0, \quad \dot{D}^i_j = 0, \quad \dot{d}^i_j = \delta^i_j.$$

The absolute energy of $\dot{F}^n$ is

$$\dot{\varepsilon}(x,y) = \dot{g}_{ij}(x,y)y^i y^j = F^2(x,y).$$

The first results on AFL spaces are the following ones.

Proposition 6.1. *i) The fundamental tensor of the AFL-space L^n coincides with the fundamental tensor $\dot{g}_{ij}$ of the associated Finsler space $\dot{F}^n$:*

$$(6.5)\quad g_{ij}(x,y) = \dot{g}_{ij}(x,y).$$

ii) The fundamental tensor $g_{ij}(x,y)$ of the AFL-space L^n is 0-homogeneous with respect to y^i.

iii) The absolute energy of the AFL-space L^n coincides with the absolute energy of the associated Finsler space $\dot{F}^n$:

$$(6.6)\quad \varepsilon(x,y) = g_{ij}(x,y)y^i y^j = \dot{\varepsilon}(x,y).$$

We shall prove that, in fact, the property ii) from the last Proposition gives locally a characterisation of AFL-spaces.

Theorem 6.1. *A Lagrange space $L^n = (M,L(x,y))$ is locally an AFL-space if and only if its fundamental tensor $g_{ij}(x,y) = \frac{1}{2}\dot{\partial}_i\dot{\partial}_j L$ is 0-homogeneous with respect to y^i.*

Proof. Assuming $g_{ij}(x,y)$ is 0-homogeneous with respect to y^i let us consider the absolute energy $\varepsilon(x,y) = g_{ij}(x,y)y^i y^j$ of the space L^n. It follows this function has all the properties of a Finsler metric function. Thus we are entitled to consider the Finsler space

$\overset{\circ}{F}{}^n = (M, \varepsilon(x,y))$. The fundamental tensor of the Finsler space $\overset{\circ}{F}{}^n$ is just $g_{ij}(x,y)$. Therefore, we have $\dot\partial_i \dot\partial_j L = \dot\partial_i \dot\partial_j \varepsilon$ or equivalently $\dot\partial_i \dot\partial_j (L - \varepsilon) = 0$. Consequently, in a coordinate system on $\tilde{TM}$ we get $L - \varepsilon = A_i(x) y^i + U(x)$. Thus the Lagrangian L is locally of the form (6.1). The converse is just ii) from the Proposition 6.1.

We shall consider in this new context the electromagnetic tensor field defined by $A_i(x)$:

$$(6.6)' \quad F_{ij}(x) = \frac{e}{2m}\left(\frac{\partial A_j}{\partial x^i} - \frac{\partial A_i}{\partial x^j}\right).$$

Then we have

Theorem 6.2. *The extremal curves of the AFL-space L^n are the solutions of the system of differential equations (2.3), where*

$$(6.7) \quad G^i = \overset{\circ}{G}{}^i - g^{ij}\left(F_{jk} y^k + \frac{\partial U}{\partial x^j}\right).$$

As to the non-linear connection of an AFL-space we get

Theorem 6.3. *The coefficients of the canonical non-linear connection of the AFL-space L^n are the following ones*

$$(6.8) \quad N^i_j = \overset{\circ}{N}{}^i_j - g^{ik} F_{kj} + 2\overset{\circ}{C}{}^{ik}_j \left(F_{ks} y^s + \frac{\partial U}{\partial x^k}\right).$$

Of course, the raising and lowering of indices is done by using $g_{ij}(x,y)$.

Let us consider the d-tensor field

$$(6.9) \quad B^i_j = g^{ik} F_{kj} - 2\overset{\circ}{C}{}^{ik}_j \left(F_{ks} y^s + \frac{1}{4}\frac{\partial U}{\partial x^k}\right).$$

Then we have

$$(6.9)' \quad B^i_j y^j = g^{ik} F_{kj} y^j.$$

Therefore (6.8) transforms into

$$(6.8)' \quad N^i_j = \overset{\circ}{N}{}^i_j - B^i_j.$$

On account on (6.9)′ one obtains

Theorem 6.4. *The autoparallel curves of the canonical non-linear connection N of the AFL-space L^n given by (6.1) are solutions of "the generalized Lorentz equations"*

$$\frac{d^2x^i}{dt^2}+\dot{\gamma}^{\,i}_{jk}(x,\dot{x})\frac{dx^j}{dt}\frac{dx^k}{dt}=g^{ij}(x,\dot{x})F_{jk}(x)\frac{dx^k}{dt}.$$

The covariant tensor $B_{ij}=g_{ik}B_j^k$ takes by (6.9) the form

$$(6.10)\quad B_{ij}=F_{ij}-2\dot{C}^k_{ij}(F_{k0}+\frac{1}{4}\frac{\partial U}{\partial x^k}).$$

This form shows us the electromagnetic tensor field $F_{ij}(x)$ is the skew-symmetric part of B_{ij}, that is,

$$(6.11)\quad F_{ij}=\frac{1}{2}(B_{ij}-B_{ji}).$$

Next we have

Proposition 6.2. *The following two assertions are equivalent:*

1° $A_i(x)$ *is a gradient,*

2° $B_{ij}(x)=-\frac{1}{2}\dot{C}^k_{ij}\frac{\partial U}{\partial x^k}.$

Proof. If A_i is a gradient then $F_{ij}=0$ and by (6.10), 2° holds. Conversely, the equation in 2° and (6.10) give $F_{ij}(x)-2C^k_{ij}F_{k0}=0$. Contracting by y^i and taking into account $\dot{C}^i_{jk}y^j=0$ one obtains $F_{ij}(x)y^j=0$ which implies $F_{ij}=0$ and by (6.6)′, A_i must be a gradient.

Theorem 6.5. *The canonical metrical d-connection* $C\Gamma=(N^i_j,L^i_{jk},C^i_{jk})$ *of an AFL-space* L^n *has the coefficients*

$$(6.12)\quad L^i_{jk}=\dot{F}^i_{jk}+\dot{C}^i_{jr}B^r_k+B^i_{jk},\quad C^i_{jk}=\dot{C}^i_{jk},$$

where the d-tensor B^i_{jk} *is given by*

(6.13) $B^i_{jk} = g^{ih}(\dot{C}_{khs}B^s_j - \dot{C}_{jks}B^s_h)$.

Proof. Inserting in (3.2) the coefficients N^i_j given by (6.8)′ and $g_{ij} = \dot{g}_{ij}$ we first get $\delta_k g_{ij} = \dot{\delta}_k g_{ij} + 2\dot{C}_{ijr}B^r_k$ and then by a direct calculation, (6.12) with B^i_{jk} as in (6.13) is obtained.

Now we are interested in finding the Einstein equations associated to the canonical metrical d-connection $C\Gamma(N)$ of an AFL-space L^n. The difficulties which are met in a direct calculation will be overcome by using certain transformations of connections. Thus we have

Theorem 6.6. *i) The canonical metrical d-connection* $C\Gamma$ *can be thought of as the composition of the d-connection* $C\dot{\Gamma} = (\dot{N}^i_k, \dot{F}^i_{jk}, \dot{C}^i_{jk})$ *and* $L\tilde{\Gamma} = (N^i_j, \tilde{L}^i_{jk}, \tilde{C}^i_{jk})$ *in a sense that*

(6.14) $N^i_j = \dot{N}^i_j - B^i_j,\quad \tilde{L}^i_{jk} = \dot{F}^i_{jk} + \dot{C}^i_{jr}B^r_k,\quad \tilde{C}^i_{jk} = \dot{C}^i_{jk},$

(6.14)′ $L^i_{jk} = \tilde{L}^i_{jk} + B^i_{jk},\quad C^i_{jk} = \tilde{C}^i_{jk},$

hold.

ii) The d-connection with the coefficients (6.14) is metrical with respect to g_{ij}.

The transformations $C\dot{\Gamma} \to L\tilde{\Gamma} \to C\Gamma$ lead to the following

Theorem 6.7. *The curvature tensors of* $C\dot{\Gamma}$ *and* $L\tilde{\Gamma}$ *are related by*

$$\begin{aligned} &\tilde{R}_j{}^i{}_{kh} = \dot{R}_j{}^i{}_{kh} + \dot{P}_j{}^i{}_{ks}B^s_h - \dot{P}_j{}^i{}_{hs}B^s_k + \dot{S}_j{}^i{}_{rs}B^r_k B^s_h, \\ (6.15)\quad &\tilde{P}_j{}^i{}_{kh} = \dot{P}_j{}^i{}_{kh} + \dot{S}_j{}^i{}_{ks}B^s_h, \\ &\tilde{S}_j{}^i{}_{kh} = \dot{S}_j{}^i{}_{kh}. \end{aligned}$$

Theorem 6.8. *The curvature tensors of* $L\tilde{\Gamma}$ *and* $C\Gamma$ *are related as follows*

$$\begin{aligned} &R_j{}^i{}_{kh} = \tilde{R}_j{}^i{}_{kh} + (B^i_{jk\,\bar{|}\,h} - B^i_{jh\,\bar{|}\,k} + B^r_{jk}B^i_{rh} - B^r_{jh}B^i_{rk} + B^i_{jr}\tilde{T}^r{}_{kh}), \\ (6.16)\quad &P_j{}^i{}_{kh} = \tilde{P}_j{}^i{}_{kh} + (B^i_{jk}|_h + B^i_{jr}C^r_{kh}), \\ &S_j{}^i{}_{kh} = \tilde{S}_j{}^i{}_{kh}, \end{aligned}$$

where $\tilde{|}$ *denotes the h-covariant derivative with respect to* $L\tilde{\Gamma}$.

It is now clear how (6.15) and (6.16) provide the curvature tensors $R_j{}^i{}_{kh}, P_j{}^i{}_{kh}$ and $S_j{}^i{}_{kh}$ of the canonical metrical d-connection $C\Gamma$.

Then the Einstein equations can be written without any difficulties.

As to the h-electromagnetic tensor field, by the Theorem 4.3 and $C_{i0r}=0$ we obtain

Theorem 6.9. *The h-electromagnetic tensor field of the AFL-space* L^n, *with respect to* $C\Gamma$ *coincides to* $F_{ij}(x)$ *given by (6.11) and it verifies the Maxwell equations*

$$F_{ij|k}+F_{jk|i}+F_{ki|j}=0,$$

$$F_{ij}\tilde{|}_k+F_{jk}\tilde{|}_i+F_{ki}\tilde{|}_j=0.$$

§7. Almost Kählerian Model of a Lagrange Space

As in the case of a Finsler space we shall see that a Lagrange space $L^n=(M,L(x,y))$ can be "lifted" to an almost Kählerian structure over the manifold $\tilde{TM}$.

Let us consider the basis $(dx^i,\delta y^i)$ which, as we know, is dual to the adapted basis $(\delta_i,\dot{\partial}_k)$. We define an almost complex structure **F** on $\tilde{TM}$ by the same way as in §7 of the previous chapter. It comes out that **F** is integrable if and only if the non-linear connection N of local coefficients given by (2.5) is integrable i.e. $R^i{}_{jk}=0$.

The metric tensor g_{ij} of L^n induces a pseudo-Riemannian structure on $\tilde{TM}$:

$$\text{(7.1)}\quad \mathbf{G}=g_{ij}(x,y)dx^i\otimes dx^j+g_{ij}(x,y)\delta y^i\otimes\delta y^j.$$

The horizontal and vertical distributions are orthogonal with respect to **G**.
We have also in this context

Theorem 7.1. *i) The pair (G,F) is an almost Hermitian structure on* $\tilde{TM}$.
ii) The 2-form associated with (G,F) is

$$\text{(7.2)}\quad \theta=g_{ij}(x,y)\delta y^i\wedge dx^j.$$

As we have seen in the Theorem 2.7, θ is a symplectic structure. Thus we have

Theorem 7.2. *The triad $K^{2n} = (\tilde{TM}, G, F)$ is an almost Kählerian space.*

The space K^{2n} will be called the almost Kählerian model of the Lagrange space L^n.

We know that K^{2n} becomes a Kähler space if the almost complex structure **F** is integrable.

Hence we have

Theorem 7.3. *The almost Kählerian model K^{2n} of the Lagrange space L^{2n} is a Kählerian space if and only if the canonical non-linear connection (2.5) is integrable.*

We shall use K^{2n} in order to characterise the canonical metrical d-connection $C\Gamma$.

As in §7 of the previous chapter we shall define a notion of almost Kählerian connection of Lagrange type as a linear connection ∇ on TM with the properties

1° ∇ preserves by parallelism the vertical distribution,

2° ∇ is compatible with the almost Kählerian structure (**G**,**F**), i.e.

(7.3) $\nabla_X G = 0, \quad \nabla_X F = 0$, hold for any $X \in \chi(\tilde{TM})$.

Thus we have

Theorem 7.4. *There exists an unique almost Kählerian connection of Lagrange type, ∇, having h(hh)- and v(vv)-torsions zero. The coefficients (L^i_{jk}, C^i_{jk}) of ∇ with respect to the basis $(\delta_i, \dot{\partial}_i)$ are just the coefficients of the canonical metrical d-connection $C\Gamma$ of L^n. Conversely, $C\Gamma$ determines on the manifold $\tilde{TM}$ an unique almost Kählerian connection of Finsler type having h(hh)- and v(vv)-torsions zero.*

It is now clear that we can use the almost Kählerian model K^{2n} in order to study:

1) The Einstein equations of the Lagrange space L^n interpreted as the Einstein equations associated to the almost Kählerian connection of Lagrange type ∇.

2) The electromagnetic field defined with the help of the Liouville vector field $y^i \dfrac{\partial}{\partial y^i}$ on $\tilde{TM}$ which does not depend on any structure on TM.

As in the case of Finsler spaces, with the help of the almost Kählerian model K^{2n} a geometrical theory of subspaces in Lagrange spaces can be developed (cf. R.Miron [106]).

Chapter X

Generalized Lagrange Spaces

A first natural generalization of the notion of Lagrange space is provided by a notion which we call a generalized Lagrange space. This notion retains from that of Lagrange space only the metric structure of it, ignoring that this metric structure could be provided by a regular Lagrangian. Thus a generalized Lagrange space GL^n was defined by the first author of this book in [107] as a pair which consists of a smooth manifold and a d-tensor field $g_{ij}(x,y)$ of rank n, symmetric and of constant signature. Of course the Finsler and Lagrange spaces are examples of generalized Lagrange spaces. But there exist generalized Lagrange spaces which are not reducible to Finsler or Lagrange spaces in a sense that their metrical structure $g_{ij}(x,y)$ can not be furnished by a regular Lagrangian. First examples are due to S. Watanabe, F. Ikeda and S. Ikeda [166]. They have considered

$$g_{ij}(x,y) = e^{2\sigma(x,y)}\gamma_{ij}(x),$$

where $\gamma_{ij}(x)$ is a Lorentz metric and σ is a function depending on x^i as well as on directional variables y^i. We shall see later these spaces provide a convenient model for an axiomatic theory of Relativity.

The geometry of the generalized Lagrange spaces can be developed by the same devices as the geometry of Finsler and Lagrange spaces. That is, one associates to GL^n a non-linear connection N - is preferred one which depends on $g_{ij}(x,y)$ only, then a canonical metrical connection is determined by help of which all important geometrical objects are expounded. The whole theory is legitimized from a geometric point of view by the fact that GL^n can be lifted to an almost Hermitian structure on $\tilde{TM}$ and then its canonical metrical d-connection appear as an almost Hermitian connection on $\tilde{TM}$.

§1. Notion of Generalized Lagrange Space

Let M be a smooth (C^∞), n-dimensional, real manifold, (TM,τ,M) its tangent bundle and $\tilde{TM} = TM - \{0\}$.

Definition 1.1. *A generalized Lagrange space is a pair $GL^n = (M, g_{ij}(x,y))$, where $g_{ij}(x,y)$ is a d-tensor field on $\tilde{TM}$, symmetric, of rank n and of constant signature on $\tilde{TM}$.*

Then n is called the dimension of GL^n and $g_{ij}(x,y)$ is called its fundamental tensor or its metric tensor.

The metric tensor has the properties

$$(1.1)\quad g_{ij}(x,y) = g_{ji}(x,y), \quad \text{rank}\, \| g_{ij}(x,y) \| = n$$

and the quadratic form

$$(1.2)\quad \Phi = g_{ij}(x,y)\xi^i\xi^j, \quad (\xi^i) \in \mathbb{R}^n,$$

has constant signature on $\tilde{TM}$.

One easily sees that any Lagrange space $L^n=(M,L(x,y))$ is a generalized Lagrange space whose fundamental tensor is $g_{ij}(x,y) = \frac{1}{2}\dot{\partial}_i\dot{\partial}_j L$. But not any GL^n-space is a L^n-space. Indeed, if $g_{ij}(x,y)$ are given, it may happen that the system of partial differential equations

$$(1.3)\quad g_{ij}(x,y) = \frac{1}{2}\dot{\partial}_i\dot{\partial}_j L(x,y)$$

does not admit any solution $L(x,y)$. Noticing that $\dot{\partial}_k g_{ij}$ is a d-tensor field on $\tilde{TM}$ we assert

Proposition 1.1. *i) A necessary condition in order that the system (1.3) to admit a solution L(x,y) is that d-tensor field $\dot{\partial}_k g_{ij}$ be completely symmetric.*

ii) If the d-tensor field $g_{ij}(x,y)$ is 0-homogeneous with respect to y^i and i) is verified then

$$(1.3)'\quad L(x,y) = g_{ij}(x,y)y^iy^j + A_i(x)y^i + U(x)$$

is a solution of (1.3). Here $A_i(x)$ is an arbitrary field of covectors and U is an arbitrary function.

Proof. If (1.3) admits a solution L then we have $\dot{\partial}_k g_{ij} = \frac{1}{2}\dot{\partial}_k\dot{\partial}_i\dot{\partial}_j L$ hence i) holds.

In account of i) and of $y^k\dot{\partial}_k g_{ij} = 0$ one verifies that $(1.3)'$ is a solution of (1.3) for any $A_i(x)$ and $U(x)$, that is, ii) holds.

Let $GL^n = (M, g_{ij}(x,y))$ be a generalized Lagrange space. If (1.3) admits a solution L we shall say GL^n is reducible to a Lagrange space. Otherwise we shall say GL^n is not reducible to a Lagrange space.

Example 1.1. The space $GL^n = (M, g_{ij}(x,y))$, with

$$(1.4)\quad g_{ij}(x,y) = e^{2\sigma(x,y)}\gamma_{ij}(x),$$

where $\frac{\partial\sigma}{\partial y^i} \neq 0$ on $\tilde{TM}$ at least for one i and $\gamma_{ij}(x)$ is a Lorentz metric, is not reducible to a Lagrange space.

A first notion geometrically associated with a GL^n-space is that of absolute energy. This is defined on $\tilde{TM}$ as

$$(1.5)\quad \varepsilon(x,y) = g_{ij}(x,y)y^i y^j.$$

Because of the lack of homogeneity of $g_{ij}(x,y)$ with respect to (y^i) we must define the action integral associated to ε for parametrized curves (Γ, c), with $c:[0,1]\to M$, $t\to x^i(t)$ of class C^∞. This action integral is

$$(1.6)\quad I(c) = \int_0^1 \varepsilon(x,\dot{x})dt,\quad \dot{x}^i = \frac{dx^i}{dt}.$$

Assuming that the usual variational problem concerning I(c) is a regular one, the Euler-Lagrange equations can be put in the form (see §2, Ch. VIII)

$$(1.7)\quad \frac{d^2x^i}{dt^2} + 2G^{*i}(x,\dot{x}) = 0,$$

where

$$(1.8)\quad G^{*i} = \frac{1}{4}g^{*ij}\left(\frac{\partial^2\varepsilon}{\partial y^j\partial x^k}y^k - \frac{\partial\varepsilon}{\partial x^j}\right).$$

Recall that the variational problem concerning I(c) is a regular one if the d-tensor field

$$(1.9)\quad g^*_{ij}(x,y) = \frac{1}{2}\dot{\partial}_i\dot{\partial}_j\varepsilon$$

has rank n. If this happen we shall say the metric tensor $g_{ij}(x,y)$ is weakly regular (R.Miron [107]).

The space GL^n with weakly regular metric tensor benefits by a non-linear connection of local coefficients

$$(1.10)\quad N^i_j(x,y) = \dot{\partial}_j G^{*i},$$

which clearly depend only on the metric tensor $g_{ij}(x,y)$.

Besides the spaces GL^n with weakly regular metric there exist and others spaces GL^n which benefit by a non-linear connection depending only on the metric tensor of the space. For instance, if we deal with the metric tensor (1.4) we can prove that $N^i_j = \{{}^{\ i}_{j\ k}\} y^k$, where $\{{}^{\ i}_{j\ k}\}$ denotes the Christoffel symbols of the Lorentz metric $\gamma_{ij}(x)$, are the local coefficients of a non-linear connection depending only on the metric tensor (1.4).

On the other hand we know if M is a paracompact manifold there exists at least a non-linear connection in the tangent bundle over it (cf. §1, Ch. II). Thus it is quite natural to fix a non-linear connection N on TM and to relate it to $g_{ij}(x,y)$ whenever this is possible. We shall denote the normal d-connection defined by using N as $D\Gamma(N) = (L^i_{jk}, C^i_{jk})$. Moreover, we shall assume that the non-linear connection N is without torsion, that is,

$$(1.11)\quad t^i_{jk} = \dot{\partial}_k N^i_j - \dot{\partial}_j N^i_k = 0.$$

A space GL^n with weakly regular metric has this property.

Of course, $(\delta_i, \dot{\partial}_i)$ with

$$(1.12)\quad \delta_i = \partial_i - N^k_i(x,y)\dot{\partial}_k,$$

is an adapted basis to the horizontal distribution N and the vertical distribution V on TM.

§2. Metrical d-Connections in a GL^n-space

In what follows we shall deal with normal d-connections $D\Gamma(N)$ which are compatible with the fundamental tensor of the generalized Lagrange space $GL^n = (M, g_{ij}(x,y))$. Recall that a normal d-connection $D\Gamma(N) = (L^i_{jk}, C^i_{jk})$ is compatible with $g_{ij}(x,y)$ or is metrical with respect to $g_{ij}(x,y)$ if

$$(2.1)\quad g_{ij|k} = 0, \quad g_{ij}|_k = 0.$$

A geometrical meaning of these later conditions may be obtained as follows. If X of components $X^i(x,y)$ is a d-vector field we define the square of its norm by

(2.2) $\| X \|^2 = g_{ij}(x,y)X^iX^j$

and we state

Proposition 2.1. *Let $\tilde{c}$ be any curve on $\tilde{TM}$ of local equations $x^i = x^i(t)$, $y^i = y^i(t)$. Then all d-vector fields $X(X^i)$ with $DX^i = 0$ have $\| X \|^2$ constant along of $\tilde{c}$ if and only if (2.1) holds.*

Proof. Recall that DX^i is given by (4.3) in Ch. III. Then along of $\tilde{c}$ we have

$$\frac{d\| X \|^2}{dt} = \frac{Dg_{ij}}{dt}X^iX^j + 2g_{ij}\frac{DX^i}{dt}X^j$$

or

$$\frac{d\| X \|^2}{dt} = \frac{Dg_{ij}}{dt}X^iX^j$$

since $DX^i = 0$. Hence $\| X \|^2$ is constant along of $\tilde{c}$ for any X, is equivalent to $\frac{Dg_{ij}}{dt} = 0$ which in turn is equivalent to

$$g_{ij|k}\frac{dx^k}{dt} + g_{ij}|_k\frac{\delta y^k}{dt} = 0.$$

The later equation is equivalent to both equations (2.1) since $\tilde{c}$ is an arbitrary curve on $\tilde{TM}$.

On account of (2.1), via the Christoffel process, we obtain

Theorem 2.1. *i) There exists an unique d-connection $C\Gamma(N)$ which is metrical with respect to $g_{ij}(x,y)$ and has the torsions $T^i{}_{jk}$ and $S^i{}_{jk}$ equal to zero.*

ii) The coefficients L^i_{jk}, C^i_{jk} of $C\Gamma(N)$ are

$$(2.3)\quad \begin{aligned} L^i_{jk} &= \frac{1}{2} g^{ih}(\delta_j g_{hk} + \delta_k g_{jh} - \delta_h g_{jk}), \\ C^i_{jk} &= \frac{1}{2} g^{ih}(\dot{\partial}_j g_{hk} + \dot{\partial}_k g_{jh} - \dot{\partial}_h g_{jk}). \end{aligned}$$

It is obvious that if GL^n is a Finsler space F^n and N is the Cartan non-linear connection of F^n, then (2.3) gives the coefficients of the Cartan connection of F^n. A similar remark applies to the Lagrange spaces.

We note that $C\Gamma(N)$ depends only on $g_{ij}(x,y)$ and $N^i_j(x,y)$. This is a reason to call it *the canonical metrical d-connection* of the generalized Lagrange space GL^n.

Using $C\Gamma(N)$ the following Ricci identities are deduced

$$(2.4)\quad \begin{aligned} &X^i_{|k|h} - X^i_{|h|k} = X^r R_r{}^i{}_{kh} - X^i|_r R^r{}_{kh}, \\ &X^i_{|k}|_h - X^i|_{k|k} = X^r P_r{}^i{}_{kh} - X^i_{|r} C^r_{kh} - X^i|_r P^r{}_{kh}, \\ &X^i|_k|_h - X^i|_h|_k = X^r S_r{}^i{}_{kh}, \end{aligned}$$

where $X^i(x,y)$ denotes a d-vector field.

Similar identities may be written for arbitrary d-tensor fields.

In (2.4) the torsions $R^i{}_{jk} = \delta_k N^i_j - \delta_j N^i_k$, $P^i{}_{jk} = \dot{\partial}_k N^i_j - L^i_{kj}$ and C^i_{jk} as well as curvatures $R_j{}^i{}_{kh}$, $P_j{}^i{}_{kh}$, $S_j{}^i{}_{kh}$ of $C\Gamma(N)$ appear.

We shall use the matrix (g_{ij}) and its inverse (g^{hk}) for lowering and raising the indices of a d-tensor field, respectively. Thus $R_{ijkh} = g_{jr} R_i{}^r{}_{hk}$, $C^{ih}_k = g^{hj} C^i_{jk}$ and so on.

The Ricci identities applied to g_{ij} lead to

Theorem 2.2. *The curvatures of* $C\Gamma(N)$ *satisfy the identities*

$$(2.5)\quad R_{ijkh} + R_{jikh} = 0,\ P_{ijkh} + P_{jikh} = 0,\ S_{ijkh} + S_{jikh} = 0.$$

Let us consider the deflection tensors of $C\Gamma(N)$:

$$(2.6)\quad D^i_j = y^i_{|j} = y^k L^i_{kj} - N^i_j,\quad d^i_j = y^i|_j = \delta^i_j + y^k C^i_{kj}$$

and $D_{ij} = g_{ik} D^k_j$, $d_{ij} = g_{ik} d^k_j$ their covariant counterparts.

The Ricci identities applied to the Liouville vector field (y^i) give

Theorem 2.3. *The following identities involving the covariant deflection tensors* D_{ij} *and* d_{ij} *of the canonical metrical d-connection* $C\Gamma(N)$ *hold good*

$$D_{ij|k}-D_{ik|j}=y^rR_{rijk}-d_{ir}R^r{}_{jk},$$

$$(2.7)\quad D_{ij}|_k-d_{ik|j}=y^rP_{rijk}-D_{ir}C^r_{jk}-d_{ir}P^r{}_{jk},$$

$$d_{ij}|_k-d_{ik}|_j=y^rS_{rijk}.$$

These identities will be useful in deriving the Maxwell equations for the electromagnetic tensor field.

Of course, metrical normal d-connections different from $C\Gamma(N)$ may be found. More precisely we have

Theorem 2.4. *i) There exists an unique normal d-connection* $D\Gamma(N)=(\bar{L}^i_{jk},\bar{C}^i_{jk})$ *which is compatible with the metric tensor* $g_{ij}(x,y)$ *of a space* GL^n *and which admits as h(hh)- and v(vv)- torsions the d-tensors* $T^i{}_{jk}=-T^i{}_{kj}$, $S^i{}_{jk}=-S^i{}_{kj}$ *a priori given.*

ii) The coefficients $\bar{L}^i_{jk}$, $\bar{C}^i_{jk}$ *of* $D\Gamma(N)$ *are the following ones*

$$(2.8)\quad \begin{aligned}\bar{L}^i_{jk}&=L^i_{jk}-\frac{1}{2}g^{ih}(g_{jr}T^r{}_{hk}+g_{kr}T^r{}_{hj}-g_{hr}T^r{}_{kj}),\\ \bar{C}^i_{jk}&=C^i_{jk}-\frac{1}{2}g^{ih}(g_{jr}S^r{}_{hk}+g_{kr}S^r{}_{hj}-g_{hr}S^r{}_{kj}),\end{aligned}$$

where $C\Gamma(N)=(L^i_{jk},C^i_{jk})$ *is the canonical metrical d-connection of* GL^n.

Another important result is given by

Theorem 2.5. *A transformation of non-linear connections* $N\to\bar{N}$ *of the form* $N^i_j\to\bar{N}^i_j=N^i_j-A^i_j$ *maps the canonical metrical d-connection* $C\Gamma(N)$ *to a normal d-connection* $D\Gamma(\bar{N})=(\tilde{L}^i_{jk},\tilde{C}^i_{jk})$, *where*

$$(2.9)\quad \tilde{L}^i_{jk}=L^i_{jk}+C^i_{jr}A^r_k,\quad \tilde{C}^i_{jk}=C^i_{jk}.$$

$D\Gamma(\bar{N})$ *is a metrical d-connection, too.*

Proof. A direct calculation shows that the transformation $N\to\bar{N}$ maps $C\Gamma(N)$ to $D\Gamma(\bar{N})$ given by (2.9). Then if $\bar{|}$ and $\tilde{|}$ denote the h- and v- covariant derivatives with respect to $D\Gamma(\bar{N})$, respectively, we have

$$g_{ij\tilde{|}k} = g_{ij|k} + A^r_k g_{ij}|_r, \quad g_{ij}\tilde{|}_k = g_{ij}|_k.$$

On account of (2.1) it comes out that $D\Gamma(N)$ is metrical.

Introducing the Obata operators Ω^{ij}_{hk} and Ω^{*ij}_{hk} as in §6, Ch. IV on a way similar to that used there one can prove

Theorem 2.6. *The set of all metrical normal d-connections from a GL^n-space is as follows*

$$\text{(2.10)} \quad \begin{aligned} \bar{N}^i_j &= N^i_j - X^i_j, \\ \bar{L}^i_{jk} &= L^i_{jk} + C^i_{jr}X^r_k + \Omega^{ir}_{sj}X^s_{rk}, \\ \bar{C}^i_{jk} &= C^i_{jk} + \Omega^{ir}_{sj}Y^s_{rk}, \end{aligned}$$

where $C\Gamma(N) = (L^i_{jk}, C^i_{jk})$ is the canonical metrical d-connection and X^i_j, X^i_{jk}, Y^i_{jk} are arbitrary d-tensor fields.

§3. Structure Equations. Parallelism

The canonical metrical d-connection $C\Gamma(N)$ satisfies certain structure equations which are obtained on a way similar to that used for a d-connection in a vector bundle (see §4, Ch. III).

Let us consider the connection 1-forms

$$\text{(3.1)} \quad \omega^i_j = L^i_{jk}dx^k + C^i_{jk}\delta y^k,$$

where $(dx^i, \delta y^k)$ is the dual basis of $(\delta_i, \dot{\partial}_i)$, hence

$$\text{(3.2)} \quad \delta y^k = dy^i + N^i_j dx^j.$$

Theorem 3.1. *The structure equations of the canonical metrical d-connection $C\Gamma(N)$ of a space GL^n are the following ones*

$$\text{(3.3)} \quad \begin{aligned} d(dx^i) - dx^h \wedge \omega^i_h &= -\Omega^i, \\ d(\delta y^i) - \delta y^h \wedge \omega^i_h &= -\tilde{\Omega}^i, \end{aligned}$$

(3.3)′ $d\omega^i_j - \omega^h_j \wedge \omega^i_h = -\Omega^i_j$,

where the torsions 2-forms Ω^i, $\tilde{\Omega}^i$ *are*

(3.4) $\Omega^i = C^i_{jk} dx^j \wedge \delta y^k, \ \tilde{\Omega}^i = \frac{1}{2} R^i{}_{jk} dx^j \wedge dx^k + P^i{}_{jk} dx^j \wedge \delta y^k$

while the curvature 2-forms Ω^i_j *are*

(3.4)′ $\Omega^i_j = \frac{1}{2} R_j{}^i{}_{kh} dx^k \wedge dx^h + P_j{}^i{}_{kh} dx^k \wedge \delta y^k + \frac{1}{2} S_j{}^i{}_{kh} \delta y^k \wedge \delta y^h$.

Exterior differentiating the equations (3.3) and (3.4) and using them again, the Bianchi identities for $C\Gamma(N)$ are obtained. These are identically to (5.6)′ - (5.6)^{iv} from Ch. VIII in which torsions and curvatures of $C\Gamma(N)$ are involved and covariant derivatives are taken with respect to $C\Gamma(N)$.

The connection 1-forms are useful in considering the covariant differentiation of d-tensor fields. For instance, the covariant differential of a d-vector field $X^i(x,y)$ is

(3.5) $DX^i = dX^i + X^j \omega^i_j$.

On account of (3.1) it becomes

(3.5)′ $DX^i = X^i_{|k} dx^k + X^i|_k \delta y^k$.

If $\tilde{c}: t \in I \to (x^i(t), y^i(t))$ is any smooth curve on $\tilde{TM}$ then along of $\tilde{c}$ we have $DX^i = \frac{DX^i}{dt} dt$ where

(3.6) $\frac{DX^i}{dt} = \frac{dX^i}{dt} + X^j \frac{\omega^i_j}{dt} = X^i_{|k} \frac{dx^k}{dt} + X^i|_k \frac{\delta y^k}{dt}$.

Such an expression was already used in §2. It is said the d-vector field X^i is parallel on $\tilde{c}$ if $\frac{DX^i}{dt} = 0$ along $\tilde{c}$. Thus the following system of differential equations is satisfied by a d-vector field X^i parallel along $\tilde{c}$:

$$(3.6)' \quad \frac{DX^i}{dt} = \frac{dX^i}{dt} + X^j \frac{\omega_j^i}{dt} = 0.$$

Hence if X_0^i in a point $(x^i(t_0), y^i(t_0))$ are given, by a general theorem on existence and uniqueness of the solutions of the (3.6)′, on a subinterval $I' \subset I$, $t_0 \in I'$ there exists an unique d-vector field $X^i(x(t),y(t))$ parallel along $\tilde{c}$ such that $X^i(x(t_0), y(t_0)) = X_0^i$.

Of course, the equations

$$(3.7) \quad X^i_{|k} = 0, \quad X^i|_k = 0$$

are sufficient conditions in order that the d-vector field (X^i) be parallel along a curve $\tilde{c}$.

If the system of differential equations (3.7) is completely integrable we shall say that the canonical metrical d-connection $C\Gamma(N)$ of the GL^n has the property of absolute parallelism of vectors.

By the Ricci identities (2.4) one obtains

Theorem 3.2. *A necessary and sufficient condition in order that $C\Gamma(N)$ to have the property of absolute parallelism of vectors is that its curvatures identically vanish.*

A curve $\tilde{c}$ on $\tilde{T}M$ is said to be an horizontal one if its tangent vector $\dot{\tilde{c}}$ belongs to the horizontal distribution. As we have

$$(3.8) \quad \dot{\tilde{c}} = \frac{dx^i}{dt}\delta_i + \frac{\delta y^i}{dt}\dot{\partial}_i$$

it results

Proposition 3.2. *A curve $\tilde{c}$ on $\tilde{T}M$ is an horizontal one if and only if*

$$(3.9) \quad \frac{\delta y^i}{dt} = \frac{dy^i}{dt} + N_j^i(x(t),y(t))\frac{dx^j}{dt} = 0, \quad t \in I.$$

We shall say that $\tilde{c}$ is an autoparallel curve with respect to the non-linear connection N if

$$(3.10) \quad y^i = \frac{dx^i}{dt}, \quad \frac{\delta}{dt}\left(\frac{dx^i}{dt}\right) = 0.$$

Hence it is obvious the following

Proposition 3.3. *The autoparallel curves of the non-linear connection N are given by*

$$(3.10)' \quad \frac{d^2x^i}{dt^2} + N^i_j(x(t), \frac{dx}{dt})\frac{dx^j}{dt} = 0, \quad y^i = \frac{dx^i}{dt}.$$

If we look at the first equation (3.10)′ the existence and uniqueness, in appropriate initial data, of $x^i = x^i(t)$ is deduced and than y^i is given by the second set of equations (3.10)′.

Definition 3.1. *A curve $\tilde{c}$ on $\tilde{TM}$ is called an h-path of $C\Gamma(N)$ if a) $\tilde{c}$ is an horizontal curve and, b) $\frac{D}{dt}(\frac{dx^i}{dt}) = 0$.*

It immediately comes out

Proposition 3.4. *The h-paths of the canonical metrical d-connection $C\Gamma(N)$ are characterized by the following system of differential equations*

$$(3.11) \quad \frac{d^2x^i}{dt^2} + L^i_{jk}(x(t), y(t))\frac{dx^j}{dt}\frac{dx^k}{dt} = 0, \quad \frac{\delta y^i}{dt} = 0.$$

It is now obvious how a theorem of existence and uniqueness for h-paths may be stated.

Definition 3.2. *A curve $\tilde{c}$ on $\tilde{TM}$ is called a v-path in the point $x_0 \in M$ if a) $\tilde{c}$ is a vertical curve i.e. $\tilde{c}$ belongs to the vertical distribution and b) $\frac{D}{dt}(\frac{\delta y^i}{dt}) = 0$.*

An easily calculation gives

Proposition 3.5. *The v-paths of $C\Gamma(N)$ are characterized by the equations*

$$(3.12) \quad x^i(t) = x^i_0, \ t \in I, \ \frac{d^2y^i}{dt^2} + C^i_{jk}(x_0, y(t))\frac{dy^j}{dt}\frac{dy^k}{dt} = 0.$$

It is obvious that there exists a v-path in any point $x_0 \in M$ and its uniqueness is assured in appropriate initial data.

Some of these special curves will be used in order to study a class of d-tensor fields which is useful in applications.

§4. On h-Covariant Constant d-Tensor Fields

The h-covariant derivative of a d-tensor field t with respect to the canonical metrical d-connection $C\Gamma(N)$ being constructed by the help of δ_i, when it is annihilated an interesting system of partial differential equations is obtained. Throughout in this section the covariant derivatives will involve $C\Gamma(N)$.

Definition 4.1. *We shall say that a d-tensor field t is h-covariant constant, shortly h-c.c., if*

$$(4.1)\quad t^{\cdots}_{\cdots|k} = 0.$$

If t reduces to a scalar $\sigma(x,y)$ we shall say that the function σ is h-constant, shortly h-c., if

$$(4.1)'\quad \sigma_{|k} = \frac{\partial\sigma}{\partial x^k} - N^j_k \frac{\partial\sigma}{\partial y^j} = 0.$$

The both equations (4.1) and (4.1)′ are geometrical in their nature i.e. do not depend on the local coordinates on TM.

The Ricci identities lead to certain differential consequences of (4.1) which in turn provide necessary conditions in order that the equation (4.1) admit solutions. We shall write these for a d-tensor field of type (1,1), but similar ones hold for any d-tensor field of type (p,q).

Theorem 4.1. *If a d-tensor field $t^i_j(x,y)$ is h-c.c., then we have*

$$(4.2)\quad \begin{aligned} 0 &= t^s_j R_s{}^i{}_{hk} - t^i_s R_j{}^s{}_{kh} - t^i_j|_s R^s{}_{kh}, \\ -t^i_j|_{h|k} &= t^s_j P_s{}^i{}_{kh} - t^i_s P_j{}^s{}_{kh} - t^i_j|_s P^s{}_{kh}, \\ t^i|_k|_h - t^i_j|_h|_k &= t^s_j S_s{}^i{}_{kh} - t^i_s S_j{}^s{}_{kh}. \end{aligned}$$

Corollary 4.1. *If $\sigma \in \mathcal{F}(\tilde{TM})$ is a h-c. function, then we have*

$$(4.3)\quad 0 = (\dot\partial_s\sigma) R^s{}_{kh},\quad \sigma|_{h|k} = (\dot\partial_s\sigma) P^s{}_{kh}.$$

A geometrical meaning of the condition h-c.c. for a d-tensor field is provided by

Theorem 4.2. *A d-tensor field* $t^{\cdots}_{\cdots}(x,y)$ *is h-c.c. if and only if its covariant differential* $Dt^{\cdots}_{\cdots}$ *vanishes on every horizontal curve in* $\tilde{TM}$.

Proof. Along of every curve $\tilde{c}$ in TM we have

$$(4.4)\quad \frac{Dt^{\cdots}_{\cdots}}{dt} = t^{\cdots}_{\cdots\,|k}\frac{dx^k}{dt} + t^{\cdots}_{\cdots}|_k\frac{\delta y^k}{dt}.$$

If $\tilde{c}$ is a horizontal curve then $\frac{\delta y^k}{dt}=0$ and $\frac{Dt^{\cdots}_{\cdots}}{dt}=0$ if and only if $t^{\cdots}_{\cdots\,|k}=0$.

Corollary 4.2. *If* $t^{\cdots}_{\cdots}(x,y)$ *is an h-c.c. tensor field, then* $Dt^{\cdots}_{\cdots}=0$ *on the autoparallel curves of the non-linear connection N as well as on the h-paths of* $C\Gamma(N)$.

Corollary 4.3. *If* $\sigma \in \mathcal{F}(\tilde{TM})$ *is an h-c. function, then* σ *is constant on the autoparallel curves of the non-linear connection N as well as on the h-paths of* $C\Gamma(N)$.

As an application let us look for conditions under which the Liouville vector field is h-c.c. On account of the definition of the deflection tensor D we immediately obtain

Proposition 4.1. *The Liouville vector field* y^i *is h-c.c. if and only if* $D^i_j=0$.

The Ricci identities applied to y^i give

Proposition 4.2. *If the Liouville vector field* y^i *is h-c.c., then* $C\Gamma(N)$ *has the following properties*

$$(4.5)\quad R_0{}^i{}_{kh} - d^i_s R^s{}_{kh} = 0,\quad -d^i_{h|k} = P_0{}^i{}_{kh} - d^s_i P^s{}_{kh},\quad d^i_k|_h - d^i_h|_k = S_0{}^i{}_{kh}.$$

Here d^i_k *denotes, as usual, the v-deflection tensor of* $C\Gamma(N)$.

Corollary 4.4. *If the Liouville vector field* y^i *is h-c.c. and* $C\Gamma(N)$ *has the property*

$$(4.6)\quad C^i_{0j}=0$$

then we have

$$(4.6)' \quad R^{i}_{0\,kh} = R^{i}{}_{kh},\ P^{i}_{0\,kh} = P^{i}{}_{kh},\ S^{i}_{0\,kh} = 0.$$

Also we have the following important facts:

Theorem 4.3. *If the Liouville vector field y^i is h-c.c. then*

i) The absolute energy of the space GL^n, $\varepsilon = g_{ij}(x,y)y^iy^j$ is h-c.

ii) The absolute energy ε is constant on the horizontal curves in $\tilde{TM}$. In particular, it is constant on the autoparallel curves of the non-linear connection N as well as on the h-paths of $C\Gamma(N)$.

Among the spaces GL^n which have the Liouville vector field h-c.c. the following ones are remarkable.

Definition 4.2. *A space GL^n is said to be a generalized locally Minkowski space, shortly $Mink^n$, if there exists a coordinate neighbourhood in every point of $\tilde{TM}$ on which g_{ij} do not depend on (x^i).*

Sufficient conditions in order that a space GL^n be a space $Mink^n$ are as follows.

Theorem 4.4. *If the canonical metrical d-connection $C\Gamma(N)$ of the space GL^n has the properties*

$$(4.7) \quad y^{i}_{|k} = 0,\ R^{i}{}_{jk} = 0,\ R_{j\ kh}^{\ i} = 0,\ P^{i}{}_{jk} = 0,\ C^{i}_{jk|h} = 0,$$

then GL^n is a generalized locally Minkowski space.

Proof. By (4.7) one gets $P_{ijkh} + P_{ikjh} = 0$ which, together with $P_{ijkh} + P_{jikh} = 0$ implies $P_{j\ kh}^{\ i} = 0$. Looking at the formula giving $P_{j\ kh}^{\ i}$ and using again (4.7), from the tensorial equation $P_{j\ kh}^{\ i} = 0$ one obtains $\dot{\partial}_k L^{i}_{jh} = 0$ which implies $\delta_k L^{i}_{jh} = 0$. Then $R_{j\ kh}^{\ i} = 0$ reads as follows

$$\frac{\partial L^{i}_{jk}}{\partial x^h} - \frac{\partial L^{i}_{jh}}{\partial x^k} + L^{s}_{jk} L^{i}_{sh} - L^{s}_{jh} L^{i}_{sk} = 0.$$

Then, just in the case of Riemannian spaces one deduces that there exists a coordinate neighbourhood in which $L^{i}_{jk} = 0$. Next $y^{i}_{|k} = L^{i}_{0k} - N^{i}_{k} = 0$ implies $N^{i}_{k} = 0$, hence

$$g_{ij|k} = \frac{\partial g_{ij}}{\partial x^k} = 0.$$

In particular if (4.6) is fulfilled, a glance at (4.6)′ gives

Corollary 4.5. *If the canonical metrical d-connection* $C\Gamma(N)$ *of the space* GL^n *has the properties*

$$(4.7)' \quad y^i_{|k} = 0,\ y^i|_k = \delta^i_k,\ R_j{}^i{}_{kh} = 0,\ P^i{}_{jk} = 0,\ C^i_{jk|h} = 0,$$

then GL^n *is a generalized locally Minkowski space.*

Remark. The later condition in (4.7) and (4.7)′ is nothing but the conditionC^i_{jk} to be h-covariant constant.

As to the h-constant functions we also have

Proposition 4.3. *Let* σ *be a function on* $\tilde{TM}$. *If the space* GL^n *admits in every point a coordinate neighbourhood in which* $N^i_j = 0$ *and* σ *depends only on* y^i *then* σ *is an h-constant function.*

Proof. On account of $N^i_j = 0$ and $\frac{\partial \sigma}{\partial x^i} = 0$, the equation (4.1) is verified.

§5. Gravitational Field

We shall think of g_{ij}, the components of the fundamental tensor of a generalized Lagrange space GL^n, as gravitational potentials and we shall admit:

The Einstein equations of the generalized Lagrange space GL^n, $(n>2)$ corresponding to the canonical metrical d-connection $C\Gamma$ are as follows

$$(5.1)\quad \begin{aligned} &R_{ij} - \frac{1}{2}R g_{ij} = \kappa \overset{H}{T}_{ij}, \quad {}'P_{ij} = \kappa \overset{1}{T}_{ij}, \\ &S_{ij} - \frac{1}{2}S g_{ij} = \kappa \overset{V}{T}_{ij}, \quad {}''P_{ij} = -\kappa \overset{2}{T}_{ij}, \end{aligned}$$

where κ is a constant and the other symbols have the known meanings (cf. §4, Ch. IX).

A support for this hypothesis will be showed in the next section where we shall see the equations (5.1) are equivalent to the Einstein equations associated to an almost Hermitian connection on $\tilde{TM}$.

The energy-momentum tensor from the right hand of the equations (5.1) must satisfy the so-called conservation laws as these are pointed out in §4, Ch. IX.

Let be the h- and v- Einstein tensors

$$(5.2)\quad \overset{H}{E}_{ij}=R_{ij}-\frac{1}{2}Rg_{ij},\ \overset{V}{E}=S_{ij}-\frac{1}{2}Sg_{ij},$$

respectively. Using the Biachi identities for $C\Gamma(N)$, one proves

Theorem 5.1. *The following identities hold good*

$$(5.3)\quad \begin{aligned}&\overset{H}{E}{}^{i}_{j|i}=-\frac{1}{2}(P^{ih}{}_{js}R^{s}{}_{hi}+2R^{s}{}_{ij}\,{}^{\prime\prime}P^{i}_{s}),\\ &\overset{V}{E}{}^{i}_{j}|_{i}=0.\end{aligned}$$

Theorem 5.2. *i) If the non-linear connection N used to construct the canonical d-connection $C\Gamma(N)$ is integrable then $\overset{H}{T}{}^{i}_{j|i}=0,\ \overset{V}{T}{}^{i}_{j}|_{i}=0$.*

ii) If $C\Gamma(N)$ has the property $P_{j}{}^{i}{}_{kh}=0$, then the Einstein equations of GL^{n} reduce to

$$(5.4)\quad R_{ij}-\frac{1}{2}Rg_{ij}=\kappa\overset{H}{T}_{ij},\ S_{ij}-\frac{1}{2}Sg_{ij}=\kappa\overset{V}{T}_{ij}$$

and "the conservation laws"

$$(5.5)\quad \overset{H}{T}{}^{i}_{j|i}=0,\ \overset{V}{T}{}^{i}_{j}|_{i}=0,$$

are verified.

If we look at the form of the curvature tensor $P_{j}{}^{i}{}_{kh}$, we can easily see several cases when we have $P_{j}{}^{i}{}_{kh}=0$. More precisely, we prove

Proposition 5.1. *If $C\Gamma(N)$ has the properties*

$$(5.6)\quad y^{i}_{|j}=0,\ C^{i}_{jk|h}=0,\ \dot{\partial}_{h}L^{i}_{jk}=0$$

then $P_{j\ kh}^{i}=0$, *hence (5.4) and (5.5) hold.*

Proof. By $y^i_{|j}=0$ and the later equation (5.6) we get $N_j^i = L_{jk}^i(x)y^k$, $P^i{}_{jk} = \dot{\partial}_k N_j^i - L_{kj}^i = 0$ and consequently $P_j{}^i{}_{kh}=0$. Thus the Theorem 5.2 applies.

A more particular situation is as follows

Proposition 5.2. *If* $C\Gamma(N)$ *has the properties (5.6) and* $R_j{}^i{}_{kh} - C_{js}^i R^s{}_{kh} = 0$ *then:*
i) GL^n *is a space* $Mink^n$,
ii) The Einstein equations are

$$(5.7)\quad S_{ij} - \frac{1}{2} S g_{ij} = \kappa \overset{V}{T}_{ij}$$

and the conservation law

$$(5.7)'\quad \overset{V}{T}{}^i_j|_i = 0,$$

is verified.

Proof. i) As we have noticed in the proof of the preceding theorem, (5.6) implies $N_j^i = L_{jk}^i(x)y^k$, $P^i{}_{jk}=0$. We shall see that $R^i{}_{jk}=0$ and $R_j{}^i{}_{kh}=0$. Indeed,

$$R_j{}^i{}_{kh} - C_{js}^i R^s{}_{kh} = \frac{\partial L_{jk}^i}{\partial x^h} - \frac{\partial L_{jh}^i}{\partial x^k} + L_{jk}^s L_{sh}^i - L_{jh}^s L_{sk}^i = 0$$

implies $R^i{}_{jk}=0$ and $R_j{}^i{}_{kh}=0$. Thus the conditions (4.7) are satisfied and the space GL^n is a $Mink^n$-space.

ii) The equations (5.4) and (5.5) reduce in this case to (5.7) and (5.7)$'$, respectively.

Of course, the Einstein equations (5.1) may be derived by using the gauge theory as it was presented in Ch. VI.

Some applications in Relativity and Relativistic optics of the theory from this section will be provided in the next chapters.

§6. Electromagnetic Fields

In a Lagrange space the electromagnetic field was introduced by extending an idea suggested by a Lagrangian from electrodynamics (see (5.1) in Ch. IX). This idea may also be applied to the generalized Lagrange spaces GL^n endowed with the canonical metrical d-connection $C\Gamma(N)$.

Thus we shall consider h- and v- covariant deflection tensor fields $D_{ij} = g_{ih}y^h_{|j}$ and $d_{ij} = g_{ih}y^h|_j$, respectively, and taking their skew symmetric parts one obtains the h-electromagnetic field F_{ij} and v-electromagnetic field f_{ij}. These fields satisfy a system of "Maxwell equations". If the space GL^n is reducible to a Lagrange space, then f_{ij} vanishes and the theory of the electromagnetic field F_{ij} expounded in §4., Ch. IX is obtained.

More precisely we set

Definition 6.1. *The d-tensor field* $F_{ij} = \frac{1}{2}(D_{ij} - D_{ji})$ *will be called the h-electromagnetic field and the d-tensor* $f_{ij} = \frac{1}{2}(d_{ij} - d_{ji})$ *will be called the v-electromagnetic field.*

It is easy to obtain

$$\text{(6.1)}\quad f_{ij} = C_{0ij} - C_{0ji} = \frac{1}{2}y^k(\dot{\partial}_i g_{kj} - \dot{\partial}_j g_{ki}).$$

Thus we have

Proposition 6.1. *If one of the following properties holds*

i) $C^i_{0j} = 0$,

ii) The space GL^n *is reducible to a Lagrange space,*
then the v-electromagnetic tensor f_{ij} *vanishes.*

On the other hand on account to the Definition 6.1 we get

Proposition 6.2. *If the Liouville vector field* y^i *is h-c.c. the h-electromagnetic tensor* F_{ij} *vanishes.*

A first important result concerning the d-tensor fields F_{ij} and f_{ij} can be stated as follows:

Theorem 6.1. *The h- and v- electromagnetic tensor fields F_{ij} and f_{ij} of the generalized Lagrange space GL^n endowed with the canonical metrical d-connection $C\Gamma(N)$ verify the equations of Maxwell type*

$$F_{ij|k}+F_{jk|i}+F_{ki|j}=-\frac{1}{2}\sum_{(ijk)}(C_{i0r}+d_{ir})R^r{}_{jk},$$

$$(6.2)\quad F_{ij}|_k+F_{jk}|_i+F_{ki}|_j=f_{ij|k}+f_{jk|i}+f_{ki|j},$$

$$f_{ij}|_k+f_{jk}|_i+f_{ki}|_j=0,$$

where $\sum_{(ijk)}$ *means cyclic sum over i,j,k.*

Proof. Starting from the first identity (2.7),

$$D_{ij|k}-D_{ik|j}=R_{0ijk}-d_{ir}R^r{}_{jk},$$

permuting in it the indices i,j,k and summing one gets

$$(*)\quad 2\sum_{(ijk)}F_{ij|k}=\sum_{(ijk)}(y^rR_{rijk}-d_{ir}R^r{}_{jk}).$$

Next the Bianchi identity $\sum_{(ijk)}(R_i{}^h{}_{jk}-R^s{}_{ij}C^h_{sk})=0$ combined with (*) yields the first equation (6.2). Analogously one finds the others equations (6.2). In finding the second equation (6.2) one uses also the identity $\sum_{(ijk)}(d_{ir}P^r{}_{jk}-d_{jr}P^r{}_{ik})=0$, a direct consequence of the equation $P^i{}_{jk}=P^i{}_{kj}$.

Remark 6.1. If GL^n is a Lagrange space L^n then the equations (6.2) reduce to the equations (4.8), Ch. IX.

Corollary 6.1. *If the non-linear connection N is integrable and $f_{ij}=0$ then*

$$\sum_{(ijk)}F_{ij|k}=0,\quad \sum_{(ijk)}F_{ij}|_k=0.$$

Corollary 6.2. *If the condition (4.6) is fulfilled then*
a) The h-electromagnetic tensor field F_{ij} vanishes,
b) The v-electromagnetic tensor field f_{ij} is h-c.c.,
c) The following Maxwell equation holds:

$$\sum_{(ijk)} f_{ij}|_k = 0.$$

Proof. The assertion a) is an immediate consequence of b). By the Theorem 4.4, the space GL^n is a $Mink^n$ space, that is, for every point of TM there exists a coordinate neighbourhood in which $N^i_k = 0$, $L^i_{jk} = 0$, $g_{ij} = g_{ij}(y)$. It follows $C_{0ij} = C_{0ij}(y)$ and $f_{ij|k} = \dfrac{\partial f_{ij}}{\partial x^k} = 0$. The assertion c) follows from the Theorem 6.1.

Definition 6.2. *The d-vector fields*

$$(6.3)\quad hJ^i = F^{ij}_{\ \ |j},\ vJ^i = F^{ij}|_j$$

are called h- and v- charge current density, respectively.

A second important result is stated as follows.

Theorem 6.2. *In the space GL^n endowed with the canonical metrical d-connection $C\Gamma(N)$, the following identities hold good:*

$$hJ^i_{\ |i} = \frac{1}{2}\{F^{ij}(R_{ij} - R_{ji}) + F^{ij}|_r R^r_{\ ij}\},$$
$$vJ^i|_i = 0.$$

Proof. A direct calculation.

Corollary 6.3. *If the non-linear connection N is integrable then the following identities hold good:*

$$hJ^i_{\ |i} = 0,\ vJ^i|_i = 0.$$

§7. Almost Hermitian Model of a GL^n Space

Extending the considerations done in order to construct the almost Kählerian model of a Lagrange space (see §7, Ch. IX) we shall lift the generalized Lagrange space GL^n to an almost Hermitian structure H^{2n} over the manifold $\widetilde{TM}$. In the case now considered the almost symplectic structure of H^{2n} is not integrable thus the symplectic geometry can not be used in order to develop those field theories which involve generalized Lagrange metrics.

In what follows we shall construct the almost Hermitian model of a generalized

Lagrange space GL^n endowed with the canonical metrical d-connection $C\Gamma(N)$.

Let $(\delta_i, \dot{\partial}_i)$ be the usual frame adapted to the horizontal and vertical distributions on TM and $(dx^i, \delta y^i)$ its dual.

Let us consider a linear operator $\mathbf{F} : \chi(\tilde{TM}) \to \chi(\tilde{TM})$ defined on $(\delta_i, \dot{\partial}_i)$ by

(7.1) $\mathbf{F}(\delta_i) = -\dot{\partial}_i, \ \mathbf{F}(\dot{\partial}_i) = \delta_i.$

Then $\mathbf{F}$ is a tensor field of type (1,1) on $\tilde{TM}$ and

(7.1)$'$ $\mathbf{F} \circ \mathbf{F} = -I,$

that is, $\mathbf{F}$ is an almost complex structure on $\tilde{TM}$. Moreover, $\mathbf{F}$ reduces to a complex structure if and only if the horizontal distribution N is integrable.

The metric tensor $g_{ij}(x,y)$ of the space GL^n induces on $\tilde{TM}$ the following metrical structure

(7.2) $\mathbf{G} = g_{ij}(x,y)dx^i \otimes dx^j + g_{ij}(x,y)\delta y^i \otimes \delta y^j.$

This is a Riemannian one if $g_{ij}(x,y)$ is positive definite.

The horizontal and vertical distributions are orthogonal with respect to $\mathbf{G}$. We have

Theorem 7.1. *i) The pair* (G, F) *is an almost Hermitian structure on* $\tilde{TM}$, *ii) The 2-form associated to* (G, F) *is*

(7.3) $\theta = g_{ij}(x,y)\,\delta y^i \wedge dx^j.$

Proof. i) In a coordinate neighbourhood, using the basis $(\delta_i, \dot{\delta}_i)$, on account of (7.1) and (7.2) one easily verifies the equality

$$\mathbf{G}(\mathbf{F}X, \mathbf{F}Y) = \mathbf{G}(X,Y), \ X,Y \in \chi(\tilde{TM}).$$

ii) In general, $\theta(X,Y) = \mathbf{G}(\mathbf{F}X,Y)$. In a coordinate system again on account of (7.1) and (7.2), θ takes the form (7.3).

The almost Hermitian space $H^{2n} = (\tilde{TM}, \mathbf{G}, \mathbf{F})$ is called the lift or the almost Hermitian model of the generalized Lagrange space GL^n.

Let us notice that the exterior differential of the 2-form θ is as follows

$$\text{(7.4)}\quad \begin{aligned} d\theta = -\frac{1}{6}\Big(\sum_{(ijk)} g_{ir}R^{r}{}_{jk}\Big)dx^i\wedge dx^j\wedge dx^k + \frac{1}{2}(g_{ij|k}-g_{ik|j})\delta y^i\wedge dx^j\wedge dx^k + \\ +\frac{1}{2}(\dot{\partial}_k g_{ij}-\dot{\partial}_i g_{kj})\delta y^k\wedge\delta y^i\wedge dx^j, \end{aligned}$$

where

$$g_{ij|k} = \delta_k g_{ij} - B^r_{ik}g_{rj} - B^r_{jk}g_{ir},\ \ B^i_{jk} = \dot{\partial}_j N^i_k.$$

It is well known that $d\theta=0$ is a necessary and sufficient condition in order that H^{2n} be an almost Kählerian space.

Thus we have

Theorem 7.2. *The space H^{2n} is almost Kählerian if and only if*

$$\text{(7.5)}\quad \sum_{(ijk)} g_{ir}R^{r}{}_{jk}=0,\ \ g_{ij|k}-g_{ik|j}=0,\ \ \dot{\partial}_k g_{ij}-\dot{\partial}_i g_{kj}=0.$$

If the generalized Lagrange space GL^n is not reducible to a Lagrange space then $\dot{\partial}_k g_{ij} \neq \dot{\partial}_j g_{ik}$. Therefore we get

Corollary 7.1. *If the space GL^n is not reducible to a Lagrange space then the almost symplectic structure θ is not integrable.*

If the almost structure **F** is integrable then H^{2n} is an Hermitian space. Thus we have

Theorem 7.3. *The space H^{2n} is Hermitian if and only if the horizontal distribution N is integrable.*

Now we characterise the canonical metrical d-connection $C\Gamma(N)$ by using the model H^{2n}.

Definition 7.1. *We shall say that a linear connection ∇ on $\widetilde{TM}$ is almost Hermitian of Lagrange type if*

1. *∇ preserves by parallelism the vertical distribution V,*
2. *∇ is compatible with the almost Hermitian structure (G,F), that is,*

(7.6) $\nabla_X G = 0, \ \nabla_X F = 0, \ X \in \chi(TM)$.

Exactly as in §7, Ch. IX one proves

Theorem 7.4. *There exists an unique almost Hermitian connection of Lagrange type* ∇ *having h(hh)- and v(vv)- torsions equal to zero. The coefficients* (L^i_{jk}, C^i_{jk}) *of* ∇ *in the adapted basis* $(\delta_i, \dot{\partial}_i)$ *are just the coefficients of the canonical metrical d-connection* $C\Gamma(N)$ *of the generalized Lagrange space* GL^n*. Conversely,* $C\Gamma(N)$ *determines on* $\tilde{TM}$ *an almost Hermitian connection of Lagrange type having h(hh)- and v(vv)- torsions equal to zero.*

The remarks given at the end of chapter IX may be repeated here word for word. For instance:

The Einstein equations associated with the almost Hermitian connections of Lagrange type ∇, in the adapted basis $(\delta_i, \dot{\partial}_i)$ are equivalent to the Einstein equations (5.1) of the generalized Lagrange space GL^n corresponding to the canonical metrical d-connection $C\Gamma(N)$.

Chapter XI

Applications of the GL^n Spaces with the Metric Tensor $e^{2\sigma(x,y)}\gamma_{ij}(x,y)$

The generalized Lagrange spaces whose metric tensor is of the form

$$\text{(I)}\quad g_{ij}(x,y) = e^{2\sigma(x,y)}\gamma_{ij}(x,y),$$

were $\gamma_{ij}(x,y)$ is the metric tensor of a Finsler space and σ is a smooth function on $\tilde{TM}$ are not reducible to Lagrange or Finsler spaces.

Their importance is mainly due to this fact since there are many problems which the Lagrangian (Finslerian or Riemannian) geometrical models are not sufficient to treat.

Following are two special cases which motivate investigation of the GL^n spaces with the metrics (I).

a) If $\gamma_{ij}(x,y) = \gamma_{ij}(x)$ is a Lorentz metric then (I) is the metric tensor of those GL^n spaces in which are fulfilled the conditions formulated by Ehlers, Pirani and Schild (EPS conditions [40]), for a constructive axiomatic theory of General relativity. These spaces were studied by R. Miron and R. Tavakol [116].

b) If $\gamma_{ij}(x,y)$ is the metric tensor of a locally Minkowski Finsler space, then (I) gives a class of GL^n spaces which generalizes the spaces applied by P.L. Antonelli [17] in ecology.

Our aim is now to study the GL^n spaces indicated in a) and b) by applying the general geometrical theory developed in the previous chapters.

§1. EPS Conditions and the Metric $e^{2\sigma(x,y)}\gamma_{ij}(x,y)$

The usual point of view that the underling geometry of space-time is Riemannian is thought of to be further strengthened by the constructive-axiomatic formulation of General Relativity due to Ehlers, Pirani and Schild [40]. Within this scheme the geometry of space-time is viewed in terms of its main substructures: conformal and projective structures, which are in turn thought of to be fixed by light propagation and freely falling non-rotating neutral test particles, respectively.

R. Tavakol and Van den Berg [154] have showed that the geometrical framework within which theories of gravity are sought can be generalized without at the same time contradicting the EPS conditions.

Let us consider a Finsler space $F^n = (M, F(x,y))$ endowed with the Cartan connection $\overset{c}{C\Gamma}(N) = (\overset{c}{L}{}^i_{jk}, \overset{c}{C}{}^i_{jk})$ (see §4, Ch. VIII) and subject to the following conditions

1. $F(x,y)$ is given by

(1.1) $F^2(x,y) = e^{2\sigma(x,y)}\gamma_{ij}(x)y^iy^j$,

where $\sigma(x,y)$ is 0-homogeneous with respect to (y^i) and smooth on $\tilde{TM}$, and $\gamma_{ij}(x)$ is a Lorentz metric on the manifold M. This assumption will ensure that the Finsler space F^n has the same causal and conformal structure as the pseudo-Riemannian space $R^n = (M, \gamma_{ij}(x))$.

2. The Finsler space F^n has the coefficients of the Berwald connection B^i_{jk} of the form

(1.2) $B^i_{jk}(x,y) = \{{}^{\,i}_{j\,k}\}(x)$,

where $\{{}^{\,i}_{j\,k}\}$ are the Christoffel symbols of the space R^n. Geometrically, (1.2) means that the Berwald connection of F^n does not depend on directional variables y^i i.e. F^n is a Berwald space, while physically it ensures that F^n has the same projective structure as R^n i.e. the autoparallel curves of the non-linear connection of F^n are the same with the autoparallel curves in R^n.

The conditions 1. and 2. together with the constancy of the norm of vectors under the parallel transport which is ensured by employing the Cartan connection are equivalent to the EPS conditions.

A main result concerning the consequences of the imposing of the EPS conditions in the Finslerian framework is the following one

Theorem 1.1. (Tavakol-Van den Bergh). *A necessary and sufficient condition for the Finsler space F^n with the metric (1.1) to satisfy the EPS conditions is for the function $\sigma(x,y)$ to satisfy*

$$(1.3)\quad \frac{\partial \sigma}{\partial x^i} - \frac{\partial \sigma}{\partial y^k} y^j \{{}^{\,k}_{i\,j}\} = 0.$$

The proof is matter of calculation in showing that (1.3) is equivalent to (1.2).

We note two important restrictions on $\sigma(x,y)$. One is its 0-homogeneity with respect to y^i and another one is to satisfy the Tavakol-Van den Bergh equation (1.3). Of course (1.3) shows that σ must be an h-constant function.

The question is if there exists a generalized Lagrange space with the metric tensor of the type (I) which satisfies the EPS-conditions without such strong conditions on $\sigma(x,y)$.

Thus we shall investigate the possibility of realising the EPS conditions by means of a special GL^n space endowed with a non-linear connection completely determined by its fundamental tensor field. Let us start with $GL^n = (M, g_{ij}(x,y))$ and let assume the axioms a.1. and a.2. from below hold.

a.1. The fundamental tensor field $g_{ij}(x,y)$ is of the form

(1.4) $g_{ij}(x,y) = e^{2\sigma(x,y)}\gamma_{ij}(x)$,

section and $\gamma_{ij}(x)$ is a given Lorentz metric on M.

a.2. The space GL^n is endowed with the non-linear connection N with the coefficients

(1.5) $N^i_j(x,y) = \{^{\ i}_{j\ k}\} y^k$.

It should be emphasised here that these assumptions eliminate the usual complications related to the causal structure of the pseudo-Riemannian space $R^n = (M, \gamma_{ij}(x))$. We note the absence of the condition of homogeneity for the function $\sigma(x,y)$ with respect to y^i.

We have the following important result

Theorem 1.2 *The axioms a.1. and a.2. are equivalent to the EPS conditions.*

Proof. The axiom a.1. asserts that the metric tensor g_{ij} of the space GL^n is con formally to the metric tensor $\gamma_{ij}(x)$ of R^n. Therefore the spaces GL^n and R^n have the same causal structure.

The axiom a.2. assures that the autoparallel curves of the non-linear connection N of the space GL^n are coincident to the autoparallel curves of the Riemannian space R^n.

We may give the following geometrical interpretation of the axiom a.1. Let (Γ,γ) be a smooth parametrized curve $\gamma: t \in I \to x(t)$ and $\Gamma = Im\gamma$. Next consider the parametrized curve $(\tilde{\Gamma},\tilde{\gamma})$ on $\tilde{TM}$ given by $\tilde{\gamma}: t \in I \to (x(t), \dot{x}(t))$ and $\tilde{\Gamma} = Im\tilde{\gamma}$. The infinitesimal arc-length of (Γ,γ) is $d_M s^2 = \gamma_{ij}(x(t))\dot{x}^i\dot{x}^j dt^2$ while that of $(\tilde{\Gamma},\tilde{\gamma})$ is $ds^2 = g_{ij}(x,\dot{x})\dot{x}^i\dot{x}^j dt^2$. By the axiom a.1. we get

(1.6.) $ds^2 = e^{2\sigma(x,\dot{x})} d_M s^2$.

Thus a kind of gauge transformation is obtained. This is well determined when σ is prescribed. Generally $\sigma(x,y)$ depends on the physical nature of the space. By the theorems in §4, Ch.XI one obtains

Theorem 1.3. *i) The function σ is h-constant if and only if it satisfies the equations (1.3).*

ii) If σ is h-constant then it is constant on the autoparallel curves of the non-linear connection N given by (1.5).

Also, it is not difficult to prove

Theorem 1.4. *i) The non-linear connection N from the axiom a.2. is completely determined by the fundamental metric tensor $g_{ij}(x,y)$ from the axiom a.1.*

ii) The horizontal distribution of the non-linear connection N from (1.5.) is integrable if and only if the pseudo-Riemannian space R^n is flat.

Theorem 1.5. (R. Miron, S. Watanabe, H. Izumi) *The generalized Lagrange*

Some examples. 1. Let $A_i(x)$ be a covector field on M and let us take

(1.7) $\sigma(x,y) = A_i(x)y^i$.

One obtains a GL^n space with some interesting properties as we shall see in a next section.

2. Another interesting form of the function σ is

(1.8) $\sigma(x,y) = \frac{1}{2}\gamma_{ij}(x)y^i y^j$.

Thus σ is half of the absolute energy of the pseudo-Riemannian space R^n.

3. If $\boldsymbol{M} = (M, V^i(x), n(x,V(x)))$ is a dispersive medium, where $V^i(x)$ is the velocity of the particle $x \in M$ and $n(x,V(x))$ is the refractive index, we may consider

(1.9) $\sigma(x,y) = \alpha(1 - \frac{1}{n^2(x,y)}), \quad \alpha \in \mathbf{R}^*_+$.

The restriction of the metric tensor $g_{ij}(x,y)$ from (1.4), with (1.9), to the cross-section $S_V(M)$ given by $x^i = x^i, y^i = V^i(x)$ of the tangent bundle (TM,τ,M) can be considered as the relativistic metric tensor of the dispersive medium $\boldsymbol{M}$. It is

(1.10) $g_{ij}(x,V(x)) = \exp[2\alpha(1 - \frac{1}{n^2(x,V(x))})]\gamma_{ij}(x)$.

§2. Canonical Metrical d-Connection

We shall determine the canonical metrical d-connection $C\Gamma(N)$ for the spaces GL^n endowed with the metric tensor g_{ij} from (1.4) and with the non-linear connection N from (1.5) by using the theory developed in §2 of Ch.X. As usual $(\delta_i, \dot\partial_i)$, were $\delta_i = \partial_i - \{^{\,h}_{i\,j}\}y^j\dot\partial_h$ is the local basis for the module of vector fields on $\tilde{TM}$ adapted to the horizontal and vertical distributions and $(dx^i, \delta y^i)$, were $\delta y^i = dy^i + \{^{\,i}_{j\,k}\}y^j dx^k$ is its dual.

By the Theorem 2.1, Ch.X we obtain

Theorem 2.1. *The canonical metrical d-connection* $C\Gamma(N)$ *of space* GL^n *has the coefficients*

$$(2.1)\quad \begin{aligned} L^i_{jk} &= \{^{\,i}_{j\,k}\} + \delta^i_j\sigma_k + \delta^i_k\sigma_j - g_{jk}\sigma^i, \\ C^i_{jk} &= \delta^i_j\dot\sigma_k + \delta^i_k\dot\sigma_j - g_{jk}\dot\sigma^i, \end{aligned}$$

$$(2.1)' \quad \sigma_i = \frac{\delta\sigma}{\delta x^i}, \quad \dot{\sigma}_i = \frac{\partial\sigma}{\partial y^i}, \quad \sigma^i = g^{ij}\sigma_j, \quad \dot{\sigma}^i = g^{ij}\dot{\sigma}_j.$$

Proof. It is easy to check the following properties hold

$$(2.2) \quad g_{ij|k}=0, \quad g_{ij}|_k=0.$$

$$(2.2)' \quad T^i{}_{jk}=0, \quad S^i{}_{jk}=0.$$

By the uniqueness of the canonical metrical d-connection the equalities (2.2) and (2.2)′ assure that indeed $C\Gamma(N)$ given by (2.1) is the canonical metrical d-connection of the space GL^n.

The torsions of $C\Gamma(N)$ are C^i_{jk} from (2.1) and

$$(2.3) \quad R^i{}_{jk} = y^h r_h{}^i{}_{jk}, \quad P^i{}_{jk} = -(\delta^i_j\dot{\sigma}_k + \delta^i_k\dot{\sigma}_j - g_{jk}\dot{\sigma}^i),$$

where $r_h{}^i{}_{jk}(x)$ is the curvature tensor of the pseudo-Riemannian space R^n.
Also we have

Theorem 2.2. *i) $C\Gamma(N)$ satisfies the equation*

$$(2.4) \quad L^i_{jk} = \{{}_j{}^i{}_k\}$$

if and only if σ is a solution of the equation (1.3) i.e. it is h-constant.
ii) $C\Gamma(N)$ verifies the equation $C^i_{jk}=0$ if and only if $\dot{\sigma}=0$.

Corollary 2.1. *If σ is h-constant then the absolute energy of the space GL^n*

$$(2.5) \quad \varepsilon(x,y) = g_{ij}(x,y)y^iy^j$$

is h-constant and, consequently, it is constant on the autoparallel curves of the non-linear connection N.

Proof. From $\sigma_k=0$, $y^i_{|k}=\{{}_j{}^i{}_k\}y^j - N^i_k=0$ and (2.5) it follows $\varepsilon_{|k}=0$.

Let us set

$$(2.6) \quad \Lambda^i_{jk} = \delta^i_j\sigma_k + \delta^i_k\sigma_j - g_{jk}\sigma^i$$

Applying the considerations from §3, Ch. X we have

Theorem 2.3. *The h-paths of $C\Gamma(N)$ are characterized by the system of differential equations*

$$\frac{dy^i}{dt}+\{_j{}^i{}_k\}y^j\frac{dx^k}{dt}=0,$$

$$\frac{d^2x^i}{dt^2}+(\{_j{}^i{}_k\}+\Lambda^i_{jk})\frac{dx^j}{dt}\frac{dx^k}{dt}=0.$$

Theorem 2.4. *The v-paths of $C\Gamma(N)$ in a point $x_0\in M$ are characterized by the system of differential equations*

$$\frac{d^2y^i}{dt^2}+C^i_{jk}(x_0,y)\frac{dy^j}{dt}\frac{dy^k}{dt}=0.$$

The Berwald connection of the non-linear connection N has the coefficients $B\Gamma(N)=(\{_j{}^i{}_k\},0)$ in the adapted basis $(\delta_i,\dot{\partial}_i)$. As usual we denote by $_|$ and $\|$ the h- and v- covariant derivatives with respect to $B\Gamma(N)$.

The following properties hold

(2.7) $\gamma_{ij|k}=0,\ \gamma_{ij}\|_k=0,$

(2.8) $g_{ij|k}=2g_{ij}\sigma_k,\ g_{ij}\|_k=2g_{ij}\dot{\sigma}_k.$

Therefore we have

Proposition 2.1. *The Berwald connection $B\Gamma(N)$ is a conformal connection with respect to the metric tensor $g_{ij}(x,y)$.*

Of course, the equations (2.1) give a transformation of d-connections $B\Gamma(N)\to C\Gamma(N)$ having as deformation tensors Λ^i_{jk} and C^i_{jk}.

Thus we get

Proposition 2.2. *The curvature tensors of $C\Gamma(N)$ have the following form*

$$R_j{}^i{}_{kh}=r_j{}^i{}_{kh}+\Lambda^i_{jk|h}-\Lambda^i_{jh|k}+\Lambda^s_{jk}\Lambda^i_{sh}-\Lambda^s_{jh}\Lambda^i_{sk}+C^i_{js}R^s{}_{kh},$$

(2.9) $P_j{}^i{}_{kh}=\Lambda^i_{jk}\|_h-C^i_{jh|k}+\Lambda^s_{jk}C^i_{sh},$

$S_j{}^i{}_{kh}$ is of the general form (with C^i_{jk} from (2.1)).

In order to find a more explicit form for these we introduce the following notations

$$
(2.10)\quad
\begin{aligned}
\sigma_{jk} &= \sigma_{j\,|\,k} - \sigma_j\sigma_k + \frac{1}{2}g_{jk}\langle\sigma,\sigma\rangle, \quad \langle\sigma,\sigma\rangle = \sigma_i\sigma^i,\\
\dot\sigma_{jk} &= \dot\sigma_j\,\|_k - \dot\sigma_j\dot\sigma_k + \frac{1}{2}g_{jk}\langle\dot\sigma,\dot\sigma\rangle, \quad \langle\dot\sigma,\dot\sigma\rangle = \dot\sigma_i\dot\sigma^i,\\
\overset{1}{\sigma}_{jk} &= \sigma_j\,\|_k - \dot\sigma_j\sigma_k + \frac{1}{2}g_{jk}\langle\sigma,\dot\sigma\rangle, \quad \langle\sigma,\dot\sigma\rangle = \sigma_i\dot\sigma^i = \sigma^i\dot\sigma_i,\\
\overset{2}{\sigma}_{jk} &= \dot\sigma_{j\,|\,k} - \sigma_j\dot\sigma_k + \frac{1}{2}g_{jk}\langle\sigma,\dot\sigma\rangle.
\end{aligned}
$$

By a direct calculation one gets

$$
(2.11)\quad \sigma_{jk} - \sigma_{kj} = -\dot\sigma_s R^s{}_{jk}, \quad \dot\sigma_{jk} - \dot\sigma_{kj} = 0, \quad \overset{1}{\sigma}_{jk} - \overset{2}{\sigma}_{kj} = 0.
$$

Now again by a direct calculation one obtains

Theorem 2.5. *The curvature tensors of the canonical matrical d-connection are given by*

$$
(2.12)\quad
\begin{aligned}
R_j{}^i{}_{kh} &= r_j{}^i{}_{kh} + \delta^i_k\sigma_{jh} - \delta^i_h\sigma_{jk} - g^{is}(g_{jk}\sigma_{sh} - g_{jh}\sigma_{sk}) + (\delta^i_s\dot\sigma_j - g_{js}\dot\sigma^i)r_0{}^s{}_{kh},\\
P_j{}^i{}_{kh} &= \delta^i_k\overset{1}{\sigma}_{jh} - \delta^i_h\overset{2}{\sigma}_{jk} - g^{is}(g_{jk}\overset{1}{\sigma}_{sh} - g_{jh}\overset{2}{\sigma}_{sk}) + g_{kh}(\sigma^i\dot\sigma_j - \dot\sigma^i\sigma_j),\\
S_j{}^i{}_{kh} &= \delta^i_k\dot\sigma_{jh} - \delta^i_h\dot\sigma_{jk} - g^{is}(g_{jk}\dot\sigma_{sh} - g_{jh}\dot\sigma_{sk}).
\end{aligned}
$$

From these formulae we deduce

Theorem 2.6. *i) If* σ *is a solution of the equation (1.3) then we have*

$$
(2.13)\quad P_j{}^i{}_{kh} = 0.
$$

ii) $\dot\sigma_k = 0$ *implies*

$$
(2.13)'\quad P_j{}^i{}_{kh} = S_j{}^i{}_{kh} = 0.
$$

Proof. By $\sigma_k = o$ it follows $\sigma_{jk} = \overset{1}{\sigma}_{jk} = \overset{2}{\sigma}_{jk} = 0$ and then by (2.12) one gets (2.13). If $\dot\sigma_k = 0$ then $\dot\sigma_{ij} = \overset{1}{\sigma}_{ij} = \overset{2}{\sigma}_{ij} = 0$ and by (2.12) one obtains (2.13)$'$.

The following properties of $C\Gamma(N)$ are found by a direct calculation.

Theorem 2.7. *The Ricci tensors of* $C\Gamma(N)$ *are the following ones*

$$
(2.14)\quad
\begin{aligned}
&R_{jk} = r_{jk} - (n-2)\sigma_{jk} - g_{jk} <\sigma> + \dot{\sigma}_j r_{0k} - r_{0jks}\gamma^{sr}\dot{\sigma}_r, \quad <\sigma> = g^{ij}\sigma_{ij},\\
&{}'P_{jk} = \overset{1}{\sigma}_{jk} - (n-1)\overset{1}{\sigma}_{kj} - g_{jk} <\overset{1}{\sigma}> + \sigma_k\dot{\sigma}_j - \sigma_j\dot{\sigma}_k, \quad <\overset{1}{\sigma}> = g^{ij}\overset{1}{\sigma}_{ij},\\
&{}''P_{jk} = -{}'P_{jk},\\
&S_{jk} = -(n-2)\dot{\sigma}_{jk} - g_{jk} <\dot{\sigma}>, \quad <\dot{\sigma}> = g^{ij}\dot{\sigma}_{ij}.
\end{aligned}
$$

Theorem 2.8. *The scalar curvatures of* $C\Gamma(N)$ *take the form*

$$
(2.15)\quad
\begin{aligned}
&R = e^{-2\sigma}(r - 2\dot{\sigma}_s r_0^s) - 2(n-1) <\sigma>\\
&S = -2(n-1) <\dot{\sigma}> .
\end{aligned}
$$

A glance at (2.12) gives

Theorem 2.9. *i) If the pseudo-Riemannian space* R^n *is flat and the d-tensor* σ_{ij} *vanishes, then the curvature tensor* $R_j{}^i{}_{kh}$ *vanishes, too.*

ii) If R^n *is flat, then* $R_{ij} = R_{ji}$.

iii) If $S_{ij} = 0$, *then* $S_j{}^i{}_{kh} = 0$ *and conversely.*

§3. Electromagnetics and Gravitational Fields

The metric tensor (1.4) can generate an electromagnetic field even in the case when the usual EPS conditions are fulfilled i.e. $\dot{\sigma}_k = 0$. In general, this field has two components F_{ij} and f_{ij} given by (cf. §6, Ch. X)

$$
(3.1)\quad F_{ij} = \frac{1}{2}(D_{ij} - D_{ji}),\quad f_{ij} = \frac{1}{2}(d_{ij} - d_{ji}),
$$

where D_{ij} and d_{ij} are the h- and v-covariant deflection tensors of the connection $C\Gamma(N)$.

It is easy to see

Proposition 3.1. *The electromagnetic tensors* F_{ij} *and* f_{ij} *have the following form*

$$
(3.2)\quad F_{ij} = g_{i0}\sigma_j - g_{j0}\sigma_i,\quad f_{ij} = g_{i0}\dot{\sigma}_j - g_{j0}\dot{\sigma}_i .
$$

Consequently, we have

Proposition 3.2. *i)* $\dot{\sigma}_k = 0$ *implies* $f_{ij} = 0$,

ii) $\sigma_k = 0$ *implies* $F_{ij} = 0$.

Hence, if $\sigma(x,y)=\sigma(x)$, the metric $e^{2\sigma(x)}\gamma_{ij}(x)$ satisfies the classical EPS conditions. However, the electromagnetic tensor F_{ij} may appear.

Next we have

Theorem 3.1. *The following Maxwell equations hold*

$$\begin{aligned} &F_{ij|k}+F_{jk|i}+F_{ki|j} = -\sum_{(ijk)} g_{i0}\dot{\sigma}_h r_{0\,jk}^{\,h}, \\ (3.3)\quad &F_{ij}|_k+F_{jk}|_i+F_{ki}|_j = -(f_{ij|k}+f_{jk|i}+f_{ki|j}), \\ &f_{ij}|_k+f_{jk}|_i+f_{ki}|_j = 0. \end{aligned}$$

Proof. By (6.2), Ch. X, noticing that

$$C_{i0k}+d_{ik} = g_{ik}+2g_{i0}\dot{\sigma}_k, \quad \sum_{(ijk)} g_{is}R^{s}{}_{jk} = 0$$

one obtains (3.3).

Proposition 3.3. *If $\sigma(x,y)$ does not depend on y^i then the electromagnetic tensor F_{ij} satisfies the Maxwell equations*

$$(3.3)' \quad F_{ij|k}+F_{jk|i}+F_{ki|j}=0, \quad F_{ij}|_k+F_{jk}|_i+F_{ki}|_j=0.$$

Proposition 3.4. *If $\sigma_k=0$ then the v-electromagnetic tensor f_{ij} verifies the equations*

$$f_{ij|k}+f_{jk|i}+f_{ki|j}=0, \quad f_{ij}|_k+f_{jk}|_i+f_{ki}|_j=0.$$

The components of the fundamental tensor field $g_{ij}(x,y)=e^{2\sigma(x,y)}\gamma_{ij}(x)$ can be considered as gravitational potentials. In order to study the gravitational field we may use the gauge theory expounded in Ch. VI. as well as the theory from Ch. X. Choosing the second way we have

Theorem 3.2. *The Einstein equations of the generalized Lagrange space GL^n, $n>2$, with the tensor (1.4) corresponding to the canonical metrical d-connection $C\Gamma(N)$ are the following ones*

$$\begin{aligned} &r_{ij}-\frac{1}{2}r\gamma_{ij}+\rho_{ij} = \kappa \overset{H}{T}_{ij}, \quad {}'P_{ij} = \kappa \overset{1}{T}_{ij}, \\ (3.4)\quad &S_{ij}-\frac{1}{2}Sg_{ij} = \kappa \overset{V}{T}_{ij}, \quad {}''P_{ij} = -\kappa \overset{2}{T}_{ij}, \end{aligned}$$

where κ, $\overset{H}{T}$, ... *have the known meanings and*

$$(3.4)' \quad \rho_{jk} = -(n-2)\sigma_{jk} + [(n-2) < \sigma > + \dot{\sigma}^s r_{s0}] g_{jk} + \dot{\sigma}_j r_{0k} - r_0{}^s{}_{jk} \dot{\sigma}_s .$$

Theorem 3.3. *If the non-linear connection N is integrable then:*

i) The Einstein equations of GL^n *are (3.4) with* $r_{ij}=0$, $r=0$ *and the additional term* ρ_{ij} *of the form*

$$(3.4)'' \quad \rho_{jk} = -(n-2)(\sigma_{jk} - < \sigma > g_{jk}), \quad \rho_{jk} = \rho_{kj} .$$

ii) The following equations hold good

$$(3.5) \quad \overset{H}{T}{}^i_{j|i} = 0, \quad \overset{V}{T}{}^i_j|_i = 0.$$

These equations show that for the general metric (1.4) the Einstein equations differ from the classical ones by the additional tensor ρ_{ij} from $(3.4)''$. This tensor does not vanish even if $R^n = (M, \gamma_{ij}(x))$ is locally Minkowski.

Combining the last two theorems with the Theorem 2.6 and 2.9 one obtains the following important results:

Theorem 3.4. *If* $\dot{\sigma}_k = 0$, *the Einstein equations of* GL^n *reduce to*

$$r_{ij} - \frac{1}{2} r \gamma_{ij}(x) + \rho_{ij} = \kappa \overset{H}{T}_{ij}$$

and the condition $\overset{H}{T}{}^i_{j|i} = 0$ *is fulfilled.*

Theorem 3.5. *If* σ *is a solution of the equation (1.3) and the pseudo-Riemannian space* R^n *is flat then*

i) The Einstein equations for GL^n *reduce to*

$$S_{ij} - \frac{1}{2} S g_{ij} = \kappa \overset{V}{T}_{ij} ,$$

ii) The equality $\overset{V}{T}{}^i_j|_i = 0$ *holds hood.*

§4. Two Particular Cases

Now we shall apply the previous theory to the metric tensor (1.4) with σ given by (1.7) and (1.8), respectively.

1) The spaces GL^n with the fundamental tensor

$$(4.1)\quad g_{ij}(x,y) = e^{2A_k(x)y^k}\gamma_{ij}(x),$$

where A_k may be thought of as electromagnetic potentials have the canonical metrical d-connection $C\Gamma(N)$ with the coefficients given by (2.1) in which

$$(4.2)\quad \sigma = A_k(x)y^k,\ \ \sigma_k = y^s A_{s\|k},\ \ \dot{\sigma} = A_k.$$

Hence we have

Proposition 4.1. *The space GL^n with the metric tensor (4.1) has the property that σ is a solution of the equation (3.1) if and only if the d-vector A_k is parallel with respect to the Berwald connection $B\Gamma(N)$. In this case the canonical metrical d-connection $C\Gamma(N)$ has the coefficients*

$$(4.3)\quad L^i_{jk} = \{^{\,i}_{j\,k}\},\ \ C^i_{jk} = \delta^i_j A_k - \delta^i_k A_j - \gamma_{jk}\gamma^{is}A_s(x).$$

We note that the coefficients of $C\Gamma(N)$ do not depend on the directional variables y^i.

Also we see that $A_{i\|j}=0$ implies the covector field A_i is a gradient. Therefore $F_{ij}=0$, $f_{ij}=y^s(g_{is}A_j - g_{js}A_i)$ and $A_{i|k}=0$, $y^i_{|k}=0$, $f_{ij|k}=0$. The Einstein equations are (3.4) with $\sigma_k=0$, $\sigma_{ij}=0$, $<\sigma>=0$, $'P_{ij}=0$, $''P_{ij}=0$ and the identities $\overset{H}{T}{}^i_{j|i}=0$, $\overset{V}{T}{}^i_j|_i=0$ are true.
If $\sigma_k \neq 0$ we have

Proposition 4.2. *The canonical metrical d-connection of the spaces GL^n with the metric tensor (4.1) is given by*

$$(4.4)\quad \begin{aligned} L^i_{jk} &= \{^{\,i}_{j\,k}\} + [\delta^i_j A_{s\|k} + \delta^i_k A_{s\|j} - \gamma_{jk}\gamma^{ih}A_{s\|h}]y^s,\\ C^i_{jk} &= \delta^i_j A_k + \delta^i_k A_j - \gamma_{jk}\gamma^{is}A_s. \end{aligned}$$

2) The space GL^n with the fundamental tensor

(4.5) $g_{ij}(x,y) = e^{2\sigma(x,y)}\gamma_{ij}(x)$, $2\sigma(x,y) = \gamma_{ij}(x)y^iy^j$.

We have

Proposition 4.3. *The function σ from (4.5) satisfies the equation (1.3).*

Indeed, it is easy to check

(4.6) $\sigma_k = 0$, $\dot{\sigma}_k = \gamma_{ki}(x)y^i$.

Proposition 4.4. *For the GL^n with the metric tensor (4.5) the d-connection $C\Gamma(N)$ has the coefficients*

(4.7) $L^i_{jk} = \{{}_j{}^i{}_k\}$, $C^i_{jk} = (\delta^i_j\gamma_{ks} + \delta^i_k\gamma_{is} - \delta^i_s\gamma_{jk})y^s$.

Proposition 4.5. *The spaces GL^n with the metric tensor (4.5) have the properties $F_{ij}=0, f_{ij}=0$.*

Proposition 4.6. *The absolute energy $\varepsilon(x,y)$ of the spaces GL^n with the metric (4.5) is h-constant and so it is constant on the autoparallel curves of the non-linear connection N.*

Proposition 4.7. *The Einstein equations of the GL^n spaces with the metric tensor (4.5) are given by (3.4) with $\sigma_k = 0$, $'P_{ij} = 0$, $''P_{ij} = 0$. The identities $\overset{H}{T}{}^i_{j|i} = 0$, $\overset{V}{T}{}^i_j|_i = 0$ hold good.*

§5. GL^n Spaces with the Metric $e^{2\sigma(x,y)}\gamma_{ij}(y)$

Let $F^n = (M, F(x,y))$ be a locally Minkowski Finsler space (see §5, Ch. VIII). Let us consider a coordinate neighbourhood on which F does not depend on x. Then the fundamental tensor of F^n is

(5.1) $\gamma_{ij}(y) = \dfrac{1}{2}\dfrac{\partial^2 F^2}{\partial y^i \partial y^j}$.

The Cartan non-linear connection of F^n has the coefficients

(5.2) $\overset{\circ}{N}{}^i_j = 0.$

It is an integrable non-linear connection i.e. $\overset{\circ}{R}{}^i{}_{jk} = 0$.

The Cartan tensor

(5.3) $\overset{\circ}{C}_{ijk} = \dfrac{1}{2}\dfrac{\partial \gamma_{ij}}{\partial y^k}$

is completely symmetric and $\overset{\circ}{C}_{0ij} = 0$.

The metrical Cartan connection $C\overset{\circ}{\Gamma}(N)$ of F^n has the coefficients

(5.4) $\overset{\circ}{L}{}^i_{jk} = 0,\ \overset{\circ}{C}{}^i_{jk} = g^{ih}\overset{\circ}{C}_{jhk}.$

Now we consider the d-tensor field

(5.5) $g_{ij}(x,y) = e^{2\sigma(x,y)}\gamma_{ij}(y),$

where σ is a smooth function on $\tilde{TM}$.

It is obvious that the pair $(M, g_{ij}(x,y))$ is a generalized Lagrange space whose associated Finsler space is just the locally Minkowski Finsler space F^n discussed in the above. The mapping $\gamma_{ij}(y) \rightarrow g_{ij}(x,y)$ given by (5.5) may be thought of as a gauge transformation.

A first result is as follows

Theorem 5.1. *The GL^n spaces with the fundamental tensor (5.5) for which $\dot{\sigma}_k \neq 0$ are not reducible to Lagrange spaces.*

Proof. By (5.5) we have $\dot{\partial}_k g_{ij} = 2(\dot{\sigma}_k g_{ij} + e^{2\sigma(x,y)}\overset{\circ}{C}_{ijk})$. Since $\dot{\sigma}_k \neq 0$ it follows $\dot{\partial}_k g_{ij} \neq \dot{\partial}_j g_{ik}$.

We admit the GL^n space with the metric (5.5) is endowed with the non-linear connection N of the associated Finsler space F^n. As the coefficients of the non-linear connection N vanish on the coordinate neighbourhood considered, the operators δ_i reduce to ∂_i and then $\sigma_k = \dfrac{\partial \sigma}{\partial x^k}$. Thus $\sigma_k = 0$ if and only if σ does not depend on x^i. Let $\varepsilon(x,y) = e^{2\sigma(x,y)} \gamma_{ij}(y) y^i y^j$ be the absolute energy of the space GL^n with the metric (5.5).

Proposition 5.1. *If* $\sigma_k = 0$ *then*

i) The absolute energy ε *is h-constant,*

ii) The absolute energy is constant on the autoparallel curves of the non-linear connection N.

Proposition 5.2. *If* $\dot{\sigma}_k = 0$ *then* GL^n *with the metric tensor (5.5) reduces to a Finsler space* $\bar{F}^n$ *whose fundamental function is obtained from that of* F^n *by the conformal transformation*

$$(5.6)\quad \bar{F}(x,y) = e^{\sigma(x)} F(y).$$

Next we have

Theorem 5.2. *The canonical metrical d-connection* $C\Gamma(\overset{\circ}{N}) = (L^i_{jk}, C^i_{jk})$ *of the space* GL^n *with the metric tensor (5.5) is given by*

$$(5.7)\quad \begin{aligned} L^i_{jk} &= \delta^i_j \sigma_k + \delta^i_k \sigma_j - g_{jk} \sigma^i, \quad \sigma^i = g^{ih} \sigma_h, \\ C^i_{jk} &= \overset{\circ}{C}{}^i_{jk} + \delta^i_j \dot{\sigma}_k + \delta^i_k \dot{\sigma}_j - g_{jk} \dot{\sigma}^i, \quad \dot{\sigma}^i = g^{ih} \dot{\sigma}_h. \end{aligned}$$

Thus we can apply the whole theory of the generalized Lagrange spaces from Ch. X.

For instance, using (5.7) we find that the covariant deflection tensors of $C\Gamma(N)$ have the form

$$D_{ij} = g_{0i} \sigma_j + g_{ij} \sigma_0 - g_{0j} \sigma_i, \quad d_{ij} = g_{ij}(1 + \dot{\sigma}_0) + g_{0i} \dot{\sigma}_j - g_{0j} \dot{\sigma}_i$$

and then the h- and v- electromagnetic tensors F_{ij} and f_{ij} are the following ones

$$(5.8)\quad F_{ij} = g_{0i} \sigma_j - g_{oj} \sigma_i, \quad f_{ij} = g_{0i} \dot{\sigma}_j - g_{oj} \dot{\sigma}_i.$$

It is clear that $\sigma_j = 0$ implies $F_{ij} = 0$ while $\dot{\sigma}_j = 0$ implies $f_{ij} = 0$. Noticing that in the associated Finsler space F^n the electromagnetic tensors disappear it results that the transformation $\gamma_{ij}(y)\to g_{ij}(x,y)$ given by (5.5) leads to the appearance of non trivial electromagnetic tensors. Taking into account the integrability of N and applying the Theorem 6.1, Ch. X one gets

Theorem 5.3. *The electromagnetic tensor fields* F_{ij} *and* f_{ij} *from (5.8) satisfy the Maxwell equations*

$$F_{ij|k}+F_{jk|i}+F_{ki|j}=0,\ f_{ij}|_k+f_{jk}|_i+f_{ki}|_j=0,$$

$$F_{ij}|_k+F_{jk}|_i+F_{ki}|_j+f_{ij|k}+f_{jk|i}+f_{ki|j}=0.$$

In particular, if $\dot{\sigma}_j=0$ we obtain

Theorem 5.4. *The conformal transformation (5.6) of the space* F^n *produces the electromagnetic tensor field* $F_{ij}=g_{0i}\sigma_i-g_{0j}\sigma_i$ *which satisfies the Maxwell equations*

$$F_{ij|k}+F_{jk|i}+F_{ki|j}=0,\ F_{ij}|_k+F_{jk}|_i+F_{ki}|_j=0.$$

Also, for $\sigma_k=0$ we deduce

Theorem 5.5. *The transformation* $\gamma_{ij}(y)\to e^{2\sigma(y)}\gamma_{ij}(y)$ *produces the electromagnetic tensor* $f_{ij}=g_{0i}\dot{\sigma}_j-g_{0j}\dot{\sigma}_i$ *which verifies the Maxwell equations*

$$f_{ij|k}+f_{jk|i}+f_{ki|j}=0,\ f_{ij}|_k+f_{jk}|_i+f_{ki}|_j=0.$$

§6. Antonelli Metrics

We shall consider a particular type of metrics (5.5) from the §5. Namely, in a preferential coordinate system we assume the metric $g_{ij}(x,y)$ is of the form

(6.1) $g_{ij}(x,y)=e^{2\sigma(x)}\gamma_{ij}(y)$

in which

(6.2) $\sigma(x)=\alpha_i x^i$ (α_i are positive constants)

and $\gamma_{ij}(y)$ is the fundamental tensor of a locally Minkowski Finsler space whose fundamental function is

(6.3) $F(y) = [(y^1)^m + \dots + (y^n)^m]^{1/m}$, $m \geq 3$, m integer.

The metrics given by (6.1)-(6.3) were introduced by P.L. Antonelli [15-17]. The GL^n spaces with Antonelli's metric provide geometrical models which are useful in ecology (cf. P.L. Antonelli [17]).

The absolute energy of the space GL^n with the metric (6.1)-(6.3) is as follows

(6.4) $\bar{F}^2(x,y) = e^{2\sigma(x)} F^2(y)$.

Thus we have

Theorem 6.1. *The space GL^n with the metric (6.1)-(6.3) is a Finsler space $\bar{F}^n = (M, \bar{F}(x,y))$ with the fundamental function (6.4) which is locally conformal to the associated Finsler space $F^n = (M, F(y))$ with the fundamental function (6.3).*

Inserting in (6.1) the $\gamma_{ij}(y)$ provided by the fundamental function (6.3) we find the following form of the metric $g_{ij}(x,y)$:

$$(6.5)\quad \begin{aligned} g_{ii} &= e^{2\alpha_k x^k}(y^i)^{m-2}F^{2-m}[m-1-(m-2)(y^i)^m F^{-m}], \text{ i is fixed and} \\ g_{ij} &= -e^{2\alpha_k x^k}(m-2)(y^i y^j)^{m-1}F^{2-m}, \ i \neq j. \end{aligned}$$

We shall apply the theory from §5 taking as non-linear connection N that of the associated Finsler space F^n. Then $\overset{\circ}{N}{}^i_j = 0$ by (5.2) and the metrical Cartan connection is $(\overset{\circ}{L}{}^i_{jk} = 0, \overset{\circ}{C}{}^i_{jk})$. Thus the Theorem 5.2 leads to

Theorem 6.2. *The canonical metrical d-connection $C\Gamma(N)$ of the space GL^n with the metric (6.1)-(6.3) has the coefficients*

$$(6.6)\quad \begin{aligned} L^i_{jk} &= \delta^i_j \alpha_k + \delta^i_k \alpha_j - g_{jk} g^{ih} \alpha_h, \\ C^i_{jk} &= \overset{\circ}{C}{}^i_{jk}. \end{aligned}$$

It follows that the d-tensor field C^i_{jk} is a conformal invariant. The following important result was obtained by a direct calculation.

Theorem 6.3.(H.Shimada [140]). *The v-curvature tensor of the space $\overline{F}^n$ is of the form*

$$\overline{S}_{j\,kh}^{\,i} = \frac{(m-2)^2}{4(m-1)\overline{F}^2}\,(\overline{h}_{jk}\delta^i_h - \overline{h}_{jh}\delta^i_k),$$

where $\overline{h}_{ij} = \overline{F}\dfrac{\partial^2 \overline{F}}{\partial y^i \partial y^j}$ is the angular metric tensor of the space $\overline{F}^n$.

We note that $\overline{h}_{ij}$ can be easily computed observing that $g_{ij} = \overline{F}\dfrac{\partial^2\overline{F}}{\partial y^i\partial y^j} + \dfrac{\partial\overline{F}}{\partial y^i}\dfrac{\partial\overline{F}}{\partial y^j}$ and then using (6.5).

By means of the Theorem 6.2, the v-Ricci tensor S_{ij} as well as the scalar curvature S can be obtained.

For applications of special interest are the geodesics of the Finsler space $\overline{F}^n$. In order to study them we compute the Christoffel symbols $\overline{\gamma}^i_{jk}$ of the Finsler space $\overline{F}^n = (M, \overline{F}(x,y))$. From $\dot{\partial}_k g_{ij} = 2\alpha_k g_{ij}$ and $g_{ij}g^{hk} = \gamma_{ij}\gamma^{hk}$ we deduce

(6.7) $\overline{\gamma}^i_{jk} = \delta^i_j\alpha_k + \delta^i_k\alpha_j - \gamma_{jk}\gamma^{ih}\alpha_h.$

Now let $c:[0,1]\to M$ be a curve represented in a coordinate system as $x^i = x^i(t)$, $t\in[0,1]$. The tangent mapping $\dot{c}:[0,1]\to TM$ is then locally represented as $x^i = x^i(t)$, $y^i = \dfrac{dx^i}{dt} = \dot{x}^i(t)$. The infinitesimal arc-length of the curve c is

$$ds^2 = g_{ij}(x,\dot{x})\frac{dx^i}{dt}\frac{dx^j}{dt} = \overline{F}(x,\dot{x})\,dt^2.$$

Taking s as parameter on c we have $\overline{F}(x, \dfrac{dx}{ds}) = 1$. In this canonical parametrization the geodesics of the Finsler space $\overline{F}^n$ are given (§3, Ch.VIII) by

$$\frac{d^2x^i}{ds^2} + \overline{\gamma}^i_{jk}\frac{dx^j}{ds}\frac{dx^k}{ds} = 0.$$

By (6.7) we get

Theorem 6.4.(Antonelli-Shimada) *The geodesics of the Finsler spaces $\overline{F}^n$ are solutions of the following system of differential equations*

$$(6.8.)\quad \frac{d^2x^i}{ds^2} + 2\alpha_k \frac{dx^i}{ds}\frac{dx^k}{ds} = \gamma^{ik}\alpha_k .$$

Until now the theory is not so simple because of many computational difficulties. If the dimension n is equal to 2 an important simplification is obtained. In this case the equations (6.8) of the geodesics reduce, putting $y^i = \frac{dx^i}{ds}$, to the equations,

$$(6.9)\quad \begin{aligned} \frac{dy^1}{ds} &= y^1(-\alpha_1 y^1 + \frac{\alpha_1}{m-1}(\frac{y^2}{y^1})^m y^1 - \frac{\alpha_2 m}{m-1} y^2) \\ \frac{dy^2}{ds} &= y^2(-\alpha_2 y^2 - \frac{\alpha_1 m}{m-1} y^1 + \frac{\alpha_2}{m-1}(\frac{y^1}{y^2})^m y^2). \end{aligned}$$

These equations describe the growth of the colonial animals like corals, Hydrozoans and Bryozoans.

§7. General Case

We shall consider now the generalized Lagrange spaces whose fundamental tensor field is (I) from the introduction of this chapter. Thus we shall analyze the space $GL^n = (M, g_{ij}(x,y))$ with

$$(7.1)\quad g_{ij}(x,y) = e^{2\sigma(x,y)}\gamma_{ij}(x,y)$$

where γ_{ij} is the metric tensor of a given Finsler space $F^n = (M, F(x,y))$:

$$(7.1)'\quad \gamma_{ij}(x,y) = \frac{1}{2}\frac{\partial^2 F^2}{\partial y^i \partial y^j}$$

and $\sigma : TM \to R$ is a smooth function on $\tilde{TM}$ and at least continuous on the image of zero section of the tangent bundle.

The space F^n will be called the associated Finsler space to GL^n. As we know, if $\dot{\sigma}_k = 0$ then the GL^n with the metric (7.1) is a Finsler space $\bar{F}^n = (M, e^{2\sigma(x)} F(x,y))$ which is conformal to the Finsler space F^n.

The GL^n spaces studied in the previous sections are spaces with particular metrics of the form (7.1).

A first result which was already proved for the considered particular cases is given by

Theorem 7.1. *The space GL^n, $n>1$, with the metric (7.1) for which $\dot\sigma_k \neq 0$ is not reducible to a Lagrange or a Finsler space.*

Proof. By (7.1) we have

$$(7.2)\quad \dot\partial_k g_{ij} = 2(\dot\sigma_k \gamma_{ij} + C_{ijk})e^{2\sigma(x,y)},$$

where

$$C_{ijk} = \frac{1}{2}\dot\partial_k \gamma_{ij} = \frac{1}{4}\dot\partial_i \dot\partial_j \dot\partial_k F^2.$$

In order that GL^n be reducible to a Lagrange or a Finsler space we must have $\dot\partial_k g_{ij} = \dot\partial_j g_{ik}$ which is equivalent to $\dot\sigma_k \gamma_{ij} + C_{ijk} = \dot\sigma_j \gamma_{ik} + C_{ikj}$ or $\dot\sigma_k \gamma_{ij} = \dot\sigma_j \gamma_{ik}$. Multiplying the last equality by γ^{ij} we find $(n-1)\dot\sigma_k = 0$, which is a contradiction.

We shall derive the canonical metrical d-connection of a space GL^n with the metric (7.1). To this aim we shall postulate:

The space GL^n with the metric (7.1) is endowed with the non-linear connection N of the associated Finsler space F^n.

The coefficients of the non-linear connection N are the following ones

$$(7.3)\quad N^i_j = \frac{1}{2}\dot\partial_j \gamma^i_{00},$$

where γ^i_{jk} are the Christoffel symbols of the metric tensor γ_{ij}.

As usual, σ_k stands for $\dfrac{\delta\sigma}{\delta x^k}$. We have

Theorem 7.2. *The canonical metrical d-connection $C\bar\Gamma(N) = (\bar L^i_{jk}, \bar C^i_{jk})$ of the space GL^n with the metric tensor (7.1) is given by*

$$(7.4)\quad \begin{aligned} \bar L^i_{jk} &= F^i_{jk} + \delta^i_j \sigma_k + \delta^i_k \sigma_j - g_{jk} g^{ih} \sigma_h, \\ \bar C^i_{jk} &= C^i_{jk} + \delta^i_j \dot\sigma_k + \delta^i_k \dot\sigma_j - g_{jk} g^{ih} \dot\sigma_h, \end{aligned}$$

where $C\Gamma(N) = (F^i_{jk}, C^i_{jk})$ is the Cartan connection of the Finsler space F^n.

Indeed, on account of the Theorem 2.1, Ch. X a direct calculation gives (7.4).

Now we are in position to apply the theory of generalized Lagrange spaces expounded in Ch. X.

Concerning to the almost Hermitian model $H^{2n} = (\tilde{TM}, \tilde{G}, F)$ of the space GL^n we note the pseudo-Riemannian structure $\tilde{G}$ is as follows

$$(7.5)\quad \tilde{G} = e^{2\sigma(x,y)}[\gamma_{ij}(x,y)dx^i \otimes dx^j + \gamma_{ij}(x,y)\delta y^i \otimes \delta y^j].$$

This is obtained from the pseudo-Riemannian structure of the almost Kählerian model $K^{2n} = (\tilde{TM}, G, F)$ of the associated Finsler space F^n by a conformal transformation $G \to e^{2\sigma}G = \tilde{G}$.

The almost symplectic structure $\tilde{\theta}$ of the space H^{2n} is obtained from the symplectic structure θ of K^{2n} by the same transformation

$$(7.6)\quad \tilde{\theta} = e^{2\sigma(x,y)}\theta.$$

We note although θ is integrable, if $\dot{\sigma}_k \neq 0$, $\tilde{\theta}$ is not integrable. Indeed, the form $d\tilde{\theta}$ does not vanishes since $\dot{\partial}_k g_{ij} - \dot{\partial}_j g_{ik} \neq 0$ as we have seen in the proof of the Theorem 7.1.

Thus we have

Theorem 7.3. *The almost Hermitian model H^{2n} of the space GL^n with the metric tensor (7.1) and $\dot{\sigma}_k \neq 0$ is not reducible to an almost Kählerian one.*

Since $\tilde{\theta}$ is not integrable, it follows that in a geometrical study of the gravitational and electromagnetic fields from the space GL^n with the metric (7.1) and $\dot{\sigma}_k \neq 0$ can not be used the symplectic techniques as those from a book by P. Liebermann and Ch. Marle, [93].

Chapter XII

Relativistic Geometrical Optics

Another application of the theory of the generalized Lagrange spaces which we shall describe is to relativistic geometrical optics. Namely, we consider the space $GL^n = (M, g_{ij}(x,y))$ with the fundamental tensor field

$$\text{(I)} \qquad g_{ij}(x,y) = \gamma_{ij}(x) + \left(1 - \frac{1}{n^2(x,y)}\right) y_i y_j ,$$

where $\gamma_{ij}(x)$ is a Riemannian or Lorentz metric tensor, $y_i = \gamma_{ij}(x) y^j$ and $n(x,y)$ is the refractive index. This metric was intensively studied by R.Miron and T.Kawaguchi [113].

The restriction of the geometry of the space GL^n with the metric tensor (I) to a cross section $S_V : M \to TM$, $x \to (x, y^i = V^i(x))$ gives a geometrical model for the relativistic optics. An electromagnetic theory for a radiation field contained within an infinite, homogeneous and isotropic dielectric medium, based on the metric (I) restricted to S_V was recently developed with many interesting results by I.Brewick [28]. If $\frac{1}{n^2} = 1 - \frac{\alpha}{c^2}$, $\alpha \in R^*_+$ and c being the light velocity, we obtain an important class of GL^n spaces, studied also by R.Miron and T.Kawaguchi [113]. For these spaces some post-Newtonian estimations were investigated by G.S. Asanov and T.Kawaguchi [20] and I.Roxburgh [113]. The general problem of a post-Newtonian estimation concerning the metric (I) was analyzed in [172].

As we shall see, the generalized Lagrange spaces GL^n with the metric tensor (I) are not reducible to Lagrange or Finsler spaces. Consequently, the problem which we solve in relativistic geometrical optics can not be treated by the methods of symplectic geometry.

In §1 we state more explicitly our object of study which contains in particular the Synge metric. A post-Newtonian estimation using a convenient form of the metric tensor (I) is discussed in §2. The geometry of GL^n spaces with the metric tensor (I) is expounded in §§3-4. The electromagnetic tensors are introduced in §5 and corresponding Maxwell equations are derived. The Einstein equations are studied in §6 and in §7 the locally Minkowski case is treated. Some special features of the almost Hermitian model are pointed out in §8 while §9 is devoted to the homogeneous case.

§1. Synge Metric in Dispersive Media

Let us assume on the base manifold M of the tangent bundle (TM,τ,M), a pseudo-Riemannian metric $\gamma_{ij}(x)$ is given. Then $\gamma_{ij}(\tau(u)))$, $u\in TM$, $\tau(u)=x$, defines a d-tensor field on TM. We shall denote $\gamma_{ij}(\tau(u))$ by $\gamma_{ij}(x)$. Thus from the Liouville vector field $y^i\partial_i$ we may derive a d-covector field on TM given by

$$(1.1)\quad y_i=\gamma_{ij}(x)y^j.$$

We set

$$(1.2)\quad \|y\|^2=\gamma_{ij}(x)y^iy^j$$

and we remark that $\|y\|^2\neq 0$ on the manifold $\tilde{TM}$.

Assuming there is given a positive function $n(x,y)$ on $\tilde{TM}$ we introduce the function

$$(1.3)\quad a(x,y)=1+(1-u^2(x,y))\|y\|^2,\quad u(x,y)=\frac{1}{n(x,y)}.$$

The function $n(x,y)$ will be called the refractive index.

Now we consider the d-tensor field

$$(1.4)\quad g_{ij}(x,y)=\gamma_{ij}(x)+(1-\frac{1}{n^2(x,y)})y_iy_j.$$

We have

Theorem 1.1. *(i) $g_{ij}(x,y)$ is a symmetric d-tensor field on $\tilde{TM}$.*
(ii) rank $\|g_{ij}(x,y)\|=n$ on $\tilde{TM}$.

Proof. (i) is obvious. To prove (ii) it is enough to show that the matrix $\|g_{ij}\|$ is invertible. By a direct calculation one may check that its inverse has the entries

$$(1.5)\quad g^{ij}(x,y)=\gamma^{ij}(x)-\frac{1}{a(x,y)}(1-u^2(x,y))y^iy^j,$$

where $(\gamma^{ij}(x))$ denotes the inverse of matrix $(\gamma_{ij}(x))$. Thus the proof is finished.

We caution there are no relations between the dimension n of the manifold M and the refractive index n(x,y).

Supposing that the quadratic form $g_{ij}(x,y)\xi^i\xi^j$ has a constant signature on $\tilde{T}M$, then $GL^n = (M, g_{ij}(x,y))$ is a generalized Lagrange space.

Let $V^i(x)$, $x \in M$ be a smooth nowhere zero vector field on M. The mapping $S_V: M \to TM$ given by

$$(1.6) \quad x^i = x^i, \; y^i = V^i(x), \; x \in M, \; i = 1,2,\dots,n$$

is a cross-section of the projection τ:TM→M. It follows $S_V(M)$ is an embedded submanifold of TM. The restriction of the fundamental tensor $g_{ij}(x,y)$ of GL^n to the submanifold $S_V(M)$ is given by

$$(1.7) \quad g_{ij}(x, V(x)) = \gamma_{ij}(x) + (1 - u^2(x, V(x)))V_i V_j,$$

where $V_i = \gamma_{ij}(x)V^j$.

The metric (1.7) was introduced by J.L. Synge [146] and was used by him in the study of the propagation of the electromagnetic waves in a medium with the index of refraction n(x,V(x)), V(x) being the velocity of medium.

Definition 1.1. *The triad* $\mathbf{M} = (M, V(x), n(x, V(x)))$ *is called a dispersive medium. If* $\dot{\partial}_i n(x,y) = 0$, *then* $\mathbf{M}$ *is called a non-dispersive medium.*

Now we can formulate

Definition 1.2. *The restriction of the geometry of the space* GL^n *with the metric tensor (1.4) to* $S_V(M)$ *will be called the geometrical theory of the dispersive medium* $\mathbf{M}$.

This is a reason to consider the geometrical theory of the fields in GL^n as the content of the relativistic geometrical optics.

§2. A Post-Newtonian Estimation

A first application of the metric (1.4) to a dispersive medium $\mathbf{M}$ with the refractive index in the form $\frac{1}{n^2} = 1 - \frac{\alpha}{c^2}$ was worked out by G.S. Asanov and T. Kawaguchi [20]. A version of this application was presented by I. Roxburgh [133]. Roxburgh's method was extended recently by Gh. Zet, M. Agop, V. Manta [172]. We shall describe in what follows their results.

Let us consider a more general metric tensor than (1.4) given by

$$(2.1)\quad g_{ij}(x,y) = \gamma_{ij}(x) + \alpha\left(1 - \frac{1}{n^2(x,y)}\right)y_i y_j,$$

where α is a positive constant.

We assume that the base manifold M is a four dimensional space-time with the local coordinates $x^i = (x^0 = ct., x^a)$, $a = 1,2,3$.

If $\gamma:[0,1]\to M$ is a smooth parametrized curve, we define the length of $\Gamma = \mathrm{Im}\gamma$ as follows

$$(2.2)\quad l(\Gamma) = \int_0^1 \sqrt{g_{ij}(x,\dot{x})}\, dt.$$

The curves for which $l(\Gamma)$ affords extreme values will be called the geodesics of the space GL^n with the metric tensor (2.1). The geodesics are the solutions of the Euler-Lagrange equations written for the Lagrangian $\sqrt{g_{ij}(x,y)y^i y^j}$, $y^i = \dot{x}^i$ The form of these equations remains unchanged if instead of this Lagrangian we consider the Lagrangian

$$(2.3)\quad L(x,y) = -m_0 c\sqrt{g_{ij}(x,\dot{x})y^i y^j}\ ,\ y^i = \dot{x}^i,$$

where m_0 is the mass of the test particle.

The Lagrangian (2.3) takes the form

$$(2.3)'\quad L(x,y) = -m_0 c\{\|y\|^2[1 + \alpha(1 - \frac{1}{n^2})\|y\|^2]\}^{1/2}.$$

It is obvious that the additional term $\alpha(1 - \frac{1}{n^2})y_i y_j$ to the Riemannian metric $\gamma_{ij}(x)$ from the generalized Lagrange metric (2.1) will influence the extremal curves of the Lagrangian (2.3)′. The associate post-Newtonian estimation will be indicate in the following, which is aimed at clarifying the restriction which would be imposed to the constant α of the metric (2.1) by post-Newtonian analysis adapted to the nowadays gravitational observations.

We shall take $n^2(x,\dot{x})$ in the form

(2.4) $$n^2(x,\dot{x}) = 1 + \varepsilon U + \delta\frac{v^2}{c^2} + \mu U\frac{v^2}{c^2} + \nu U^2 + \sigma\frac{v^4}{c^2},$$

where $U = \frac{Gm}{rc^2}$ is the gravitational potential (G is the gravitational constant and m the mass of the source of the gravitational field),

$$r = (\delta_{ab}x^a x^b)^{1/2},\ v^2 = \delta_{ab}v^a v^b,$$

ε, δ, μ, ν and σ are the post-Newtonian parameters.

This choice of n^2 is in accordance with the known results (see Fock, [43], pg.340) obtained by studying the deviation of a light ray which propagates in a gravitational field. Let us note the post-Newtonian tests for the solar system impose some restrictions on the parameters $\varepsilon,\mu,\nu,\sigma$, on the constant α as well as on the post-Newtonian standard parameter β of a Riemannian metric.

The nondispersive case, which is frequently met in applications, impose the choice $\delta = \mu = \sigma = 0$ which simplifies considerable our analysis. Neglecting the terms of the higher order in V^a we may set

(2.5) $$\|v\|^2 = (\gamma_{00} + 2\gamma_{0a}\frac{v^a}{c} + \gamma_{ab}\frac{v^a v^b}{c^2})c^2.$$

Then the Lagrangian (2.3)′ takes the form

$$L(x,\dot{x}) = -m_0 c^2\{\gamma_{00} + 2\gamma_{0a}\frac{v^a}{c} + \gamma_{ab}\frac{v^a v^b}{c^2} + \alpha(1 - \frac{1}{n^2})[\gamma_{00}^2 +$$

(2.6) $$+4\gamma_{0a}\gamma_{ab}\frac{v^a v^b}{c^2} + \gamma_{ab}\gamma_{cd}\frac{v^a v^b v^c v^d}{c^4} + 4\gamma_{00}\gamma_{0a}\frac{v^a}{c} + 2\gamma_{00}\gamma_{ab}\frac{v^a v^b}{c^2}$$

$$+4\gamma_{0a}\gamma_{bc}\frac{v^a v^b v^c}{c^3}]\}^{1/2}.$$

In the static case, using the spherical coordinates, the post-Newtonian approximation of the Riemannian metric $\gamma_{ij}(x)$ will be of the form

(2.7) $$\begin{aligned}\gamma_{00} &= 1 - 2U + 2\beta U^2,\\ \gamma_{ab} &= -\delta_{ab}(1 + 2\gamma U),\\ \gamma_{0a} &= 0,\end{aligned}$$

where β and γ are the so called post-Newtonian parameters. On account of (2.7), the Lagrangian (2.6) becomes

$$(2.6)' \quad \begin{aligned} L(x,\dot{x}) = -m_0c^2\{1+\alpha(1-\frac{1}{n^2}) - 2U[1+2\alpha(1-\frac{1}{n^2})] - \frac{v^2}{c^2}[1+2\alpha(1-\frac{1}{n^2})] \\ -2U\frac{v^2}{c^2}[\gamma+2\alpha(\gamma-1)(1-\frac{1}{n^2})] + 2U^2[\beta+2\alpha(1+\beta)(1-\frac{1}{n^2})] + \alpha\frac{v^4}{c^4}(1-\frac{1}{n^2})\}^{1/2}. \end{aligned}$$

Expanding the square root from the right hand of (2.6)′ in powers of U and $\frac{v^2}{c^2}$ and neglecting the terms of order higher then three one gets

$$(2.8) \quad \frac{L(x,\dot{x})}{m_0\lambda_0} = \frac{1}{2}v^2 + Uc^2\frac{\lambda_1}{\lambda_0} + U^2c^2\frac{\lambda_2}{\lambda_0} + U^2v^2\frac{\lambda_3}{\lambda_0} + \frac{v^4}{c^2}\frac{\lambda_4}{\lambda_0},$$

where

$$(2.9) \quad \begin{aligned} &\lambda_0 = 1 - \alpha\delta \neq 0, \ \lambda_1 = 1 - \frac{\alpha\varepsilon^2}{2}, \\ &\lambda_2 = -\beta + \frac{1}{2} + \frac{3\alpha\varepsilon}{2} - \frac{\nu\alpha}{2} + \alpha\varepsilon^2 + \frac{\alpha^2\varepsilon^2}{8}, \\ &\lambda_3 = \gamma + \frac{1}{2} + \frac{3\alpha\varepsilon}{4} + \frac{3\alpha\delta}{2} - \frac{\alpha\mu}{2} + \alpha\varepsilon\delta + \frac{\alpha^2\varepsilon\delta}{4}, \\ &\lambda_4 = \frac{1}{8} + \frac{3\alpha\delta}{4} - \frac{\alpha\sigma}{2} + \frac{\alpha\delta^2}{2} + \frac{\alpha^2\delta^2}{8}. \end{aligned}$$

Note that the right hand of (2.8) approximates L up to a constant factor. This factor does not influence the Euler-Lagrange equations verified by L.

The approximation of order zero of the Lagrangian $\frac{L(x,\dot{x})}{m_0\lambda_0}$ must give the Newtonian case. Thus we have to assume $\lambda_1 = \lambda_0$ and it follows

$$(2.10) \quad \varepsilon = 2\delta.$$

Next in order that (2.8) be useful for the test regarding the shift of perihelion of planets we must accept the condition

$$(2.10)' \quad \Delta = \frac{\lambda_2}{\lambda_0} + 2\frac{\lambda_3}{\lambda_0} + 4\frac{\lambda_4}{\lambda_0} = 3 \pm 10^{-2}.$$

Finally, the theory is self contained if the metric (2.1) verifies the condition of propagation of light:

$$\gamma_{ij}\frac{dx^i}{dt}\frac{dx^j}{dt}[1+\alpha(1-\frac{1}{n^2})\gamma_{hk}\frac{dx^h}{dt}\frac{dx^k}{dt}]=0.$$

This is a reason to choose

$(2.10)''$ $\gamma=1.$

Thus we may summarise the discussions in the above as follows: *the metric (2.1) from the relativistic optics is compatible with the results of the gravitational tests (experiments) if the parameter* β, *the constant* α *and the parameters* ε, δ, μ, ν, σ *from (2.4) satisfy* $(2.10)'$ *with* $\gamma=1$ and $\varepsilon=2\sigma$.

A case which occurs in practice is that of nondispersive media. Taking $\delta=\mu=\sigma=0$ in $(2.10)'$ one deduces by (2.10) that $\varepsilon=0$. On account of $(2.10)''$ we get:

In order that the metric (2.1) from the relativistic optics of the nondispersive media to be compatible with the results of the gravitational experiments it is necessary that the post-Newtonian parameters β, ν *and the constant* α *to satisfy the restriction*

(2.11) $4-\beta-\frac{\nu\alpha}{2}=3\pm 10^{-2}.$

The analysis done in the above shows that the dispersive media with the index of refraction $n(x,\dot{x})$, the post-Newtonian solar system tests do not require a restriction on the constant α but only on a combination of α with the post-Newtonian parameter β and with the parameters introduced in the expansion (2.4) of $n^2(x,\dot{x})$. This restriction is $(2.10)'$ with $\varepsilon=2\delta$, $\gamma=1$. It is very complicated since the choosing of the parameters in (2.4) is dependent of the nature of the space. An important simplification appeared for nondispersive media is manifested by (2.11).

To end, we remark that the additional term $(1-\frac{1}{n^2})y_iy_j$ in the Riemannian metric $\gamma_{ij}(x)$ may influence not only the geodesics of the space, but many other aspects of the behaviour of the light signals, the Doppler effect etc.

§3. A Non-linear Connection

A first important property of the spaces GL^n endowed with the metric tensor (1.4) is given by

Theorem 3.1. *The generalized Lagrange space* GL^n *with the metric tensor (1.4) is not reducible to a Lagrange space.*

Proof. Using the notation (1.3) we get

$$\frac{\partial g_{ij}}{\partial y^k} = (1-u^2)(\gamma_{ik}y_j + \gamma_{jk}y_i) - 2u\frac{\partial u}{\partial y^k}y_i y_j. \tag{3.1}$$

Then keeping in mind the Proposition 1.1, Ch.X, it comes out that $\frac{\partial g_{ij}}{\partial y^k} = \frac{\partial g_{kj}}{\partial y^i}$ implies

$$(1-u^2)(\gamma_{jk}y_i - \gamma_{ji}y_k) - 2uy_j(y_i\frac{\partial u}{\partial y^k} - y_k\frac{\partial u}{\partial y^i}) = 0.$$

Contracting this equation by y^i one deduces

$$y_i\frac{\partial u}{\partial y^k} - y_k\frac{\partial u}{\partial y^i} = 0, \quad \gamma_{jk}y_i - \gamma_{ji}\gamma_k = 0.$$

Multiplying the last equation by γ^{jk} one obtains $(n-1)y_i = 0$ which implies $y_i = 0$ on $\tilde{TM}$. This contradiction proves the theorem.

By the Theorem 3.1, the relativistic geometrical optics can not be studied by means of methods of symplectic geometry since as we shall see later the almost Hermitian model of the space GL^n is not reducible to an almost Kählerian one. It means its almost symplectic structure is not integrable. In the following we shall use the d-tensor fields

$$\begin{aligned} \dot{C}_{ijk} &= (1-u^2)\gamma_{ik}y_j, \\ \overset{1}{C}_{ijk} &= -u(y_i y_j\frac{\partial u}{\partial y^k} + y_j y_k\frac{\partial u}{\partial y^i} - y_k y_i\frac{\partial u}{\partial y^j}). \end{aligned} \tag{3.2}$$

Clearly, we have

$$\begin{aligned} &\dot{C}_{ijk} = \dot{C}_{kji}, \quad \overset{1}{C}_{ijk} = \overset{1}{C}_{kji}, \\ &\dot{C}_{ijk} \neq 0, \quad \dot{C}_{ijk} \neq \dot{C}_{ikj}. \end{aligned} \tag{3.2'}$$

Now it is easy to prove

Proposition 3.1. *The medium M is nondispersive if and only if the d-tensor field* $\overset{1}{C}_{ijk}$ *vanishes.*

Let us note that if a dispersive medium M is given, then the fundamental tensor field $g_{ij}(x,y)$ from (1.4) of the space GL^n is completely determined by the pseudo-Riemannian metric $\gamma_{ij}(x)$ only. Therefore we may introduce the following:

Postulate. The space GL^n is endowed with the non-linear connection

(3.3) $$N^i_j = \left\{ {i \atop jk} \right\} y^k,$$

where $\left\{ {i \atop jk} \right\}$ are the Cristoffel symbols of the metric $\gamma_{ij}(x)$.

At least three arguments support this Postulate:

1° The autoparallel curves of the non-linear connection (3.3) are given by

(3.4) $$\frac{d^2x^i}{dt^2} + \left\{ {i \atop jk} \right\} \frac{dx^j}{dt}\frac{dx^k}{dt} = 0.$$

This is the first EPS (Ehlers, Pirani, Schild) condition in a constructive-axiomatic formulation of General Relativity.

2° Considering the absolute energy of the space GL^n

(3.5) $$\varepsilon(x,y) = g_{ij}(x,y)y^i y^j = a(x,y)\|y\|^2$$

we get the following interesting result:

Theorem 3.2. *(J.L. Synge). If M is a nondispersive medium, then the extremal curves* $c:[0,1]\to M$ *of the action integral*

$$I(c) = \int_0^1 \varepsilon\left(x, \frac{dx}{dt}\right) dt$$

which have the property

(*) $$g_{ij}\left(x, \frac{dx}{dt}\right) \frac{dx^i}{dt}\frac{dx^j}{dt} = 0$$

coincide to the curves which are solution of the equations (3.4).

Proof. The extremal curves of I(c) are the solutions of the Euler-Lagrange equations

(**) $$\frac{d}{dt}\left(\frac{\partial \varepsilon}{\partial y^i}\right) - \frac{\partial \varepsilon}{\partial x^i} = 0, \quad y^i = \frac{dx^i}{dt}.$$

On account of (*) and of $\frac{\partial u}{\partial y^i} = 0$, the system of equations (**) reduces to (3.4).

3° In the case $\frac{1}{n^2(x,y)} = 1 - \frac{1}{c^2}$, where c is the light velocity, the space GL^n is weakly regular and then its canonical non-linear connection is just that given by (3.3).

Let $\delta_i = \partial_i - \left\{ {j \atop ik} \right\} y^k \dot{\partial}_j$ be the local basis adapted to the horizontal distribution determined by the non-linear connection (3.3). The dual basis of $(\delta_i, \dot{\partial}_i)$ is $(dx^i, \delta y^i)$, $\delta y^i = dy^i + \left\{ {i \atop jk} \right\} y^k dx^j$.

The refractive index n(x,y) is an h-constant function if it satisfies the Tavakol - Van den Bergh equation (cf. §4, Ch.X):

(3.6) $$\frac{\partial n(x,y)}{\partial x^k} - \left\{ {i \atop kj} \right\} y^j \frac{\partial n(x,y)}{\partial y^i} = 0.$$

When **M** is a nondispersive medium, (3.6) is verified if and only if n(x,y) is a constant.

Using the h-covariant derivative of the refractive index, the following analogon of the d-tensor $\overset{1}{C}_{ijk}$ is constructed:

(3.7) $$\Lambda_{ijk} = -u\left(y_i y_j \frac{\delta u}{\delta x^k} + y_j y_k \frac{\delta u}{\delta x^i} - y_k y_i \frac{\delta u}{\delta x^j}\right).$$

Clearly we have

1° Λ_{ijk} is a d-tensor field on $\tilde{T}M$,

2° $\Lambda_{ijk} = \Lambda_{kji}$,

as well as

Proposition 3.2. *The refractive index n(x,y) is h-constant if and only if* Λ_{ijk} *vanishes.*

Now let us consider the d-connection $L\dot{\Gamma}(N) = \left(\left\{ {i \atop jk} \right\}, \dot{C}^i_{jk}\right)$ where $\dot{C}^i_{jk} = g^{ih}\dot{C}_{jhk}$.

Denoting by $\overset{\circ}{|}$ and $|$ the h- and v- covariant derivatives with respect to the d-connection $L\overset{*}{\Gamma}(N)$ we state

Theorem 3.3. *The d-tensor fields* Λ_{ijk} and $\overset{1}{C}_{jik}$ *take the following form*

$$(3.8)\quad \begin{aligned} \Lambda_{jik} &= \frac{1}{2}(g_{ji\overset{\circ}{|}k} + g_{ik\overset{\circ}{|}j} - g_{jk\overset{\circ}{|}i}), \\ \overset{1}{C}_{ijk} &= \frac{1}{2}(g_{ji}\overset{\circ}{|}_k + g_{ik}\overset{\circ}{|}_j - g_{jk}\overset{\circ}{|}_i). \end{aligned}$$

Proof. Taking into account (3.2), (3.7) as well as

$$(3.9)\quad g_{ij\overset{\circ}{|}k} = -2uy_iy_j\frac{\delta u}{\partial x^k},\quad g_{ij}\overset{\circ}{|}_k = -2uy_iy_j\frac{\partial u}{\partial y^k}$$

one gets (3.8).

By (3.9) it follows without difficulties

Theorem 3.4. *If the refractive index is h-constant then*

i) The absolute energy $\varepsilon(x,y)$ *is h-constant.*

ii) The functions n(x,y) and $\varepsilon(x,y)$ *are constant on the autoparallel curves of the non-linear connection (3.3).*

§4. Canonical Metrical d-Connection

Now we are in position to determine the canonical metrical d-connection $C\Gamma(N) = (L^i_{jk}, C^i_{jk})$ of the generalized Lagrange space GL^n endowed with the metric tensor (1.4).

Applying the theory developed in §2 of Ch.X we obtain

Theorem 4.1. *The canonical metrical d-connection* $C\Gamma(N)$ *of the space* GL^n *with the fundamental tensor* $g_{ij}(x,y)$ *from (1.4) has the coefficients*

$$(4.1)\quad L^i_{jk} = \left\{ {i \atop jk} \right\} + \Lambda^i_{jk},\quad C^i_{jk} = \overset{*}{C}{}^i_{jk} + \overset{1}{C}{}^i_{jk},$$

where the d-tensor fields Λ^i_{jk} and $\overset{1}{C}{}^i_{jk}$ *are given by*

$$(4.1)' \quad \Lambda^i_{jk} = g^{ih}\Lambda_{jhk}, \quad \overset{1}{C}{}^i_{jk} = g^{ih}\overset{1}{C}_{jhk}.$$

Proof. Using (3.8) and the identities

$$(4.2) \quad g^{ij}y_j = \frac{1}{2}y^i, \quad g_{ij}y^j = ay_i,$$

we get

$$(4.3) \quad \dot{C}^i_{jk} = \frac{1}{a}(1-u^2)\gamma_{jk}y^i.$$

Then, a direct calculation in which (2.3) of Ch.X is used gives us (4.1).

The form (4.1) of the coefficients of the canonical metrical d-connection $C\Gamma(N)$ allows us to write easily the differential equations of the h-paths by particularising the Propositions 3.3 and 3.4 from Ch. X.

Of course, the canonical metrical d-connection has the properties

$$(4.4) \quad g_{ij|k} = 0, \quad g_{ij}|_k = 0,$$

$$(4.5) \quad T^i{}_{jk} = 0, \quad S^i{}_{jk} = 0.$$

The other torsions of $C\Gamma(N)$ are as follows

$$(4.6) \quad R^i{}_{jk} = r_0{}^i{}_{jk}, \quad C^i_{jk} = \dot{C}^i_{jk} + \overset{1}{C}{}^i_{jk}, \quad P^i{}_{jk} = -\Lambda^i_{jk},$$

where $r_h{}^i{}_{jk}$ is the Riemannian curvature tensor of the Levi-Civita connection $\left\{ {i \atop jk} \right\}$. Therefore we get

Theorem 4.2. *The non-linear connection N with the coefficients (3.3) is integrable if and only if the pseudo-Riemannian metric $\gamma_{ij}(x)$ is flat.*

The Theorem 4.1 shows that the mapping $L\mathring{\Gamma}(N) \to C\Gamma(N)$ is a transformation of d-connections with the deformation tensors Λ^i_{jk} and $\overset{1}{C}{}^i_{jk}$ from (4.1)′.

We shall study the effects of this transformation on the curvature tensors of $C\Gamma(N)$. Indicating by a diacritic "$\circ$" the geometrical objects associated to the d-connection

$L\dot{\Gamma}(N) = (\left\{ {i \atop jk} \right\}, \dot{C}^i_{jk})$, by a direct calculation one obtains

Proposition 4.1. *The d-connection* $L\dot{\Gamma}(N)$ *has the torsions*

$$\dot{T}^i{}_{jk} = \dot{S}^i{}_{jk} = 0, \quad \dot{R}^i{}_{jk} = R^i{}_{jk}, \quad \dot{C}^i_{jk}, \quad \dot{P}^i{}_{jk} = 0. \tag{4.7}$$

Proposition 4.2. *The curvature of the d-connection* $L\dot{\Gamma}(N)$ *are given by*

$$\begin{aligned} &\dot{R}_j{}^i{}_{kh} = g^{is} r_{jskh}, \quad \dot{P}_j{}^i{}_{kh} = -\dot{C}^i{}_{jk \overset{\circ}{|} h}, \\ &\dot{S}_j{}^i{}_{kh} = \frac{1}{a}(1-u^2) g^{is} [\gamma_{jk}\gamma_{sh} - \gamma_{jh}\gamma_{sk} - \frac{2u}{a^2} y^i (\gamma_{jk}\frac{\partial u}{\partial y^h} - \gamma_{jh}\frac{\partial u}{\partial y^k})]. \end{aligned} \tag{4.8}$$

An immediate consequence of (4.8) is

Proposition 4.3. *The d-tensor field* $\dot{P}_j{}^i{}_{kh}$ *vanishes if and only if the d-tensor field* $\dot{C}^i_{jk}$ *is h-covariant constant with respect to the d-connection* $L\dot{\Gamma}(N)$.

Now applying the formulae from §6 in Ch. IX one gets

Theorem 4.3. *The curvatures of the canonical metrical d-connection* $C\Gamma(N)$ *have the form*

$$\begin{aligned} &R_j{}^i{}_{kh} = \dot{R}_j{}^i{}_{kh} + \rho_j{}^i{}_{kh}, \quad P_j{}^i{}_{kh} = \dot{P}_j{}^i{}_{kh} + \Pi_j{}^i{}_{kh}, \\ &S_j{}^i{}_{kh} = \dot{S}_j{}^i{}_{kh} + \sigma_j{}^i{}_{kh}, \end{aligned} \tag{4.9}$$

where

$$\begin{aligned} &\rho_j{}^i{}_{kh} = \Lambda^i{}_{jk \overset{\circ}{|} h} - \Lambda^i{}_{jh \overset{\circ}{|} k} + \Lambda^s_{jk}\Lambda^i_{sh} - \Lambda^s_{jh}\Lambda^i_{sk} + \overset{1}{C}{}^i_{js} R^s{}_{kh}, \\ &\Pi_j{}^i{}_{kh} = -\overset{1}{C}{}^i{}_{jh \overset{\circ}{|} k} + \Lambda^i_{jk} \overset{\circ}{|}_h + \Lambda^s_{jk} \overset{1}{C}{}^i{}_{sh} - \overset{1}{C}{}^s_{jh} \Lambda^i_{sk} + \Lambda^i_{js} \dot{C}^s_{kh}, \\ &\sigma_j{}^i{}_{kh} = \overset{1}{C}{}^i_{jk} \overset{\circ}{|}_k - \overset{1}{C}{}^i_{jh} \overset{\circ}{|}_k + \overset{1}{C}{}^s_{jk} \overset{1}{C}{}^i{}_{sh} - \overset{1}{C}{}^s_{jh} \overset{1}{C}{}^i{}_{sk}. \end{aligned} \tag{4.10}$$

Proposition 4.4. *If* M *is a nondispersive medium, the curvature tensors of* $C\Gamma(N)$ *are given by (4.9) and (4.10) with* $\overset{1}{C}{}^i_{jk} = 0$.

Proposition 4.5. *If the refractive index n(x,y) verifies the differential equation (3.6) i.e. it is h-covariant constant, then the curvatures of the canonical metrical d-connection* $C\Gamma(N)$ *are given by (4.9) and (4.10) with* $\Lambda^i_{jk}=0$.

Proposition 4.6. *If the function n(x,y) is h-covariant constant and the d-tensor fields* $\overset{\circ}{C}{}^i_{jk}$, $\overset{1}{C}{}^i_{jk}$ *are h-covariant constant with respect to* $L\overset{\circ}{\Gamma}(N)$ *i.e.*

$$(4.11)\quad \delta_k n = 0, \quad \overset{\circ}{C}{}^i_{jk\overset{\circ}{|}h} = 0, \quad \overset{1}{C}{}^i_{jk\overset{\circ}{|}h} = 0,$$

then we have

$$(4.12)\quad P_j{}^i{}_{kh} = 0.$$

§5. Electromagnetic Tensors

For the nondispersive media there exists an electromagnetic tensor field given by the skew-symmetric part of the h-covariant deflection tensor. In general, for a dispersive medium we may consider two electromagnetic tensors which satisfy a sort of generalized Maxwell equations.

Definition 5.1. *We call the h- and v- electromagnetic tensors of the generalized Lagrange space* GL^n *with the metric tensor (1.4) the d-tensors*

$$(5.1)\quad F_{ij} = \frac{1}{2}(D_{ij} - D_{ji}), \quad f_{ij} = \frac{1}{2}(d_{ij} - d_{ji}).$$

Of course, in (5.1), D_{ij} and d_{ij} are the covariant h- and v- deflection tensors of the canonical metrical d-connection. Therefore we have

Theorem 5.1. *The h- and v- electromagnetic tensors of the space* GL^n *with the metric tensor (1.4) have the following expressions*

$$(5.2)\quad \begin{aligned} F_{ij} &= u\,\|y\|^2(y_j\delta_i u - y_i\delta_j u), \\ f_{ij} &= u\,\|y\|^2(y_j\dot{\partial}_i u - y_i\dot{\partial}_j u). \end{aligned}$$

Proof. Indeed, we get

(5.3) $D_{ij} = \Lambda_{0ij}, \quad d_{ij} = g_{ij} + C_{0ij}$

which, together with (3.2) and (3.7) lead to (5.2).

Corollary 5.1. *i) If* $\delta_k n = 0$, *then* $F_{ij}(x,y) = 0$,
ii) If $\dot{\partial}_k n = 0$ *then* $f_{ij}(x,y) = 0$.

In conformity with the theory presented in §6, Ch.X the electromagnetic tensors F_{ij} and f_{ij} of a space GL^n satisfy the Maxwell equations (6.2), Ch.X. In our case, it is easy to prove

Theorem 5.2. *The electromagnetic tensor fields* F_{ij} *and* f_{ij} *of the space* GL^n *with the metric tensor (1.4) satisfy the following generalized Maxwell equations*

$$(5.4)\quad \begin{aligned} &F_{ij|k} + F_{jk|i} + F_{ki|j} = -\frac{1}{2}\sum_{(ijk)} (C_{ios} + d_{is}) r_0{}^s{}_{jk}, \\ &F_{ij}|_k + F_{jk}|_i + F_{ki}|_j = -(f_{ij|k} + f_{jk|i} + f_{ki|j}), \\ &f_{ij}|_k + f_{jk}|_i + f_{ki}|_j = 0. \end{aligned}$$

Theorem 5.3. *If the medium M is nondispersive, then the h-electromagnetic tensor field* F_{ij} *of the generalized space* GL^n *with the metric tensor (1.4) satisfies the Maxwell equations*

$$(5.4)'\quad \begin{aligned} &F_{ij|k} + F_{jk|i} + F_{ki|j} = 0, \\ &F_{ij}|_k + F_{jk}|_i + F_{ki}|_j = 0, \end{aligned}$$

and f_{ij} *vanishes.*

Proof. If the medium **M** is nondispersive i.e. $\dot{\partial}_i n = 0$ it results

$$f_{ij} = 0 \text{ and } \sum_{(ijk)} (C_{ios} + d_{is}) r_0{}^s{}_{jk} = a y^s \sum_{(ijk)} r_{sijk} + 2(1 - u^2) y^p y^q \sum_{(ijk)} y_i r_{pqjk} = 0.$$

Therefore (5.4) reduce to (5.4)$'$.

Theorem 5.4. *If the refractive index n(x,y) is h-covariant constant then the h-electromagnetic tensor* f_{ij} *of the space* GL^n *with the metric tensor (1.4) satisfies the Maxwell equations*

$$(5.4)'' \quad \begin{aligned} & f_{ij|k} + f_{jk|i} + f_{ki|j} = 0\,, \\ & f_{ij}|_k + f_{jk}|_i + f_{ki}|_j = 0\,. \end{aligned}$$

Theorem 5.5. *If the pseudo-Riemannian space $(M, \gamma_{ij}(x))$ is locally flat, then the electromagnetic fields F_{ij} and f_{ij} of the space GL^n with the metric tensor (1.4) satisfy the generalized Maxwell equations*

$$(5.4)''' \quad \begin{aligned} & F_{ij|k} + F_{jk|i} + F_{ki|j} = 0,\ f_{ij}|_k + f_{jk}|_i + f_{ki}|_j = 0\,, \\ & F_{ij}|_k + F_{jk}|_i + F_{ki}|_j = -(f_{ij|k} + f_{jk|i} + f_{ki|j})\,. \end{aligned}$$

Now the reader could write the vector fields hJ and vJ and then to apply the Theorem 6.2 Ch.X in order to derive the conservation laws in this theory of electromagnetism.

§6. Einstein Equations

The Einstein equations (5.1) from Ch.X can be written, in particular, for the generalized Lagrange spaces GL^n whose fundamental tensor is of the form (1.4). In order to do this, the canonical metrical d-connection $C\Gamma(N)$ described in the above is used. The restrictions of these equations to the cross-section $S_V(M)$ give the Einstein equations in a nondispersive medium endowed with the Synge metric (1.8). The Ricci tensors of $C\Gamma(N)$ are obtained from (4.8) by using convenient contractions. Thus we have

$$(6.1) \quad \begin{aligned} & R_{ij} = \dot{R}_{ij} + \rho_{ij},\quad {}'P_{ij} = {}'\dot{P}_{ij} + \overset{1}{\Pi}_{ij}\,, \\ & S_{ij} = \dot{S}_{ij} + \sigma_{ij},\quad {}''P_{ij} = {}''\dot{P}_{ij} + \overset{2}{\Pi}_{ij}\,, \end{aligned}$$

where

$$(6.1)' \quad \begin{aligned} & \rho_{ij} = \rho_i{}^s{}_{js},\ \sigma_{ij} = \sigma_i{}^s{}_{js},\ \overset{1}{\Pi}_{ij} = \Pi_i{}^s{}_{js}\,, \\ & \overset{2}{\Pi}_{ij} = \Pi_i{}^s{}_{sj},\ {}'\dot{P}_{ij} = \dot{P}_i{}^s{}_{js},\ {}''\dot{P}_{ij} = \dot{P}_i{}^s{}_{sj}\,. \end{aligned}$$

Also we shall set

$$(6.2) \quad \rho = g^{ij}\rho_{ij},\quad \sigma = g^{ij}\sigma_{ij}\,.$$

Now we can state

Theorem 6.1. *The Einstein equations of the generalized Lagrange space GL^n with the metric (1.4) are the following ones*

$$
(6.3)\qquad
\begin{aligned}
&\dot{R}_{ij}+\rho_{ij}-\frac{1}{2}(\dot{R}+\rho)g_{ij}=\kappa \overset{H}{T}_{ij}, \quad {}'\dot{P}_{ij}+\overset{1}{\Pi}_{ij}=\kappa \overset{1}{T}_{ij},\\
&\dot{S}_{ij}+\sigma_{ij}-\frac{1}{2}(\dot{S}+\sigma)g_{ij}=\kappa \overset{V}{T}_{ij}, \quad {}''P_{ij}+\overset{2}{\Pi}_{ij}=-\kappa \overset{2}{T}_{ij}.
\end{aligned}
$$

Proof. On account of (6.1), (6.1)$'$ and (6.2) we get

$$
(6.4)\qquad \overset{H}{E}_{ij}=\dot{R}_{ij}+\rho_{ij}-\frac{1}{2}(\dot{R}+\rho)g_{ij}, \quad \overset{V}{E}_{ij}=\dot{S}_{ij}+\sigma_{ij}-(\dot{S}+\sigma)g_{ij}
$$

and then the Theorem 5.1 from Ch.X applies.

The same Theorem 5.1 from Ch.X leads to

Theorem 6.2. *The following identities hold:*

$$
(6.5)\qquad
\begin{aligned}
&\overset{H}{E}{}^{i}_{j|i}=-\frac{1}{2}\{(\dot{P}^{ih}{}_{js}+\Pi^{ih}{}_{js})R^{s}{}_{hi}+2R^{s}{}_{ij}({}''\dot{P}^{i}_{s}+\overset{2}{\Pi}{}^{i}_{s})\},\\
&\overset{V}{E}{}^{i}_{j}|_{i}=0.
\end{aligned}
$$

Also we have

Theorem 6.3. *If the refractive index n(x,y) is constant (for example $\frac{1}{n^2}=1-\frac{1}{c^2}$), then the Einstein equations of the space GL^n with the metric (1.4) endowed with the canonical metrical d-connection $C\Gamma(N)=L\dot{\Gamma}(N)$ are the following ones*

$$
(6.6)\qquad
\begin{aligned}
&\dot{R}_{ij}-\frac{1}{2}\dot{R}g_{ij}=\kappa \overset{H}{T}_{ij},\\
&\dot{S}_{ij}-\frac{1}{2}\dot{S}g_{ij}=\kappa \overset{V}{T}_{ij},
\end{aligned}
$$

and the conditions

(6.6)′ $\overset{H}{T}{}^{i}_{j|i}=0,\quad \overset{V}{T}{}^{i}_{j}|_{i}=0,$

are verified.

Proof. Indeed, if n(x,y) is constant we have $\Lambda^{i}_{jk}=0$, $\overset{1}{C}{}^{i}_{jk}=0$ and $\dot{C}^{i}{}_{jk\,|\,h}^{\;\circ}=0$. By the Proposition 4.6, $P_{j}{}^{i}{}_{kh}=0$. Also we have $\rho^{i}_{jkh}=\Pi^{i}_{jkh}=\sigma^{i}_{jkh}=0$. The equations (6.3) reduce to (6.6) and so (6.6)′ holds.

Another remarkable case is treated by

Theorem 6.4. *If M is a nondispersive medium, then the Einstein equations of the space GL^n with the metric (1.4) endowed with the d-connection $C\Gamma(N)$ are given by (6.3) with $\overset{1}{C}{}^{i}_{jk}=0$ $(\sigma_{ij}=0,\ \sigma=0)$.*

In particular, we have

Theorem 6.5. *If the refractive index of a nondispersive medium M is h-covariant constant, then the Einstein equations of the space GL^n with the metric (1.4) endowed with the d-connection $C\Gamma(N)$ are given by (6.3) with $\Lambda^{i}_{jk}=0$.*

By the Proposition 4.6 we get

Theorem 6.6. *If the refractive index of a dispersive medium M is h-covariant constant and the d-tensors $\dot{C}^{i}_{jk}$, $\overset{1}{C}{}^{i}_{jk}$ are h-covariant constant with respect to $L\dot{\Gamma}(N)$, then the Einstein equations of GL^n with the metric (1.4) endowed with the d-connection $C\Gamma(N)$ are the following ones*

$$
(6.7)\quad
\begin{aligned}
&\dot{R}_{ij}+\rho_{ij}-\frac{1}{2}(\dot{R}+\rho)g_{ij}=\kappa\overset{H}{T}_{ij},\\
&\dot{S}_{ij}+\sigma_{ij}-\frac{1}{2}(\dot{S}+\sigma)g_{ij}=\kappa\overset{V}{T}_{ij},
\end{aligned}
$$

and the conditions

(6.7)′ $\overset{H}{T}{}^{i}_{j|i}=0,\quad \overset{V}{T}{}^{i}_{j}|_{i}=0,$

are fulfilled.

§7. Locally Minkowski GL^n Spaces

In many applications, we are interested in the previous theory when it is considered for the case of locally Minkowski spaces. According to the Definition 4.2 Ch.X the space GL^n endowed with the metric (1.4) is locally Minkowski if in every point $(x,y) \in TM$, there exists a preferential coordinate system such that $\frac{\partial n}{\partial x^i}=0$, $\frac{\partial \gamma_{ij}}{\partial x^i}=0$. Thus the pseudo-Riemannian space (M,γ_{ij}) is locally Minkowski one.

Assuming that GL^n with the metric (1.4) is locally Minkowski and considering a preferential coordinate system we shall determine the canonical metrical d-connection of this space, as well as the electromagnetic tensors. The Maxwell and Einstein equations will be derived, too.

To begin with, the coefficients of the associated non-linear connection N are given (due to (3.3)) by:

(7.1) $N_j^i=0.$

Hence the non-linear connection N is integrable and the autoparallel curves of N are characterized by $\frac{d^2x^i}{dt^2}=0$

Consequently, the refractive index n(y) satisfies the equation (3.6). We state this fact in

Theorem 7.1. *The refractive index n(y) of a locally Minkowski space GL^n with the metric (1.4) is an h-covariant constant function:*

(7.2) $\delta_k n=0.$

Proposition 7.1. *For a locally Minkowski space GL^n with the metric (1.4) the medium M is nondispersive if and only if the refractive index n(y) is constant.*

The d-connection $L\mathring{\Gamma}(N)$ has the coefficients

(7.3) $\left\{ {i \atop jk} \right\}=0,\ \dot{C}^i_{jk}=\frac{1}{a}(1-u^2(y))\gamma_{jk}y^i.$

Proposition 7.2. *The locally Minkowski space GL^n with the metric (1.4) has the canonical metrical d-connection*

$$(7.4)\quad L^i_{jk}=0,\quad C^i_{jk}=\mathring{C}^i_{jk}+\overset{1}{C}{}^i_{jk},$$

where $\mathring{C}^i_{jk}$ is in (7.3) and $\overset{1}{C}{}^i_{jk}$ is as follows

$$(7.4)'\quad \overset{1}{C}{}^i_{jk}=\frac{1}{2}g^{ih}(g_{hk}\overset{\circ}{|}_j+g_{ih}\overset{\circ}{|}_k-g_{jk}\overset{\circ}{|}_h).$$

On account of (7.2) and (7.3) we get

Proposition 7.3. *The following equations hold*

$$(7.5)\quad \begin{aligned} &y^i_{|k}=0,\ \ \|y\|^2_{|k}=0,\ \ a(y)_{|k}=0,\ \ \varepsilon(y)_{|k}=0,\\ &\mathring{C}^i_{jk|h}=0. \end{aligned}$$

Proof. The first four equations are immediate. For the last one we remark that $\varepsilon(y)$ depending only on the directional argument y^i verifies the equations (3.6) as we can see by using (7.1).

The previous considerations lead to

Theorem 7.2. *In a locally Minkowski space GL^n with the metric (1.4) we have:*
i) The absolute energy $\varepsilon(y)$ is h-covariant constant.
ii) The absolute energy is constant on the autoparallel curves of the non-linear connection N.

Proposition 7.4. *i) The h-paths of $C\Gamma(N)$ are given by*

$$\frac{d^2x^i}{dt^2}=0,\ \ \frac{dy^i}{dt}=0.$$

ii) The v-paths of $C\Gamma(N)$ are characterized by

$$x^i=x^i_0,\ \ \frac{d^2y^i}{dt^2}+[\frac{1}{a(y)}(1-u^2(y))\gamma_{jk}y^i+\overset{1}{C}{}^i_{jk}(y)]\frac{dy^j}{dt}\frac{dy^k}{dt}=0.$$

Proposition 7.5. *The torsions of $C\Gamma(N)$ are*

(7.6) $T^i{}_{jk}=0,\ S^i{}_{jk}=0,\ R^i{}_{jk}=0,\ P^i{}_{jk}=0$ *and* C^i_{jk} *as in (7.4).*

Proposition 7.6. *i) The curvatures of* $C\Gamma(N)$ *are*

(7.7) $R^{\,i}_{j\,kh}=0,\ P^{\,i}_{j\,kh}=-\overset{1}{C}{}^i_{jk|h},\ S_j{}^i{}_{kh}=\dot{S}_j{}^i{}_{kh}+\sigma_j{}^i{}_{kh}.$

ii) Also we have

(7.7)′ $\dot{R}_j{}^i{}_{kh}=\dot{P}_j{}^i{}_{kh}=0.$

Concerning the h- and v- electromagnetic tensor fields we have:

Proposition 7.7. *The electromagnetic tensors of the locally Minkowski space* GL^n *with the metric (1.4) are the following ones*

(7.8) $F_{ij}=0,\ f_{ij}=u\|y\|^2(y_j\dfrac{\partial u}{\partial y^i}-y_i\dfrac{\partial u}{\partial y^j})$

and f_{ij} *is h-covariant constant:*

(7.8)′ $f_{ij|K}=0.$

Therefore

Theorem 7.3. *The following Maxwell equation holds*

$$f_{ij}|_k+f_{jk}|_i+f_{ki}|_j=0.$$

Using the Proposition 7.6 we deduce

Theorem 7.4. *The Einstein equations of the locally Minkowski space* GL^n *with the metric (1.4) endowed with the canonical metrical d-connection* $C\Gamma(N)$ *are the following ones:*

$$\dot{S}_{ij}+\sigma_{ij}-\frac{1}{2}(\dot{S}+\sigma)g_{ij}=\kappa\overset{V}{T}_{ij};\quad -\overset{1}{C}{}^h_{ij|h}=\kappa\overset{1}{T}_{ij};\quad \overset{1}{C}{}^h_{ih|j}=\kappa\overset{2}{T}_{ij}$$

and the following equation holds: $\overset{V}{T}{}^i_j|_i=0.$

If **M** is a nondispersive medium, then n(y) is constant, $\overset{1}{C}{}^{i}_{jk}=0$ and

$$\dot{S}_{j\ kh}^{\ i} = \frac{1}{a}(1-u^2)g^{is}(\gamma_{jk}\gamma_{sh}-\gamma_{jh}\gamma_{sk}).$$

Therefore we have

Theorem 7.5. *The Einstein equations of the locally Minkowski space GL^n with the metric (1.4) endowed with the canonical metrical d-connection $C\Gamma(N)$, for a nondispersive medium are the following ones*

$$\dot{S}_{ij} - \frac{1}{2}\dot{S}g_{ij} = \kappa \overset{V}{T}_{ij}$$

and the condition $\overset{V}{T}{}^{i}_{j}|_{i} = 0$ *is fulfilled.*

§8. Almost Hermitian Model

The previous theory receives a convenient geometrical meaning on the almost Hermitian model of the generalized Lagrange space GL^n endowed with the metric tensor $g_{ij}(x,y)$ from (1.4) and with the non-linear connection N of coefficients (3.3).

Following the construction from the last section of the Chapter X we have the local basis $(\delta_i, \dot{\partial}_i)$ adapted to the horizontal distribution N and to the vertical distribution V on the manifold TM. Its dual basis is $(dx^i, \delta y^i)$ with

(8.1) $\delta y^i = dy^i + \begin{Bmatrix} i \\ jk \end{Bmatrix} y^j dx^k.$

Then the almost complex structure **F**, defined on the vector fields $(\delta_i, \dot{\partial}_i)$ by

(8.2) $F(\delta_i) = -\dot{\partial}_i,\ F(\dot{\partial}_i) = \delta_i,\ i = 1, 2, \dots, n,$

together with the metrical structure

(8.3) $G = g_{ij}(x,y)dx^i \otimes dx^j + g_{ij}(x,y)\delta y^i \otimes \delta y^j$

give us an almost Hermitian structure **(G,F)**.

It follows

Theorem 8.1. *The structure (G,F) is an Hermitian one if and only if the pseudo-Riemannian space* $(M, \gamma_{ij}(x))$ *is flat.*

Proof. As we know, the almost complex structure **F** is integrable if and only if the horizontal distribution N is integrable. But this property is equivalent to $R^i{}_{jk} = r_0{}^i{}_{jk}$ on $\tilde{TM}$. The last equation is equivalent to $r_h{}^i{}_{jk} = 0$.

The 2-form

$$(8.4)\quad \theta = [\gamma_{ij}(x) + (1 - u^2)y_i y_j]\delta y^i \wedge dx^j$$

is associated with the almost Hermitian structure **(G,F)** i.e.

$$\theta(X,Y) = G(FX, Y),\ X, Y \in \chi(TM).$$

We have

Theorem 8.2. *i)* θ *is an almost symplectic structure on* $\tilde{TM}$,
ii) θ *is not integrable.*

Proof. i) is immediate. ii) The exterior differential $d\theta$ of the 2-form θ in the adapted cobasis $(dx^i, \delta y^i)$ is given by (7.4), Ch.X. Taking into account the Theorem 3.1 it follows $\frac{\partial g_{ij}}{\partial y^k} \neq \frac{\partial g_{kj}}{\partial y^i}$. Therefore $d\theta \neq 0$.

Corollary 8.1. *The almost Hermitian space* $H^{2n} = (\tilde{TM}, G, F)$ *is not reducible to an almost Kählerian space.*

Therefore we cannot apply the methods of symplectic geometry (see P.Liebermann and Ch.Marle [93]) in order to study the dispersive media (or nondispersive media) endowed with the Synge metric (1.5). This fact was already recognized by Synge himself (see J.L. Synge, [146] p.311).

Now we can introduce (according to the Definition 7.1 in Ch.X) the notion of almost Hermitian connection of Lagrange type ∇ and we can prove that ∇ characterises the canonical metrical d-connection of the generalized Lagrange space GL^n. The Einstein equations of the space GL^n endowed with the d-connection $C\Gamma(N)$ are equivalent to the Einstein equations associated to the connection ∇.

§9. A Finslerian Approach to the Relativistic Optics

The relativistic optics based on the generalized metric tensor (1.4) has some clear advantages from a physical point of view. The particular case of locally Minkowski space, studied in the last section is however a restrictive one since it asks for the condition that the Lorentz metric tensor $\gamma_{ij}(x)$ has constant components. The locally Minkowski spaces are much more interesting in the category of Finsler spaces.

This fact suggests us to study the generalized Lagrange spaces GL^n , whose fundamental tensor field $g_{ij}(x,y)$ is of the form

$$(9.1)\quad g_{ij}(x,y)=\gamma_{ij}(x,y)+(1-\frac{1}{n^2(x,y)})y_i y_j\,,$$

where $\gamma_{ij}(x,y)$ is the fundamental tensor of a Finsler space $F^n=(M,F(x,y))$ and

$$(9.2)\quad y_i=\gamma_{ij}(x,y)y^j=\frac{1}{2}\frac{\partial F^2}{\partial y^i}\,.$$

In this case we have

$$(9.3)\quad \|y\|^2=F^2(x,y),\quad a=1+(1-\frac{1}{n^2(x,y)})F^2$$

and

$$(9.1)'\quad g^{ij}(x,y)=\gamma^{ij}(x,y)-\frac{1}{a(x,y)}(1-\frac{1}{n^2(x,y)})y^i y^j.$$

Of course, the refractive index n(x,y) can be strongly related to the fundamental function F(x,y) of the space F^n, but here it is considered as an apriori given positive function on the manifold $\tilde{TM}$.

On a cross-section $S_V(M)$ the metric tensor (9.1) gives us

$$(9.1)''\quad g_{ij}(x,V(x))=\gamma_{ij}(x,V(x))+(1-\frac{1}{n^2(x,V(x))})V_i V_j$$

and $V_i=\dfrac{1}{2}\dfrac{\partial F^2}{\partial y^i}$, $y^i=V^i$.

This metric is obviously a direct generalization of the Synge metric (1.8). It include some interesting particular cases which can not be studied by means of Synge metric.

For example, taking in a preferential local coordinate system the Finslerian metric function

$$F(y) = \sqrt[m]{(y^1)^m + \dots + (y^n)^m}$$

(which leads to the Antonelli metric studied in Ch.XI), from (9.1)″ we obtain a new metric, useful in applications.

A first result about the generalized Lagrange space GL^n with the fundamental tensor (9.1) is stated as follows:

Theorem 9.1. *The space GL^n, $n>1$, with the metric tensor (9.1) is not reducible to a Lagrange (or a Finsler) space.*

Proof. Applying the Proposition 1.1, Ch.X, from (9.2), first we have

$$\frac{\partial g_{ij}}{\partial y^k} = \frac{\partial \gamma_{ij}}{\partial y^k} + (1-u^2)(\gamma_{ik}y_j + \gamma_{jk}y_i) - 2u\frac{\partial u}{\partial y^k}y_i y_j. \tag{9.4}$$

This implies (because $\frac{\partial \gamma_{ij}}{\partial y^k}$ is totally symmetric)

$$0 = \frac{\partial g_{ij}}{\partial y^k} - \frac{\partial g_{kj}}{\partial y^i} = (1-u^2)[\gamma_{jk}y_i - \gamma_{ji}y_k - 2uy_j(\frac{\partial u}{\partial y^k}y_i - \frac{\partial u}{\partial y^i}y_k)]$$

if and only if

$$\gamma_{jk}y_i - \gamma_{ji}y_k = 0, \; y_i\frac{\partial u}{\partial y^k} - y_k\frac{\partial u}{\partial y^i} = 0.$$

Multiplying the first equation by γ^{jk} we get $(n-1)y_i = 0$ or $y^i = 0$ on $\tilde{TM}$. This is a contradiction.

Now we may assume the following:

Postulate. The space GL^n is endowed with the Cartan non-linear connection N of the Finsler space F^n:

$$N^i_j = \frac{1}{2}\frac{\partial \gamma^i_{00}}{\partial y^j}. \tag{9.5}$$

If F^n is a pseudo-Riemannian space, then N^i_j given by (9.5) is coincident to the non-linear connection (3.3).

We can determine the canonical metrical d-connection of the space GL^n with the metric tensor (9.1).

Let us consider the Cartan connection $\overset{c}{C\Gamma}(N) = (\overset{c}{L}{}^i_{jk}, \overset{c}{C}{}^i_{jk})$ of the Finsler space F^n and the d-tensor fields $\Lambda_{ijk}, \overset{1}{C}_{ijk}$ given by (3.7).

Therefore we have:

Theorem 9.2. *The canonical metrical d-connection $C\Gamma(N)$ of the generalized Lagrange space GL^n, with the metric tensor (9.1), endowed with the non-linear connection (9.5), has the coefficients:*

$$\text{(9.7)}\quad \begin{aligned} L^i_{jk} &= \overset{c}{L}{}^i_{jk} + \Lambda^i_{jk}, \\ C^i_{jk} &= \overset{c}{C}{}^i_{jk} + \dot{C}^i_{jk} + \overset{1}{C}{}^i_{jk}, \end{aligned}$$

where $\overset{c}{C\Gamma}(N) = (\overset{c}{L}{}^i_{jk}, \overset{c}{C}{}^i_{jk})$ is the Cartan connection of the Finsler space $F^n = (M, F(x,y))$, $\dot{C}^i_{jk} = \frac{1}{a}(1 - \frac{1}{n^2(x,y)})\gamma_{jk}y^i$ and Λ^i_{jk}, C^i_{jk} are given by $g^{ih}\Lambda_{jhk}$, $g^{ih}\overset{1}{C}_{jhk}$, respectively.

Proof. The conditions $L^i_{jk} = L^i_{kj}$, $C^i_{jk} = C^i_{kj}$ are obvious. It remains to verify that $g_{ij|k} = 0$, $g_{ij}|_k = 0$. These equations will be verified by a direct calculation. Indeed, we have

$$(*)\ g_{ij|k} = g_{ij\overset{c}{|}k} - (\Lambda_{ijk} + \Lambda_{ijk}); \quad g_{ij}|_k = g_{ij}\overset{c}{|}_k - (\dot{C}_{ijk} + \dot{C}_{jik} + \overset{1}{C}_{ijk} + \overset{1}{C}_{jik}),$$

But

$$g_{ij\overset{c}{|}k} = -2u(\delta_k u)y_i y_j,$$

$$g_{ij}\overset{c}{|}_k = -2u(\dot{\partial}_k u)y_i y_j + (1 - u^2)(\gamma_{ik}y_j + \gamma_{jk}y_i),$$

$$\Lambda_{ijk} + \Lambda_{jik} = -2u(\delta_k u)y_i y_j, \quad \overset{1}{C}_{ijk} + \overset{1}{C}_{jki} = -u(\dot{\partial}_k u)y_i y_j,$$

$$\dot{C}_{ijk} + \dot{C}_{jik} = (1 - u^2)(\gamma_{ik}y_j + \gamma_{jk}y_i).$$

Consequently, from (*) we have $g_{ij|k}=0,\ g_{ij}|_k=0$.

Now, based on the canonical metrical d-connection (9.7) we may develop, by the same devices as in the previous sections of this chapter the whole geometrical theory of the relativistic optics in a Finslerian framework.

Chapter XIII

Geometry of Time Dependent Lagrangians

A time dependent Lagrangian on a smooth n-dimensional manifold M is a smooth real function on R x TM.

There are many problems in Physics and Variational calculus in which time dependent Lagrangians are involved.

It seems that an explicit dependence on time in problems of dynamics was considered for the first time by A. Wundheiler [167] who called the geometrical background related to it "Rheonomic Geometry". This is an explanation for using the term "Rheonomic Lagrange geometry" for the geometrical theory of time dependent Lagrangians.

In this chapter we shall present a geometrization of time dependent Lagrangians by using as a pattern the geometry of Lagrange spaces and as a main tool the geometry of the total space of a vector bundle developed in the chapters II and III.

In order to do this we shall consider the manifold E = R x TM fibred over R x M by $\pi(t,v)=(t,\tau(v))$, $(t,v)\in E$, where $\tau: TM \to M$ is the usual projection. The reader is invited to compare this geometrization to that contained in a book by M. de Leon and P.R. Rodrigues, [92], as well as to that from the papers [12-14].

In §1 we point out the particular features of the non-linear connections in $\xi=(R\times TM,\pi,R\times M)$. A canonical non-linear connection is derived from a time dependent Lagrangian in §2. The relationship between non-linear connections and semisprays is discussed in §3. Normal d-connections on R x TM are introduced in §4 and a canonical metrical normal d-connection is showed in §5. The next section is devoted to homogeneous time dependent Lagrangians. The another important time dependent Lagrangians are treated in §7. A metrical almost contact structure described in §8 will serve as a model for further geometrical developments. A generalization of rheonomic Lagrange geometry is sketched in §9.

§1. Non-linear Connections in $\xi=(R\times TM,\pi,R\times M)$

As usual, the manifold M is of finite dimension n and the indices i,j,k,... run over 1,...,n. We denote by $(t,x^i)\equiv(t,x)$, the local coordinates on R x M. The coordinates in fibres of ξ are (y^i) defined by $u_{(t,x)} \equiv (t, v_x) \equiv (t, y^i \frac{\partial}{\partial x^i}) \in \pi^{-1}(t,x)$. Thus the local coordinates on $E = R x TM$ are $(t, x^i, y^i) \equiv (t, x, y)$. A change of local coordinates on E has the following form

$$(1.1)\ \tilde{t} = t,\ \tilde{x}^i = \tilde{x}^i(x^1,\ldots,x^n),\ \tilde{y}^i = \frac{\partial \tilde{x}^i}{\partial x^k} y^k,\ \mathrm{rank}\left(\frac{\partial \tilde{x}^i}{\partial x^k}\right) = n.$$

The projection π has the local form: $(t,x,y) \to (t,x)$.

The kernel of the mapping $\pi^T : TE \to T(R x M)$ supplies a vector subbundle of the tangent bundle $\tau_E : TE \to E$ which is called the vertical subbundle and is denoted by VE. The local vector fields $(\frac{\partial}{\partial y^1}, \ldots, \frac{\partial}{\partial y^n})$ define a local basis of $V_u E$, $u \in E$.

The natural basis $(\frac{\partial}{\partial t}, \frac{\partial}{\partial x^i}, \frac{\partial}{\partial y^i})$ changes under a coordinate transformation (1.1) as follows

$$(1.2)\quad \begin{aligned} \frac{\partial}{\partial t} &= \frac{\partial}{\partial \tilde{t}}, \\ \frac{\partial}{\partial x^j} &= \frac{\partial \tilde{x}^i}{\partial x^j}\frac{\partial}{\partial \tilde{x}^i} + \frac{\partial \tilde{y}^i}{\partial x^j}\frac{\partial}{\partial \tilde{y}^i}, \\ \frac{\partial}{\partial y^j} &= \frac{\partial \tilde{y}^i}{\partial y^j}\frac{\partial}{\partial \tilde{y}^i}. \end{aligned}$$

Of course, $\frac{\partial \tilde{y}^i}{\partial y^j} = \frac{\partial \tilde{x}^i}{\partial x^j}$.

Recall that TM admits a natural tangent structure J given by (1.9), Ch. VII. In order to prolong it to an almost tangent structure J on RxTM we must define conveniently $J(\frac{\partial}{\partial t})$. The condition $J^2=0$ asks for $J(\frac{\partial}{\partial t}) = a^i \frac{\partial}{\partial y^i}$, where (a^i) define a d-vector field on TM. We shall use two prolongations of J corresponding to $a^i=0$ and $a^i=-y^i$, respectively. In the former case we shall keep the notation J for the prolongation, too. In the later case the prolongation will be denoted by $\tilde{J}$. Thus we have

$$(1.3)\ \tilde{J}\left(\frac{\partial}{\partial t}\right) = -y^i \frac{\partial}{\partial y^i} = -\mathbf{C},\ \tilde{J}\left(\frac{\partial}{\partial x^i}\right) = \frac{\partial}{\partial y^i},\ \tilde{J}\left(\frac{\partial}{\partial y^i}\right) = 0.$$

By a direct calculation we find that the adjoint $\tilde{J}^*$ of $\tilde{J}$ is given by

(1.4) $\tilde{J}^*(dt)=0,\ \tilde{J}^*(dx^i)=0,\ \tilde{J}^*(dy^i)=dx^i-y^i dt.$

The 1-forms dx^i-$y^i dt$ are frequently met in the study of time dependent Lagrangian systems. By a direct calculation it comes out that $\tilde{J}$ is not integrable.

The homogeneity with respect to (y^i) of various geometrical objects on E may be expressed by using again $\mathbb{C}=y^i\dfrac{\partial}{\partial y^i}$. We notice that in contrast with J, the almost tangent structure $\tilde{J}$ is not homogeneous.

In the following we shall put $t=x^0$ and we shall introduce the indices $\alpha,\beta,\gamma,\ldots$ ranging over $0,1,2,\ldots,n$. We set

$$\frac{\partial\tilde{x}^\alpha}{\partial x^\beta}=\begin{pmatrix}1 & 0\ \ldots\ 0\\ 0 & \dfrac{\partial\tilde{x}^i}{\partial x^k}\end{pmatrix}$$

Now let us consider a non-linear connection in ξ, that is, a distribution $u\to H_uE$ on E which is supplementary to the vertical distribution $u\to V_uE$. As we have seen in Ch. II, the horizontal distribution $u\to H_uE$ may be given by $n+1$ local vector fields,

(1.5) $\delta_\alpha=\dfrac{\partial}{\partial x^\alpha}-N^i_\alpha(t,x,y)\dfrac{\partial}{\partial y^i},$

where N^i_α are real functions, locally defined on E and subject to the following transformation rule under (1.1):

(1.6) $N^i_\alpha\dfrac{\partial\tilde{x}^\alpha}{\partial x^\beta}=\dfrac{\partial\tilde{x}^i}{\partial x^k}N^k_\beta-\dfrac{\partial^2\tilde{x}^i}{\partial x^\beta\partial x^k}y^k.$

Let us rewrite (1.5) in the form

(1.7) $\delta_0=\dfrac{\partial}{\partial t}-N^i_0(t,x,y)\dfrac{\partial}{\partial y^i}\ ,\ \delta_i=\dfrac{\partial}{\partial x^i}-N^k_i(t,x,y)\dfrac{\partial}{\partial y^k}.$

We have

Theorem 1.1. *To give a non-linear connection in* ξ *is equivalent to giving a set of local coefficients* (N^i_0, N^k_i) *which transform under (1.1) as follows:*

$$(1.8)\quad \tilde{N}_0^i = \frac{\partial \tilde{x}^i}{\partial x^k} N_0^k ,$$

$$(1.9)\quad N_i^k \frac{\partial \tilde{x}^i}{\partial x^j} = \frac{\partial \tilde{x}^k}{\partial x^i} N_j^i - \frac{\partial \tilde{y}^k}{\partial x^i}.$$

Proof. As we know from Ch.II, to give a non-linear connection in ξ is equivalent to giving a set of local coefficients (N_α^i) which transform under (1.1) according to (1.6). But (1.6) is equivalent to (1.8) and (1.9) if we keep in mind (1.7).

Remark 1.1. The local coefficients $(N_0^i(t,x,y))$ transform like the components of a vector field on M, although they depend on t, x and y. The local coefficients $N_i^k(t,x,y)$ transform like those of a non-linear connection on TM.

Let P be the almost product structure on E defined by a non-linear connection $N = (N_0^i, N_k^i)$ in ξ according to

$$(1.10)\quad P(\delta_\alpha) = \delta_\alpha \ , \quad P(\dot{\partial}_i) = -\dot{\partial}_i ,$$

where $\dot{\partial}_i$ stands for $\dfrac{\partial}{\partial y^i}$.

A direct calculation gives (see also the Theorem 1.4, Ch.VII) the following

Theorem 1.2. *To give a non-linear connection on E is equivalent to giving a morphism* $\Gamma : \chi(E) \to \chi(E)$ *satisfying the conditions*

$$(1.11)\quad \tilde{J}\Gamma = \tilde{J} \ , \ J\Gamma = J \ , \ \Gamma J = -J .$$

Remark 1.2. In (1.11), the last equality may be replaced by $\Gamma\tilde{J} = -\tilde{J}$.

Remark 1.3. As we know from Ch.II, a non-linear connection N on E always exists if M is paracompact.

As for any vector bundle, a non-linear connection N on E induces a linear connection $\overset{\circ}{D}$ in the vertical subbundle, called the Berwald connection associated with N. It is defined as follows

(1.12) $\overset{\circ}{D}_{\frac{\partial}{\partial x^\alpha}} \frac{\partial}{\partial y^j} = B^k_{j\alpha} \frac{\partial}{\partial y^k} \ , \ \overset{\circ}{D}_{\frac{\partial}{\partial y^i}} \frac{\partial}{\partial y^i} = 0,$

where $B^k_{j\alpha} = \frac{\partial N^k_\alpha}{\partial y^j}$.

Denoting by ∇ the operator of absolute covariant differentiation of vertical valued q-forms (q=1,2,...,n) the curvature of N is given by $\Omega = -\nabla v$, where v is the vertical projector of N, regarded as a vertical valued 1-form. Expliciting ∇ we immediately obtain

(1.13) $\Omega(X,Y) = -v[hX, hY] \ , \ X, Y \in \chi(E).$

Let h be the horizontal projector associated to N. We have

Theorem 1.3. *The following structure equation holds*

(1.14) $\Omega = N_h,$

where N_h is the Nijenhuis tensor field associated with h.

The local form of Ω is obtained from (1.13) as follows

$$(1.15) \quad \begin{aligned} &\Omega\left(\frac{\partial}{\partial x^\alpha}, \frac{\partial}{\partial x^\beta}\right) = \Omega^i_{\alpha\beta} \frac{\partial}{\partial y^i} \ , \ \Omega\left(\frac{\partial}{\partial x^\alpha}, \frac{\partial}{\partial y^i}\right) = 0 \ , \\ &\Omega\left(\frac{\partial}{\partial y^i}, \frac{\partial}{\partial y^j}\right) = 0, \end{aligned}$$

where $\Omega^i_{\alpha\beta} = \delta_\alpha N^i_\beta - \delta_\beta N^i_\alpha$. On the other hand we have

(1.16) $[\delta_\alpha, \delta_\beta] = R^i_{\alpha\beta} \dot{\partial}_i \ , \ R^i_{\alpha\beta} = -\Omega^i_{\alpha\beta}.$

Thus the horizontal distribution on E is integrable if and only if $\Omega = 0$.

The following d-tensor field

(1.17) $t^k_{ij} = B^k_{ij} - B^k_{ji}$

will be called the torsion of N.

As for any vector bundle, a non-linear connection N in ξ is homogeneous (resp. linear) if the functions $N^i_0(t,x,y)$ and $N^i_k(t,x,y)$ are 1-homogeneous (resp. linear) with respect to (y^i).

If we define the tension H of N as $H = \frac{1}{2} L_C P$, where L_C means the Lie derivative, we obtain that N is homogeneous if and only if $H=0$. Of course, in order to speak of homogeneous connections we must delete from E the image of the null-section in ξ.

Remark 1.4. If N becomes a linear connection, then from $N_0^i(t,x,y) = K_j^i(t,x)y^j$ and $N_j^i(t,x,y) = \Gamma_{jk}^i(t,x)y^k$ one obtains a pair $(K_j^i(t,x)\ ,\ \Gamma_{jk}^i(t,x))$ which may be thought of as a generalized affine connection on R x M in a sense from [89, p. 127].

§2. Time Dependent Lagrangians

In this section we shall derive a non-linear connection on E = R x TM from the Euler-Lagrange equations associated with a regular time dependent Lagrangian.

Definition 2.1. *A time dependent Lagrangian is a function* $L: R \times TM \to R\ ,\ (t,v) \to L(t,v)$.

Definition 2.2. *A time dependent Lagrangian L is said to be regular if the matrix with the entries*

$$(2.1)\quad g_{ij}(t,x,y) = \frac{1}{2}\frac{\partial^2 L}{\partial y^i \partial y^j}$$

is of rank n on E.

The condition for L to be regular is independent on the choice of local chart.

Definition 2.3. *A pair RL^n = $(M, L(t,x,y))$, where L is a regular time dependent Lagrangian such that the quadratic form with the coefficients $g_{ij}(t,x,y)$ from (2.1) has constant signature will be called a rheonomic Lagrange space.*

Let $c:[0,1] \to M$ be a curve on M and $\dot{c}:[0,1] \to TM$ be the tangent vector field to c. If a local representation of c is $x^i = x^i(\tau)\ ,\ \tau \in [0,1]$, then $\dot{c}(\tau)$ is represented as $(x^i(\tau), \dot{x}^i = \frac{dx^i}{d\tau})$. The time dependent Lagrangian L defines a function

$\mathcal{L}:c\to\mathcal{L}(c)=\int_0^1 L(\tau,x(\tau),\dot{x}(\tau))d\tau$, from which it follows the variational problem: find those curves which afford extreme values for $\mathcal{L}$. Looking for such an extremal curve in the set of all curves with fixed endpoints, one finds from the first variational formula that it is among the curves which satisfy the Euler-Lagrange equations (cf. R. Hermann, [56], p. 153):

$$(2.2)\quad \frac{\partial L}{\partial x^i}-\frac{d}{d\tau}\left(\frac{\partial L}{\partial \dot{x}^i}\right)=0.$$

Using the differentiation on composite functions and (2.1), the system of differential equations (2.2) may be rewritten as

$$(2.3)\quad 2g_{ij}\frac{d^2x^j}{d\tau^2}+\left(\frac{\partial^2 L}{\partial \dot{x}^i\partial x^j}\dot{x}^j-\frac{\partial L}{\partial x^i}\right)+\frac{\partial^2 L}{\partial t\partial \dot{x}^i}=0.$$

Transvecting (2.3) by g^{ki}, the entries of the inverse of the matrix (g_{ij}), one obtains

$$(2.4)\quad \frac{d^2x^k}{d\tau^2}+2G^k(\tau,x,\dot{x})+N_0^k(\tau,x,\dot{x})=0,$$

where we have put

$$(2.5)\quad N_0^k(t,x,y)=\frac{1}{2}g^{ki}\frac{\partial^2 L}{\partial t\partial y^i},$$

$$(2.6)\quad G^k(t,x,y)=\frac{1}{4}g^{ki}\left(\frac{\partial^2 L}{\partial y^i\partial x^j}y^j-\frac{\partial L}{\partial x^i}\right)\ ,\ y^i=\dot{x}^i.$$

Since $\frac{\partial L}{\partial t}$ is invariant under the coordinate transformation (1.1), it follows from (1.2) that the functions N_0^k given by (2.5) are the components of a d-vector field on E and thus they may be taken as first local coefficients of a non-linear connection on E. We assert that the next n^2 coefficients can be taken as

$$(2.7)\quad N_j^i(t,x,y)=\frac{\partial G^i(t,x,y)}{\partial y^j}$$

such that $N_L=(N_0^i,N_j^i)$ given by (2.5) and (2.7) is a non-linear connection on E, completely determined by the time dependent Lagrangian L. This N_L will be called the canonical non-linear connection of the rheonomic Lagrange space RL^n.

It remains to show that (N_j^i) from (2.7) satisfies the transformation rule (1.9). Under the coordinates transformation $(t,x,y)\to(\tilde{t},\tilde{x},\tilde{y})$ given by (1.1), L remains invariant while its partial derivatives take the following change:

$$\frac{\partial L}{\partial x^i}=\frac{\partial L}{\partial \tilde{x}^k}\frac{\partial \tilde{x}^k}{\partial x^i}+\frac{\partial L}{\partial \tilde{y}^k}\frac{\partial^2 \tilde{x}^k}{\partial x^i \partial x^j}y^j,$$

$$\frac{\partial^2 L}{\partial y^i \partial x^k}=\frac{\partial^2 L}{\partial \tilde{y}^j \partial \tilde{x}^h}\frac{\partial \tilde{x}^j}{\partial x^i}\frac{\partial \tilde{x}^h}{\partial x^k}+\frac{\partial L}{\partial \tilde{y}^j}\frac{\partial^2 \tilde{x}^j}{\partial x^i \partial x^k}+2\tilde{g}_{jh}\frac{\partial \tilde{x}^j}{\partial x^i}\frac{\partial^2 \tilde{x}^h}{\partial x^k \partial x^s}y^s.$$

Multiplying the second equality by y^k, we introduce the result in the following form of (2.6):

$$4g_{ij}G^j=\frac{\partial^2 L}{\partial y^i \partial x^k}y^k-\frac{\partial L}{\partial x^i}$$

which thus becomes

$$4g_{ij}G^j=\left(\frac{\partial^2 L}{\partial \tilde{y}^j \partial \tilde{x}^k}\tilde{y}^k-\frac{\partial L}{\partial \tilde{x}^j}\right)\frac{\partial \tilde{x}^j}{\partial x^i}+2\tilde{g}_{jk}\frac{\partial \tilde{x}^j}{\partial x^i}\frac{\partial^2 \tilde{x}^k}{\partial x^r \partial x^s}y^r y^s,$$

because two terms cancel each other. Hence we have the change rule of G^i as follows:

$$4\tilde{g}_{jk}\frac{\partial \tilde{x}^j}{\partial x^i}\frac{\partial \tilde{x}^k}{\partial x^h}G^h=4\tilde{g}_{jk}\tilde{G}^k\frac{\partial \tilde{x}^j}{\partial x^i}+2\tilde{g}_{jk}\frac{\partial \tilde{x}^j}{\partial x^i}\frac{\partial^2 \tilde{x}^k}{\partial x^r \partial x^s}y^r y^s.$$

On account of rank $(g_{ij})=\text{rank}\left(\frac{\partial \tilde{x}^j}{\partial x^i}\right)=n$, the last equality is equivalent to

$$\frac{\partial \tilde{x}^k}{\partial x^i}G^i=\tilde{G}^k+\frac{1}{2}\frac{\partial^2 \tilde{x}^k}{\partial x^i \partial x^j}y^i y^j.$$

Differentiating the both sides of this equality with respect to y^j, one obtains the transformation rule of N_j^i from N_L in the form (1.9).

Remark 2.1. The non-linear connection N_L is without torsion.

Let us consider an 1-form on E: $\omega = \frac{\partial L}{\partial y^i} dx^i + (L - \frac{\partial L}{\partial y^i} y^i) dt$ and let $\theta = d\omega = [d(\frac{\partial L}{\partial y^i}) - \frac{\partial L}{\partial x^i} dt] \wedge (dx^i - y^i dt)$.

In general, a vector field X on E is said to be characteristic for θ if the inner product of θ by X vanishes i.e. $X \rfloor \theta = 0$.

A curve on E is said to be characteristic for θ if its tangent vector field is characteristic for θ.

Now let $c: R \to M$ be a curve on M. Then $\sigma(t) = (t, \dot{c}(t))$, $t \in R$, defines a curve on E and $\dot{\sigma}(t) = \frac{\partial}{\partial t} + \frac{dx^i}{dt}\frac{\partial}{\partial x^i} + \frac{d^2x^i}{dt^2}\frac{\partial}{\partial y^i}$. We have

Theorem 2.1. *If a curve c on M is an extremal curve for $\mathcal{L}$, then the curve $\sigma(t) = (t, \dot{c}(t))$ is a characteristic curve for θ.*

Proof. Putting $\varphi_i = d(\frac{\partial L}{\partial y^i}) - \frac{\partial L}{\partial x^i} dt$ and $\psi^i = dx^i - y^i dt$ we have $\theta = \varphi_i \wedge \psi^i$ and $(\dot{\sigma}(t) \rfloor \theta)(Y) = \varphi_i(\dot{\sigma}(t))\psi^i(Y) - \varphi_i(Y)\psi^i(\dot{\sigma}(t))$ for any $Y \in \chi(E)$. But $\psi^i(\dot{\sigma}(t)) = 0$ since along σ we have $y^i = \frac{dx^i}{dt}$ and $\varphi_i(\dot{\sigma}(t)) = 0$ in virtue of the Euler-Lagrange equations. Thus $\dot{\sigma}(t) \rfloor \theta = 0$.

§3. Non-linear Connections and Semisprays

The relationship between non-linear connections and semisprays which holds for tangent bundle (see §2, Ch. VII) can be partly recovered in this new framework.

The local form of $\dot{\sigma}(t)$ mentioned above suggests to set

Definition 3.1. *A vector field S on $E = R \times TM$ is said to be a semispray if in every local chart it takes the form*

$$(3.1)\quad S = \frac{\partial}{\partial t} + y^i \frac{\partial}{\partial x^i} + S^i(t,x,y)\frac{\partial}{\partial y^i},$$

where S^i are smooth real functions.

Remark 3.1. 1) If $\tilde{S}^i$ denotes the functions defining S in another local chart, then on overlap we must have

(3.2) $\tilde{S}^i = \frac{\partial \tilde{x}^i}{\partial x^k} S^k + \frac{\partial^2 \tilde{x}^i}{\partial x^j \partial x^k} y^j y^k.$

2) A set of smooth real functions (S^i) defined for every local chart satisfying (3.2) on overlaps defines via (3.1) a semispray on E.

A direct calculation leads to

Theorem 3.1. *Let S be a vector field on E. Then we have*
(i) S is a semispray if and only if $\mathring{J}S = 0,\ JS = \mathbb{C}.$
(ii) S is a semispray if and only if $\psi^i(S) = 0,\ dt(S) = 1,$ *where* $\psi^i = dx^i - y^i dt.$

We show there exist two remarkable semisprays defined by a given non-linear connection on E.

Theorem 3.2. *If N is a non-linear connection on E given by the local coefficients* $(N^i_0(t,x,y), N^i_k(t,x,y))$ *then* $S_0 = \frac{\partial}{\partial t} + y^i \frac{\partial}{\partial x^i} - N^i_k y^k \frac{\partial}{\partial y^i}$ *as well as* $S_1 = \frac{\partial}{\partial t} + y^i \frac{\partial}{\partial x^i} - (N^i_0 + N^i_k y^k) \frac{\partial}{\partial y^i}$ *are semisprays on E.*

Proof. One verifies with the help of (1.8) and (1.9) that $S^i_0 = -N^i_k y^k$ as well as $S^i_1 = -(N^i_0 + N^i_k y^k)$ satisfies (3.2).

Conversely, we have

Theorem 3.3. *If S is a semispray on E defined locally by* $(S^i(t,x,y))$ *then the pairs of functions* $(0, -\frac{1}{2}\frac{\partial S^i}{\partial y^j}),\ (\frac{\partial S^i}{\partial t}, -\frac{1}{2}\frac{\partial S^i}{\partial y^j})$ *define two non-linear connections on E.*

Proof. One verifies by a direct calculation that (1.8) and (1.9) hold good for the mentioned pairs of functions.

Now, if S_0 and $(0, -\frac{1}{2}\frac{\partial S^i}{\partial y^j})$ are taken into considerations and the notion of homogeneous semispray is introduced, the interplay between semisprays and non-linear connections can be much more refined and the results similar with those from the Theorems 2.1 and 2.2, Ch. VII could be obtained.

However, we leave this line and we shall use S_1 in order to get a new characterisation of the extremal curves of $\mathcal{L}$.

First, we set

Definition 3.2. *A curve* $t \to x^i(t)$, $t \in \mathbf{R}$, *on M is said to be a path for a non-linear connection N on E if the curve* $\sigma: t \to (t, x^i(t))$ *on E is horizontal with respect to N i.e. its tangent vector field belongs to the horizontal distribution.*

The dual basis of the adapted basis $(\delta_\alpha, \dot{\partial}_i)$ is $(dx^\alpha, \delta y^i)$, where $\delta y^i = dy^i + N^i_\alpha dx^\alpha$. It is obvious that the curve σ is horizontal if and only if $\delta y^i(\dot{\sigma}) = 0$, $i = 1,2,\ldots,n$. Writing down these equations one obtains

Theorem 3.4. *A curve* $t \to x^i(t)$, $t \in \mathbf{R}$, *on M is a path for a non-linear connection given by* (N^i_0, N^i_j) *if and only if*

$$(3.3)\quad \frac{d^2x^i}{dt^2} + N^i_j(t,x,x)\frac{dx^j}{dt} + N^i_0(t,x,x) = 0.$$

If we look at the local form of S_l we immediately obtain

Theorem 3.5. *The paths of a non-linear connection N coincide with the integral curves of the semispray* S_l *associated with N.*

Now, if we compare (3.3) with (2.4), we see that if the functions G^k given by (2.6) are 2-homogeneous with respect to (y^i), then by the Euler theorem we have $2G^k(t,x,y) = N^k_i(t,x,y)y^i$ and thus (2.4) reduces to (3.3).

Therefore, we have proved

Theorem 3.6. *If the functions* G^k *given by (2.6) are 2-homogeneous with respect to* (y^i), *then the extremal curves of* $\mathcal{L}$ *coincide with the paths of the non-linear connection* N_L.

We shall meet bellow a condition on L which implies that the functions G^k are 2-homogeneous with respect to (y^i).

§4. Normal d-Connections on R x TM

As we have seen, for rheonomic Lagrange space $RL^n = (M, L(t,x,y))$ a canonical non-linear connection N_L may be considered and thus the local basis $(\delta_\alpha, \dot{\partial}_i)$ adapted to the decomposition $T_uE = H_uE \oplus V_uE$, $u \in E$, is obtained. Considering also the dual basis $(dx^\alpha, \delta y^i)$, it comes out that $(1, \delta_\alpha, \dot{\partial}_i, dx^\alpha, \delta y^i)$ span a subalgebra $\mathcal{T}(E)$

of the tensorial algebra over E. This will be called the algebra of d-tensor fields on E. For instance, $t_j^i\delta_i\otimes\delta y^j$, $t_{0j}^i dt\otimes\delta_i\otimes\delta y^j$, $t_j^{0i}\delta_0\otimes\delta_i\otimes\delta y^j$ are elements of this subalgebra and t_j^i, t_{0j}^i, t_j^{0i} are their local components called also d-tensor fields on E. Notice that the local components of a d-tensor field on E change under a coordinate transformation on E like the components of a tensor field on M (ignoring the index 0) although they depend on t, x and y.

As in the general setting of vector bundles, we may consider d-connections on E. Recall these are linear connections which preserve by parallelism the horizontal and vertical distributions. On the other hand, for tangent bundle a particular class of d-connections, called normal d-connections, we have considered. Our bundle $\xi=(E,\pi,M)$ is very close to the tangent bundle, a fact which suggests to set

Definition 4.1. *A linear connection D and E said to be a normal d-connection, shortly an nd-connection, if in the adapted basis* $(\delta_0,\delta_i,\dot\partial_j)$ *it is given as follows*

$$
\begin{aligned}
&D_X\delta_0=0\ ,\ X\in\chi(E),\\
(4.1)\quad &D_{\delta_0}\delta_j=L_{j0}^i\delta_i\ ,\ D_{\delta_k}\delta_j=L_{jk}^i\delta_i\ ,\ D_{\dot\partial_k}\delta_j=C_{jk}^i\delta_i\ ,\\
&D_{\delta_0}\dot\partial_j=L_{j0}^i\dot\partial_i\ ,\ D_{\delta_k}\dot\partial_j=L_{jk}^i\dot\partial_i\ ,D_{\dot\partial_k}\dot\partial_j=C_{jk}^i\dot\partial_i\ .
\end{aligned}
$$

In order this definition be correct, $L_{jk}^i(t,x,y)$ must change under a coordinate transformation on E as the local coefficients of a linear connection on M and L_{j0}^i, C_{jk}^i must define d-tensor fields. Conversely, a set of functions $D\Gamma=(L_{j0}^i,L_{jk}^i,C_{jk}^i)$ which change as we have mentioned above under a coordinate transformation on E defines via (4.1) an nd-connection on E. Thus we have

Theorem 4.1. *To give an nd-connection on E is equivalent to giving* $D\Gamma=(L_{j0}^i,L_{jk}^i,C_{jk}^i)$.

Remark 4.1. It is obvious that the horizontal distribution $u\to H_uE$ splits into two distributions $H_uE=H_0\oplus H_1$, where H_0 is spanned by $\{\delta_0\}$ and H_1 is spanned by $\{\delta_i\}$. From (4.1) it comes out that H_0 is parallel with respect to an nd-connection $D\Gamma$ and H_1, V_uE are preserved by the operator D_X, $X\in\chi(E)$.

An nd-connection $D\Gamma$ on E induces an algorithm of h-covariant derivation as well as an algorithm of v-covariant derivation in the algebra of d-tensor fields on E. These will be denoted by $_{|\alpha}$ and $|k$, respectively. For instance

$$
\begin{aligned}
s^{\;i}_{j0\,|\,0} &= \delta_0 s^{\;i}_{j0} + L^{\;i}_{k0} s^{\;k}_{j0} - L^{\;k}_{j0} s^{\;i}_{k0}, \\
s^{\;i}_{j0\,|\,h} &= \delta_h s^{\;i}_{j0} + L^{\;i}_{kh} s^{\;k}_{j0} - L^{\;k}_{jh} s^{\;i}_{k0}, \\
s^{\;i}_{j0}\,\big|_h &= \dot\partial_h s^{\;i}_{j0} + C^{\;i}_{kh} s^{\;k}_{j0} - C^{\;k}_{jh} s^{\;i}_{k0}.
\end{aligned}
$$

These rules are general ones.

Particularising the formulae (3.7), Ch.III one finds that the torsions of an nd-connection $D\Gamma$ are the following ones

$$
\begin{aligned}
&T^{i}{}_{jk} = L^{i}{}_{jk} - L^{i}{}_{kj},\ L^{i}{}_{k0},\ R^{i}{}_{\alpha\beta} = \delta_\beta N^i_\alpha - \delta_\alpha N^i_\beta,\ C^{i}{}_{jk}, \\
&P^{i}{}_{0j} = \dot\partial_j N^i_0 - L^{i}{}_{j0},\ P^{i}{}_{kj} = \dot\partial_j N^i_k - L^{i}{}_{jk},\ S^{i}{}_{jk} = C^i_{jk} - C^i_{kj}.
\end{aligned}
\tag{4.2}
$$

Analyzing the formulae (3.13), Ch.III it comes out that the curvatures of an nd-connection are the following ones

$$
\begin{aligned}
&R_j{}^i{}_{\alpha\beta} = \delta_\beta L^{\;i}_{j\alpha} - \delta_\alpha L^{\;i}_{j\beta} + L^{\;h}_{j\alpha} L^{\;i}_{h\beta} - L^{\;h}_{j\beta} L^{\;i}_{h\alpha} + C^{\;i}_{jh} R^{h}{}_{\alpha\beta}, \\
&P_j{}^i{}_{0k} = \dot\partial_k L^{\;i}_{j0} - C^{\;i}_{jk|0} + C^{\;i}_{jh} P^{h}{}_{0k}, \\
&P_j{}^i{}_{hk} = \dot\partial_k L^{\;i}_{jh} - C^{\;i}_{jh|k} + C^{\;i}_{js} P^{s}{}_{hk}, \\
&S_j{}^i{}_{hk} = \dot\partial_k C^{\;i}_{jh} - \dot\partial_h C^{\;i}_{jk} + C^{\;i}_{jh} C^{\;i}_{sk} - C^{\;s}_{jk} C^{\;i}_{sh}.
\end{aligned}
\tag{4.3}
$$

We shall say that $T^{i}{}_{jk}$ is h(hh)-torsion of $D\Gamma$ and $S^{i}{}_{jk}$ is v(vv)-torsion of $D\Gamma$. The torsions and curvatures of $D\Gamma$ satisfy a number of Bianchi identities which could be established by particularizing the formulae (3.17)-(3.20), Ch.III. We do not write down them.

§5. Metrical Normal d-Connections on R x TM

Let us come back to the functions $g_{ij}(t,x,y)$ given by (2.1). As it is easy to check, these functions transform under (1.1) as the components of a tensor field on M. Thus they define a d-tensor field of type (0,2) which, moreover, is symmetric and nondegenerate $(\det(g_{ij}) \neq 0)$. Assume the quadratic form $g_{ij}(t,x,y)\xi^i\xi^j$ has a constant signature. We shall say that $g_{ij}(t,x,y)$ is the metric tensor or the fundamental tensor of the rheonomic Lagrange space $RL^n = (M, L(t,x,y))$.

As usual we shall denote by $(g^{ij}(t,x,y))$ the inverse of the matrix $(g_{ij}(t,x,y))$. The functions $g^{jk}(t,x,y)$ define a d-tensor field of type (0,2) on E.

Definition 5.1. *An nd-connection* $D\Gamma = (L^i_{j0}, L^i_{jk}, C^i_{jk})$ *is said to be metrical if*

(5.1) $g_{ij|0}=0\ ,\ g_{ij|k}=0\ ,\ g_{ij}|_k=0\ .$

Concerning the existence of metrical nd-connections we have

Theorem 5.1. *Let* $RL^n=(M,L(t,x,y))$ *be a rheonomic Lagrange space. If* $T^i{}_{jk}$ *and* $S^i{}_{jk}$ *are two arbitrary skew-symmetric d-tensor fields on* $E=R\times TM$*, then there exists a set of metrical nd-connections on E such that each of them has* $T^i{}_{jk}$ *and* $S^i{}_{jk}$ *as h(hh)- and v(vv)- torsions, respectively. The local coefficients of a connection from this set are given as follows:*

$$L^k_{i0}=\frac{1}{2}g^{kh}\delta_0 g_{ih}+O^{jk}_{ih}X^h_{j0}\ ,$$

(5.2)
$$L^k_{ij}=\frac{1}{2}g^{kh}(\delta_i g_{hj}+\delta_j g_{ih}-\delta_h g_{ij}+g_{is}T^s{}_{jk}+g_{js}T^s{}_{ih}+g_{hs}T^s{}_{ij})\ ,$$

$$C^k_{ij}=\frac{1}{2}g^{kh}(\dot\partial_i g_{hj}+\dot\partial_j g_{ih}-\dot\partial_h g_{ij}+g_{is}S^s{}_{jk}+g_{js}S^s{}_{ih}+g_{hs}S^s{}_{ij})\ ,$$

where X^h_{j0} *is an arbitrary d-tensor field on E and* O^{jk}_{ih} *denotes the Obata operator*

(5.3) $O^{jk}_{ih}=\frac{1}{2}(\delta^j_i\delta^k_h-g_{ih}g^{jk}).$

Proof. The condition $g_{ij|k}=0$ is equivalent to the equality $\delta_k g_{ij}=L^h_{ik}g_{hj}+L^h_{jk}g_{ih}$. Applying the Christoffel process and putting $T^i{}_{jk}=L^i_{jk}-L^i_{kj}$ one obtains L^k_{ij} as in (5.2). We may proceed analogously in order to obtain C^k_{ij} as in (5.2), using $g_{ij}|_k=0$ and setting $C^i_{jk}-C^i_{kj}=S^i{}_{jk}$. Next it is easy to check that $L^k_{i0}=\frac{1}{2}g^{kh}\delta_0 g_{ih}$ are solutions of the equations $g_{ij|0}=\delta_0 g_{ij}-L^k_{i0}g_{kj}-L^k_{j0}g_{ik}=0$ in the unknowns L^k_{i0}.

Now if L^k_{i0} are any solutions of the equations $g_{ij|0}=0$, then $B^k_{i0}=L^k_{i0}-\frac{1}{2}g^{kh}\delta_0 g_{ih}$ verify the equations $g_{ki}B^k_{j0}+g_{jk}B^k_{i0}=0$. The general solutions of these equations are $B^k_{i0}=O^{jk}_{ih}X^h_{j0}$, where X^h_{j0} is an arbitrary d-tensor field. Thus L^k_{i0} have the form given in (5.2).

Taking $X^i_{j0}=0$ in (5.2) one obtains

Corollary 5.1. *Let $RL^n=(M,L(t,x,y))$ be a rheonomic Lagrange space and $T^i{}_{jk}$, $S^i{}_{jk}$ be two arbitrary skew-symmetric d-tensor fields on E. Then there exists an unique metrical d-connection $D\Gamma=(\frac{1}{2}g^{kh}\delta_0 g_{ih}, L^k_{ij}, C^k_{ij})$, whose h(hh)-torsion is T^i_{jk} and v(vv)-torsion is S^i_{jk}. The coefficients $L^i{}_{jk}$ and $C^i{}_{jk}$ are given by (5.2).*

Proof. The uniqueness of $L^i{}_{jk}$ and $C^i{}_{jk}$ from (5.2) follows by contradiction.

In particular, taking $T^i{}_{jk}=S^i{}_{jk}=0$ one obtains

Corollary 5.2. *Let $RL^n=(M,L(t,x,y))$ be a rheonomic Lagrange space. Then there exists a set of metrical nd-connections such that each of them has the vanishing h(hh)- and v(vv)- torsion. The local coefficients of any connection from this set are given by*

$$
(5.4)\quad
\begin{aligned}
L^k_{i0} &= \frac{1}{2}g^{kh}\delta_0 g_{hi} + O^{jk}_{ih}X^h_{j0}\ ,\\
L^k_{ij} &= \frac{1}{2}g^{kh}(\delta_i g_{hj} + \delta_j g_{hi} - \delta_h g_{ij})\ ,\\
C^k_{ij} &= \frac{1}{2}g^{kh}(\dot{\partial}_i g_{hj} + \dot{\partial}_j g_{hi} - \dot{\partial}_h g_{ij})\ ,
\end{aligned}
$$

where X^h_{j0} is an arbitrary d-tensor field on E.

Definition 5.2. *The metrical nd-connection whose local coefficients are given by (5.4) with $X^h_{j0}=0$ will be called the canonical metrical nd-connection on E. It will be denoted by $\overset{c}{D}\Gamma$.*

The nd-connection $\overset{c}{D}\Gamma$ is completely determined by the time dependent Lagrangian L. Thus $\overset{c}{D}\Gamma$ is similar to the connection $C\Gamma$ in Lagrange spaces (see Theorem 3.1, Ch. IX).

The h- and v-covariant of $\mathbb{C}=y^i\dot{\partial}_i$ with respect to $\overset{c}{D}\Gamma$ lead us to introduce the following deflexion tensor fields for $\overset{c}{D}\Gamma$:

(5.5) $D^i_0 = y^i_{|0}$, $D^i_k = y^i_{|k}$, $d^i_k = y^i|_k$.

Setting $D_{k0} = g_{ki} y^i_{|0}$, $D_{kj} = g_{ki} y^i_{|j}$, $d_{kj} = g_{ki} y^i|_j$, using the Ricci identities found in §3, Ch. III, and taking into account $\overset{c}{D\Gamma}$ is metrical, one gets

$$
\begin{aligned}
&D_{i0|k} - D_{ik|0} = R_{ji0k} y^j - d_{ih} R^h{}_{0k},\\
&D_{hi|k} - D_{hk|i} = R_{jhik} y^j - d_{hs} R^s{}_{ik},\\
(5.6)\quad &D_{h0}|_k - d_{hk|0} = P_{jh0k} y^j - d_{kj} P^j{}_{0k},\\
&D_{hi}|_k - d_{hk|i} = P_{jhik} y^j - D_{hj} C^j_{ik} - d_{hj} P^j{}_{ik},\\
&d_{ik}|_h - d_{ih}|_k = S_{jikh} y^j.
\end{aligned}
$$

We may also introduce the h- and v- electromagnetic tensor fields, respectively,

(5.7) $F_{ij} = \frac{1}{2}(D_{ij} - D_{ji})$, $f_{ij} = \frac{1}{2}(d_{ij} - d_{ji})$.

As it is easy to check, $f_{ij} = 0$.

As to regard F_{ij}, a direct calculation leads to

Theorem 5.2. *The tensor field F_{ij} given by (5.7) satisfies the Maxwell equations:*

$$
(5.8)\quad
\begin{aligned}
&F_{ij|k} + F_{jk|i} + F_{ki|j} = \sum_{(ijk)} R^h{}_{jk} C_{ish} y^s,\\
&F_{ij}|_k + F_{jk}|_i + F_{ki}|_j = 0.
\end{aligned}
$$

§6. Rheonomic Finsler Spaces

Now we shall deal with certain particular rheonomic Lagrange spaces which are rheonomic counterpart of the Finsler spaces.

Definition 6.1. *We shall say that a pair $RF^n = (M, F(t,x,y))$ is a rheonomic Finsler space if the following properties are verified:*

1. F is a real function on $R \times TM$, of class C^∞ on $\tilde{E} = R \times \widetilde{TM}$ and only continuous in the points $(t,x,0)$,

2. F is 1-homogeneous with respect to $y = (y^i)$,

3. The matrix with the elements

(6.1) $g_{ij}(t,x,y) = \frac{1}{2}\dot{\partial}_i\dot{\partial}_j F^2$, *has rank n on E.*

4. The quadratic form $g_{ij}\xi^i\xi^j$ $(\xi = (\xi^i) \in \mathbf{R}^n)$ *has a constant signature on E.*

We notice that the functions $g_{ij}(t,x,y)$ given by (6.1) defines a d-tensor field of type (0,2), symmetric and nondegenerate. It will be called the metric tensor or fundamental tensor of the space RF^n.

Theorem 6.1. *Every RF^n space is an RL^n space.*

Proof. Taking $L(t,x,y)=F^2(t,x,y)$ one sees by the previous definition that L is a regular Lagrangian and thus (M,L) is an RL^n space.

Notice that the Lagrangian F^2 is 2-homogeneous with respect to $y=(y^i)$. Taking into account this property of F^2, by the Euler theorem on homogeneous functions one gets

Theorem 6.2. *Let $RF^n = (M, F(t,x,y))$ be a rheonomic Finsler space. Then we have*

(6.2) $F^2(t,x,y) = g_{ij}(t,x,y)y^iy^j$.

Thus we may call F^2 the absolute energy of the space RF^n.

The following d-tensor fields

(6.3) $C_{ijk} = \frac{1}{2}\dot{\partial}_i g_{jk}$, $C_{ij0} = \frac{1}{2}\partial_0 g_{ij}$, $\partial_0 := \frac{\partial}{\partial t}$,

will be called the Cartan tensor fields of RF^n. These d-tensor fields are totally symmetric and

(6.4) $\dot{\partial}_k C_{ij0} = \partial_0 C_{ijk}$.

We set $\partial_0 C_{ijk} = C_{0ijk}$ and notice that the d-tensor field C_{0ijk} is also totally symmetric. Warning here the index "0" does not mean a contraction by y^i as till now.

The homogeneity of F implies

Theorem 6.3. *The following equations hold good*

$$(6.5)\quad \begin{aligned} y^i C_{ijk} &= y^i C_{jik} = y^i C_{jki} = 0, \\ y^i C_{0ijk} &= y^i C_{0jik} = y^i C_{0jki} = 0. \end{aligned}$$

In order to apply to RF^n spaces the theory of RL^n spaces we set

Definition 6.2. *The canonical non-linear connection (metrical nd-connection) of the space* $RL^n = (M, F^2(t,x,y))$ *will be called the Cartan non-linear connection (resp. Cartan metrical connection) of the space* $RF^n = (M, F(t,x,y))$.

Introducing the usual Christoffel symbols

$$\gamma^i_{jk} = \frac{1}{2} g^{ih}\left(\frac{\partial g_{hk}}{\partial x^j} + \frac{\partial g_{jh}}{\partial x^k} - \frac{\partial g_{jk}}{\partial x^h}\right)$$

we may state

Theorem 6.4. *The local coefficients of the Cartan non-linear connection are as follows*

$$(6.6)\quad N^i_0 = \frac{1}{2} g^{ik} \partial_0 \dot{\partial}_k F^2, \quad N^i_k = \dot{\partial}_k G^i,$$

where

$$(6.7)\quad G^i = \frac{1}{2} \gamma^i_{jk} y^j y^k.$$

Proof. By a direct calculation one finds that G^k given by (2.6) with $L = F^2$, takes the form (6.7).

Theorem 6.5. *The Cartan metrical connection* $\overset{c}{F}\Gamma = (\overset{c}{F}{}^i_{j0}, \overset{c}{F}{}^i_{jk}, \overset{c}{C}{}^i_{jk})$ *of the space* RF^n *has the coefficients*

$$(6.8)\quad \begin{aligned} \overset{c}{F}{}^i_{j0} &= \frac{1}{2} g^{ik} \delta_0 g_{kj}, \\ \overset{c}{F}{}^i_{jk} &= \frac{1}{2} g^{ih}(\delta_j g_{hk} + \delta_k g_{hj} - \delta_h g_{jk}), \\ \overset{c}{C}{}^i_{jk} &= g^{ih} C_{hjk}, \end{aligned}$$

where δ_j is constructed by the help of (6.6).

By a direct calculation one finds

(6.9) $N_0^i y^i = \partial_0 F^2$,

where $y^i = g_{ih} y^h$.

§7. Remarkable Time Dependent Lagrangians

Now we shall consider a class of rheonomic Lagrange spaces whose Lagrangians are counterparts of Lagrangians from electrodynamics studied in Ch. IX.

Coming back again to the function $g_{ij}(t,x,y)$ given by (2.1) we state

Theorem 7.1. *If the functions $g_{ij}(t,x,y)$ in (2.1) do not depend on t and are 0-homogeneous with respect to (y^i), then L has the general form*

(7.1) $L(t,x,y) = \overset{\circ}{L}(x,y) + A_i(t,x)y^i + U(t,x)$,

where $\overset{\circ}{L}$ is a regular time independent Lagrangian which is 2-homogeneous with respect to (y^i), A_i is a d-covector field and U is a real function on E.

Proof. Let us put $\overset{\circ}{L}(x,y) = g_{ij}(x,y)y^i y^j$. Then by 0-homogeneity of g_{ij} it follows $\dot{\partial}_i\dot{\partial}_j(L - \overset{\circ}{L}) = 0$, which implies (7.1).

We may say that in (7.1), $\overset{\circ}{L}$ comes out from gravitational potentials (g_{ij}), A_i are electromagnetic potentials and U is a potential function.

We notice the following particular form of the Lagrangian (7.1):

(7.2) $L'(t,x,y) = \gamma_{ij}(x)y^i y^j + \alpha_i(t,x)y^i + U(t,x)$,

where (γ_{ij}) is a Riemannian metric.

Now we shall apply our previous theory to the rheonomic Lagrange space (M,L), where L is given by (7.1) as well as by L′ from (7.2).

We call this space a RDL space and we stress that $\overset{\circ}{L}$ in (7.1) is 2-homogeneous with respect to (y^i).

The metric tensor field of a RDL space is

$$(7.3)\quad \overset{\circ}{g}_{ij}(x,y)=\frac{1}{2}\dot{\partial}_i\dot{\partial}_j\overset{\circ}{L}(x,y).$$

As to the coefficients of the non-linear connection derived from L we have

$$(7.4)\quad N_0^k(t,x,y)=\frac{1}{2}\overset{\circ}{g}{}^{ki}\partial_0 A_i,$$

$$(7.5)\quad N_h^k(t,x,y)=\overset{\circ}{N}{}_h^k(x,y)-B_h^k(t,x,y),$$

where $\overset{\circ}{N}{}_h^k$ are the coefficients of the non-linear connection derived from $\overset{\circ}{L}$ and

$$(7.6)\quad B_h^k(t,x,y)=\overset{\circ}{g}{}^{ks}\overset{c}{C}{}^i_{sh}E_{ij}y^j+\frac{1}{2}\overset{\circ}{g}{}^{ki}E_{hi}-\frac{1}{2}g^{ks}\overset{c}{C}{}^i_{sh}\frac{\partial U}{\partial x^i},$$

$$(7.7)\quad E_{ij}=\frac{1}{2}\left(\frac{\partial A_i}{\partial x^j}-\frac{\partial A_j}{\partial x^i}\right).$$

The local coefficients $(\tilde{L}^h_{i0},\tilde{L}^h_{ik},\tilde{C}^h_{ik})$ of the cannonical metrical nd-connection of a RDL space are as follows

$$(7.8)\quad \begin{aligned} \tilde{L}^h_{i0}&=\frac{1}{2}\overset{\circ}{g}{}^{ks}\overset{c}{C}{}^h_{si}\partial_0 A_k,\\ \tilde{L}^h_{ik}&=\frac{1}{2}\overset{\circ}{g}{}^{hs}(\delta_i\overset{\circ}{g}_{sk}+\delta_k\overset{\circ}{g}_{is}-\delta_s\overset{\circ}{g}_{ik}),\\ \tilde{C}^h_{ik}&=\overset{c}{C}{}^h_{ik}, \end{aligned}$$

where δ_i are constructed with N_h^k from (7.5). If we replace L from (7.1) with L' from (7.2) the above formulae become much simpler. Thus $\overset{\circ}{g}_{ij}(x,y)=\gamma_{ij}(x)$ and we have

$$(7.9)\quad \begin{aligned} N_0^k(t,x,y) &= \frac{1}{2}\gamma^{ki}\partial_0 A_i\,, \\ N_h^k(t,x,y) &= \gamma_{hj}^k(x)y^j - \frac{1}{2}E_h^k\,, \end{aligned}$$

where $E_h^k = \gamma^{ki}E_{hi}$ and γ_{hj}^k are the Christoffel symbols constructed with γ_{ij}.

The canonical metrical nd-connection reduces to $(0, \gamma_{ik}^h, 0)$. Its covariant deflection tensor fields are $D_{ij} = \frac{1}{2}E_{ji}$, $d_{ij} = \gamma_{ji}$, and thus $2F_{ij} = E_{ji}$, a equality which supports the term of h-electromagnetic tensor field for F_{ij} when (7.7) is taking into account.

The following time dependent Lagrangian

$$(7.10)\quad L(t,x,y) = a_{ij}(t,x)y^i y^j + 2\alpha_i(t,x)y^i + U(t,x),$$

where a_{ij} is a time dependent Riemannian metric and α_i is a covector field was considered by N. Oshima ([122]) in treating of some problems of dynamics.

We shall apply our theory to the Lagrangian (7.10). The fundamental metric tensor reduces to $a_{ij}(t,x)$. If we denote by $a_{jk}^i(t,x)$ the Christoffel symbols constructed from $a_{ij}(t,x)$, it comes out that the non-linear connection associated to this L is

$$(7.11)\quad \begin{aligned} N_0^k(t,x,y) &= a^{kh}\partial_0 g_{hi}y^i + \partial_0\alpha(t,x) \\ N_h^k(t,x,y) &= a_{hi}^k(t,x)y^i - a^{kj}A_{hj} \end{aligned}$$

where

$$(7.12)\quad A_{jh} = \frac{1}{2}\left(\frac{\partial\alpha_j}{\partial x^h} - \frac{\partial\alpha_h}{\partial x^j}\right).$$

The canonical metrical nd-connection is given by $(0, a_{jk}^i(t,x), 0)$ and its covariant deflexion tensors are as follows: $D_{ij} = A_{ij}$, $d_{ij} = \gamma_{ij}$. It comes out that $F_{ij} = A_{ij}$ and the term of h-electromagnetic term for F_{ij} is supported again if we look at (7.12) and think α_i as electromagnetic potentials. We notice that for the Lagrangians (7.2) and (7.10) the Maxwell equations (5.8) reduce to classical ones.

§8. Metrical Almost Contact Model of a Rheonomic Lagrange Space

Let $RL^n = (M, L(t,x,y))$ be a rheonomic Lagrange space and N_L its non-linear connection. Thus we have a decomposition of the tangent bundle TE over $E = R\times TM$ as a sum $TE = HE \oplus VE$. Let $(\delta_0, \delta_i, \dot{\partial}_i)$ be the local frame adapted to this decomposition and $(dt, dx^i, \delta y^i)$ is dual.

Let us consider the $\mathcal{F}(E)$-linear mapping F given by

$$(8.1)\quad F(\delta_0) = 0\ ,\ F(\delta_i) = -\dot{\partial}_i\ ,\ F(\dot{\partial}_i) = \delta_i .$$

It follows F in an well-defined tensor field of type (1,1) on E. It is obvious that rank F = 2n and an easy calculation gives $F^3 + F = 0$. Thus F defines an f(3,1)-structure on E, (see [92]).

Theorem 8.1. *Let* $(M, L(t,x,y))$ *be a rheonomic Lagrange space. Then the manifold* $E = R\times TM$ *carries an almost contact structure* (F, δ_0, dt).

Proof. We have $dt(\delta_0) = 1$ and by (8.1) we find $F^2(\delta_i) = -\delta_i\ ,\ F^2(\dot{\partial}_i) = -\dot{\partial}_i$. Thus it follows that $F^2 = -I + \delta_0 \otimes dt$.

Remark 8.1. The almost contact structure (F, δ_0, dt) is just that induced by the f(3,1)-structure F according to a general method, described for instance in [92].

The torsion tensor field of the almost contact structure (F, δ_0, dt) (cf. [138]) reduces to the Nijenhuis tensor field N_F of F. Thus the almost contact structure(F, δ_0, dt) is normal if and only if $N_F = 0$.

Evaluating N_F in the frame $(\delta_0, \delta_i, \dot{\partial}_i)$ one gets

Theorem 8.2. *The almost contact structure* (F, δ_0, dt) *is normal if and only if*

1. N_L *is without curvature,*
2. $\dot{\partial}_i N_0^k = 0$.

Now we shall consider the following metrical structure on E, induced by $g_{ij}(t,x,y)$.

$$(8.2)\quad G = dt \otimes dt + g_{ij} dx^i \otimes dx^j + g_{ij} \delta y^i \otimes \delta y^j .$$

Notice that if (g_{ij}) is positive definite then G is a Riemannian metric on E.

Remark 8.2. The horizontal and vertical distributions are orthogonal with respect to G.

Theorem 8.3. *The metrical structure G satisfies the following equations*

$$(8.3)\quad G(FX,FY)=G(X,Y)-dt(X)dt(Y)\ ,\ dt(X)=G(\delta_0,X)\ ,\ X,Y\in\chi(E).$$

Proof. A direct calculation in the adapted local frame $(\delta_0,\delta_i,\dot{\partial}_i)$.

Definition 8.1. *The manifold $E=R\times TM$ fibred over $R\times M$ by $(t,v)\to(t,\tau(v))$, endowed with the metrical almost contact structure (F,δ_0,dt,G) will be called the metrical almost contact model of the rheonomic Lagrange space $RL^n=(M,L(t,x,y))$.*

The main facts from the geometry of a rheonomic Lagrange space may be stated in its metrical almost contact model.

For instance, let us look for what corresponds in this model to the metrical nd-connections introduced in §3.

Definition 8.2. *A linear connection D on E said to be an almost contact connection of Lagrange type, shortly an acL-connection, if*

1. D preserves by parallelism the vertical distribution,
2. $D_XG=0\ ,\ D_XF=0\ ,\ D_X\delta_0=0\ ,X\in\chi(E)$.

It is clear that an acL-connection D preserves also the horizontal distribution and $D_Xdt=0\ ,\ X\in\chi(E)$. Moreover, we have

Theorem 8.4. *To give a metrical nd-connection for a RL^n space is equivalent to giving an acL-connection on E.*

Proof. If a metrical nd-connection is given, then (3.1) and (4.1) show that D satisfies 1. and 2. from the Definition 8.2. Thus D is an acL-connection. Conversely, if D is an acL-connection, writing down the conditions 1. and 2. from the Definition 8.2 in the frame $(\delta_0,\delta_i,\dot{\partial}_i)$ one finds D is given by (3.1) with the constraints (4.1). Thus D is an nd-connection.

Thus we have

Theorem 8.5. *There exists a set of almost contact connections of Lagrange type on E having h(hh)- and v(vv)- torsions given a priori.*

§9. Generalized Rheonomic Lagrange Spaces

As we have done before in order to obtain a notion of generalized Lagrange space, we may ignore that the metric tensor $g_{ij}(t,x,y)$ of a rheonomic Lagrange space is provided by a regular time dependent Lagrangian and to study the geometry of a pair $(M, g_{ij}(t,x,y))$ i.e. of a generalized rheonomic Lagrange space.

Definition 9.1. *A pair $(M, g_{ij}(t,x,y))$ which consists of a smooth n-dimensional real manifold M and a d-tensor field $g_{ij}(t,x,y)$ is said to be a generalized rheonomic Lagrange space, shortly a GRL^n space, if $g_{ij}(t,x,y)$ is symmetric, nondegenerate and the quadratic form $g_{ij}(t,x,y)\xi^i\xi^j$, $(\xi^i)\in R^n$, has constant signature.*

Now we sketch only the theory of GRL^n spaces, since it is similar to the theory of GL^n spaces.

An important problem here is to find conditions in which the metric tensor $g_{ij}(t,x,y)$ of a GRL^n space determines a non-linear connection.

Let us consider the absolute energy of a GRL^n space

$$(9.1)\quad \varepsilon = g_{ij}(t,x,y)y^i y^j.$$

Definition 9.2. *We shall say that a GRL^n space $(M, g_{ij}(t,x,y))$ is with weakly regular metric if its absolute energy ε is a regular time dependent Lagrangian.*

Assuming that a GRL^n space is with weakly regular metric, then the metric tensor $g^*_{ij} = \frac{1}{2}\dot{\partial}_i\dot{\partial}_j\varepsilon$ admits a reciprocal $\overset{*}{g}{}^{ij}$. The canonical non-linear connection N_ε determined by ε has the coefficients

$$(9.2)\quad N_j^{*i} = \dot{\partial}_j G^{*i}\ , \quad N_0^{*k} = \frac{1}{2} g^{*ki}\partial_0\dot{\partial}_i\varepsilon,$$

where we have put

$$(9.3)\quad G^{*i}(t,x,y) = \frac{1}{4}\overset{*}{g}{}^{ik}\left(\frac{\partial^2\varepsilon}{\partial y^k \partial x^j}y^j - \frac{\partial\varepsilon}{\partial x^k}\right).$$

Analyzing the structure of these coefficients we obtain

Theorem 9.1. *The non-linear connection whose coefficients are given by (9.2) and (9.3) is completely determined by the metric tensor $g_{ij}(t,x,y)$ of the GRL^n space $(M, g_{ij}(t,x,y))$ with weakly regular metric.*

Using the non-linear connection N_ε or an arbitrary one if the space GRL^n is not with weakly regular metric we may consider nd-connections on E as in the Definition 3.1.

Next we introduce the metrical nd-connections.

Definition 9.3. *A nd-connection in a GRL^n space is said to be metrical if*

$$(9.4)\quad g_{ij|0}=0\ ,\ g_{ij|k}=0\ ,\ g_{ij}|_k=0,$$

hold good.

Concerning the existence of metrical nd-connections we have

Theorem 9.2. *Let $(M, g_{ij}(t,x,y))$ be a GRL^n space with a fixed non-linear connection N. If $T^i{}_{jk}$ and $S^i{}_{jk}$ are two arbitrary skew-symmetric d-tensor fields, then there exists a set of metrical nd-connections such that each of them has T^i_{jk} and S^i_{jk} as h(hh)- and v(vv)- torsion, respectively.*

The local coefficients of any nd-connection from this set are given by (5.2).

Proof. One repeats the proof of the Theorem 5.1 with some minor ajustements.

Corollary 9.1. *Let $(M, g_{ij}(t,x,y))$ be a GRL^n space endowed with a non-linear connection and $T^i{}_{jk}$, $S^i{}_{jk}$ be two arbitrary skew-symmetric d-tensor fields. There exists an unique metrical nd-connection $(\frac{1}{2}g^{kh}\delta_0 g_{ih}, L^k_{ij}, C^k_{ij})$ whose h(hh)-torsion is $T^i{}_{jk}$ and v(vv)-torsion is $S^i{}_{jk}$. The coefficients $L^k{}_{ij}$ and $C^k{}_{ij}$ are given by the last two lines in (5.2).*

Taking $T^i{}_{jk} = S^i{}_{jk} = 0$ one obtains

Corollary 9.2. *Let $(M, g_{ij}(t,x,y))$ be a GRL^n space endowed with a non-linear connection. There exists an unique metrical nd-connection $\overset{c}{D}\Gamma$ whose h(hh)- and v(vv)-*

torsions vanish and $\overset{c}{L}{}^{k}_{i0} = \frac{1}{2} g^{kh} \delta_0 g_{ih}$. *The other coefficients are as follows*

$$
(9.5) \qquad
\begin{aligned}
\overset{c}{L}{}^{k}_{ij} &= \frac{1}{2} g^{kh} (\delta_i g_{hj} + \delta_j g_{ih} - \delta_h g_{ij}) \,, \\
\overset{c}{C}{}^{k}_{ij} &= \frac{1}{2} g^{kh} (\dot{\partial}_i g_{hj} + \dot{\partial}_j g_{ih} - \dot{\partial}_h g_{ij}) \,.
\end{aligned}
$$

The metrical nd-connection $\overset{c}{D}\Gamma$ could be considered as canonical for a GRL^n space.

Bibliography

1. Abraham R., Marsden J., *Foundations of Mechanics*, Benjamin, New York, 1978.
2. Aikou T., Differential Geometry of Finslerian Vector Bundles, Ph. D. Thesis, Iaşi, Romania, 1991.
3. Aikou T., Hashiguchi M., On Generalized Berwald Connections, Rep. Fac. Sci. Kagoshima Univ., 17 (1984), 9-13.
4. Akbar-Zadeh H., Les espaces de Finsler et certain de leurs generalisations, Ann. Sci. Ecole Norm. Sup., (3), 80 (1963), 1-79.
5. Albu I.D., Opriş D., Finsler Hyperconnections, An. şt. Univ. "Al.I. Cuza" Iaşi, s I a, Mat. 30 (1984), 5-10.
6. Amici O., Casciario B., An Example of Lagrangian Theory of Relativity Satisfying the Conservation laws, Proc. Nat. Sem. Finsler and Lagrange Spaces, University of Braşov, Romania, 1986, 37-42.
7. Anastasiei M., Some Existence Theorems in Finsler Geometry, An. şt. Univ. "Al.I. Cuza" Iaşi, 29, s I a (1983), 79-84.
8. Anastasiei M., On Metrical Finsler Connections, Proc. Nat. Sem. Finsler and Lagrange Spaces, Braşov, Romania, 1984, 19-28.
9. Anastasiei M., On Subspaces in Finsler Spaces, Proc. Romanian-Japanese Coll. on Finsler Geom., Braşov, Romania, 1984, 17-30.
10. Anastasiei M., Vector Bundles. Einstein Equations, An. şt. Univ. Iaşi, 32, s I a Mat., 1986, 17-24.
11. Anastasiei M., Conservation Laws in the (V,H)-Bundle Model of Relativity, Tensor, N.S., 46 (1987), 323-328.
12. Anastasiei M., Kawaguchi H., A Geometrical Theory of Time Dependent Lagrangians I, Non-linear Connections, Tensor, N.S., 48 (1989), 273-282.
13. Anastasiei M., Kawaguchi H., A Geometrical Theory of Time Dependent Lagrangians II, M-connections, Tensor, N.S., 48 (1989), 283-293.
14. Anastasiei M., Kawaguchi H., A Geometrical Theory of Time Dependent Lagrangians III, Applications, Tensor, N.S., 49 (1990), 296-304.
15. Antonelli P.L., On y-Berwald Connections and Hutchinson's Ecology of Social Interactions, Tensor, N. S., 51 (1992), 223-234.
16. Antonelli P.L., Volterra-Hamilton Systems of Finsler Type with Applications to Coral Reef Ecology, Acta Cien Venezolana, 6 (1988), 231-246.
17. Antonelli P.L., Finsler Volterra-Hamilton Systems in Ecology, Tensor, N.S., 50 (1991), 22-31.
18. Aringazin A.K., Asanov G.S., Problems of Finslerian Theory of Gauge Fields and Gravitation, Rep. Math. Phys., 25 (2), 1988, 35-93.
19. Asanov G.S., Finslerian Fibred Extension of Gauge Fields Theory: Lie-invariance in Fibres and Effective Lagrangian, Rep. Math. Phys., 26 (3), 1988, 367-399.
20. Asanov G.S., *Fibred Generalization of the Gauge Field Theory. Finslerian and Jet Gauge Fields*, Moskow Univ., 1989.
21. Asanov G. S., *Finsler Geometry, Relativity and Gauge Theories*, D. Reidel, Dordrecht, 1985.

22. Atanasiu Gh., Metrical Kawaguchi Connections on the Total Space of a Vector Bundle and Some Special Subordinated Cases, Tensor, N.S., 46 (1987), 248-264.
23. Atanasiu Gh., Hashiguchi M., Miron R., Supergeneralized Finsler Spaces, Rep. Fac. Sci. Kagoshima Univ., 18 (1985), 19-35.
24. Barbilian D., Natural Finslerian J-metrics (in Romanian), Stud. Cerc. Mat., 11 (1960), 7-47.
25. Barthel W., Nichtlineare Zusammenhange und Deren Holonomie Gruppen, J. Reine Angew. Math., 212 (1963), 120-149.
26. Beil R.G., Electrodynamics from a Metric, Int. J. Theor. Phys., 26 (1987), 189-197.
27. Beil R.G., New Class of Finsler Metrics, Int. J. Theor. Phys., 28 (1989), 659-667.
28. Beil R.G., Finsler Gauge Transformations and General Relativity, Int. J. Theor. Phys., 31 (1992), 1025-1044.
28a. Beil R.G., Finsler and Kaluza-Klein Gauge Theories, to appear.
29. Buchner K., Kräftige Losungen der Einsteinschen Gleichungen, Tensor, N.S., 38 (1982), 65-68.
30. Buchner K., Roşca R., On Real Left Kählerian Space-times, Abh. Math. Sem. Univ. Hamburg, 51 (1981), 5-12.
31. Capursi M., Palombella A., On the Almost Hermitian Structures Associated to a Generalized Lagrange Space, Mem. Sect. St. Acad. Românǎ, VIII (1985), 57-64.
32. Carathéodory C., *Variationsrechnung*, Teubner, Leipzig, 1935.
33. Cartan E., *Sur les equations de la gravitation d'Einstein*, Gautier Villars, Paris, 1922.
34. Cartan E., *Les espaces de Finsler*, Hermann, Paris, 1934.
35. Comič I., Curvature Theory of Generalized Miron's d-Connections, Differential Geometry and its Applications, Proceedings Conf. Brno, 1989, 17-26.
36. Cruceanu V., Sur la structure presque-produit associee a une connexion sur un espace fibre, An. şt. Univ. "Al.I. Cuza" Iaşi, s I a Mat., 15 (1969), 159-167.
37. Dodson C.T.J., *Categories, Bundles and Space Time Topology*, Shiva Publ., England, 1980.
38. Dombrowski P., On the Geometry of the Tangent Bundles, J. Reine Angew Math., 210 (1962), 73-88.
39. Duc T.V., Sur la géométrie differentielle des fibres vectoriels, Kodai Math. Sem. Rep., 26 (1975), 349-408.
40. Ehlers J., Pirani F.A.F., Schild A., in General Relativity, ed. Raifeartaigh, L.O. Oxford, 1972.
41. Eliopoulos H.A., Subspaces of a Generalized Metric Spaces, Canad. J. Math., 11 (1959), 235-255.
42. Finsler P., Uber Kurven und Flächen in Allgemeinen Räumen, Dissertation, Gottingen, 1918, Birkhäuser, Basel, 1951.
43. Fock V.A., *Theory of Space, Time and Gavitation*, (in Romanian), Ed. Acad. Românǎ, Bucureşti, 1962.
44. Fukui M., Udrişte C., Properties of v-Curvature Tensors in Finsler Manifolds, Tensor, N.S., 1987, 52-57.

45. Gheorghiev Gh., Sur les espaces de Riemann-Einserhart (R-E), Tensor, N.S., 46 (1987), 271-277.
46. Ghinea I., Finsler Connections Compatible with Certain Geometrical Structures (in Romanian), Ph.D. Thesis, Cluj-Napoca, 1978.
47. Grifone J., Structures presque-tangente et connexions, I, II, Ann. Inst. Fourier, Grenoble, 22 (1972), 287-334; 22, 3 (1972), 291-338.
48. Haimovici A., Sur les espaces à connection non linéaire à parallelisme absolu, Rev. Roum. Math. Pures Appl., 10 (1965), 1121-1128.
49. Haimovici M., Les formule fundamentales des la théories des hypersurfaces d'un espace general, Ann. Sci. Univ. Jassy, 20 (1935), 39-58.
50. Haimovici M., Variétés totalement extrémales et variétés totalement géodésique dans les espaces de Finsler, Ann. Sci. Univ. Jassy, 29 (1939), 559-644.
51. Hangan T., On Totally Geodesic Distributions of Planes, Coll. Math. Soc. Janos Bolyai, 46, North-Holland, 1988, 519-530.
52. Hashiguchi M., On Determinations of Finsler Connections by Deflection Tensor Fields, Rep. Fac. Sci. Kagoshima Univ. (Math., Phys., Chem.), 2 (1969), 29-39.
53. Hashiguchi M., On Conformal Transformations of Finsler Metrics, J. Math. Kyoto Univ. 16 (1976), 25-50.
54. Hashiguchi M., On Wagner's Generalized Berwald Space, J. Korean Math. Soc., 12 (1975), 51-61.
55. Hassan B.T.M., The Theory of Geodesics in Finsler Spaces, Ph.D. Thesis, Southampton, 1967.
56. Hermann R., *Differential Geometry and the Calculus of Variations*, Academic Press, New York, 1968.
57. Hojo S., On Some Generalized Connections and their Applications, An. St. Univ. "Al.I. Cuza" Iaşi, s I a Mat., 30 (1984), 23-26.
58. Horvath J.I., New Geometrical Methods of the Theory of Physical Fielfs, Nuovo Cimento, 10 (1958), 444-496.
59. Hrimiuc D., On the Differential Geometry of Infinite Dimensional Lagrange Spaces, An. şt. Univ. "Al.I. Cuza" Iaşi, s I a Mat., 36 (1990), 75-86.
60. Iacob C., *Theoretical Mechanics* (in Romanian), Ed. Didactică şi Pedagogică, Bucureşti, 1980.
61. Ianuş S., *Differential Geometry and Applications to Relativity* (in Romanian), Ed. Academiei Române, Bucureşti, 1983.
62. Ichijio Y., Finsler Manifold with a Linear Connection, J. Math. Tokushima Univ. 10 (1976), 1-11.
63. Ichijio Y., On Some G-structures defined on Tangent Bundles, Tensor, N.S., 42 (1985), 179-190.
64. Ichijio Y., Miron R., On Some Structures Defined on Dominant Vector Bundles, J. Math. Tokushima Univ., 20 (1986), 13-26.
65. Igarashi T., On Characteristic Bivectors in Submetric Areal Spaces, Tensor, N.S., 45 (1987), 70-73.

66. Ikeda F., On Finsler Spaces Satisfying the Condition $L^2C^2 = f(x)$, An. şt. Univ. "Al.I. Cuza" Iaşi, s I a Mat., 30 (1984), 23-26.
67. Ikeda S., Some Remarks on the Lagrabgian Theory of Electromagnetism, Tensor, N.S., 49 (1990), 204-208.
68. Ikeda S., On the Theory of Gravitational Field in Finsler Spaces, III, An. Phys., 47 (1990), 1-5.
69. Ingarden R.S., Differential Geometry and Physics, Tensor, N.S., 30 (1976), 201-209.
70. Ingarden R.S., On Physical Interpretations of Finsler and Kawaguchi Geometries, Tensor, N.S., 46 (1987), 354-360.
71. Ionescu-Pallas N., *General Relativity and Cosmology* (in Romanian), Ed. Ştiinţifică, Bucureşti, 1981.
72. Ishikawa H., Note on Finslerian Relativity, J. Math. Phys., 22 (1981), 995-1004.
73. Ishikawa H., Einstein Equations in Lifted Finsler Spaces, Nuovo Cimento, vol. 56 B, 252-262.
74. Izumi H., Non-holonomic Frames in a Finsler Space, An. şt. Univ. "Al.I. Cuza" Iaşi, s I a Mat., 30 (1984), 39-44.
75. Izumi H., Yoshida M., On the Geometry of Generalized Metric Spaces I, II, Publ. Math. Debrecen, 39 (1991), 113-134, 185-197.
76. Kawaguchi A., On the Theory of Non-linear Connections, I, II, Tensor, N.S., 2 (1952), 123-142; 6 (1956), 165-199.
77. Kawaguchi A., Beziehung zwischen einer metrischen linearen Ubertragung und iener nicht-metrischen in einem allgemeinen metrischen Raume, Akad. Wetensch. Amsterdam, Proc. 40 (1937), 596-601.
78. Kawaguchi H., Some Characterisations of the Vector Field A_i Proper to Finsler Spaces, Tensor, N.S., 34 (1980), 367-372.
79. Kawaguchi H., The d-Connections in Lagrange Geometry, Differential Geometry and Appl., Proc. Conf. 1989, Brno, Czechoslovakia World Scientific, Singapore, 1990, 230-235.
80. Kawaguchi T., On the Applications of Finsler Geometry to Engineering Dynamical Systems, Periodica Math. Hung., 8 (1977), 281-289.
81. Kawaguchi T., Miron R., A Lagrangian Model for Gravitation and Electromagnetism, Tensor, N.S., 48 (1989), 153-168.
82. Kern J., Lagrange Geometry, Arch. Math., 25 (1974), 438-443.
83. Khu Quac Anh, Geometry of Myller Configurations in Finsler Spaces (in Romanian), Ph.D. Thesis, Iaşi, Romania, 1977.
84. Kikuchi S., On the Theory of Subspaces in Finsler Spaces, Tensor, N.S., 2 (1952), 67-79.
85. Kikuchi S., Theory of Minkowski Spaces and Non-linear Connection in a Finsler Space, Tensor, N.S., 12 (1962), 47-60.
86. Kirkovits M., Calculus of Variations in Lagrange Spaces, Ph.D. Thesis, Iaşi, Romania, 1989.
87. Klein J., Espaces variationels et mecanique, Ann. Inst. Fourier, Grenoble, 13 (1962), 1-124.

88. Klepp F., Remarkable Finsler Structures and the Finsler Geometry of Vector Bundles (in Romanian), Ph.D. Thesis, Iaşi, Romania, 1982.
89. Kobayashi S., Nomizu K., *Foundations of Differential Geometry*, I,II, Wiley (Interscience), New York, 1963 and 1969.
90. Kondo K., Kawaguchi T., On the Origin of the Hysteresis in the Finslerian Magnetic Dynamical System, RAAG Memoirs, Vol. 4, B-IX (1969), 215-222.
91. Kozma L., On the Classical Conditions for Finsler Type Connections, Proc. National Sem on Finsler and Lagrange Spaces, Braşov, Romania, 1988, 207-216.
92. Leon de M., Rodrigues P.R., *Methods of Differential Geometry in Analytical Mechanics*, North-Holland, Amsterdam, 1989.
93. Libermann P., Marle Ch.-M., *Symplectic Geometry and Analytical Mechanics*, D. Reidel Publ. Comp., 1987.
94. Lichnerowitz A., *Theories relativistes de la gravitation et de l'électromagnetisme*, Masson, Paris, 1955.
95. Lovelock D., Rund H., *Tensors, Differential Forms and Variational Principles*, Wiley-Interscience, New York, 1975.
96. Matsumoto K., Conformal Killing Vector Fields in a P-Sasakian Manifold, J. Korean Math. Soc., 14 (1977), 135-142.
97. Matsumoto M., *Foundations of Finsler Geometry and Special Finsler Spaces*, Kaiseisha Press, Otsu, Japan, 1986.
98. Matsumoto M., On Einstein's Gravitational Field Equation in a Tangent Riemannian Space of a Finsler Space, Rep. Math. Phys., 8 (1875), 103-108.
99. Matsumoto M., The Induced and Intrinsic Finsler Connections of a Hypersurface and Finslerian Projective Geometry, J. Math. Kyoto Univ., 25 (1985), 107-144.
100. Matsumoto M., The inverse Problem of Variation Calculus in Two-dimensional Finsler Space, J. Math. Kyoto Univ., (1989), 489-496.
101. Matsumoto M., Miron R., On an Invariant Theory of the Finsler Spaces, Periodica Math. Hung., 8 (1977), 73-82.
102. Munteanu Gh., Lagrange Geometry of Second Order, Proc. National Seminar on Finsler and Lagrange spaces, Braşov, Romania, 1988, 285-297.
103. Miron R., Introduction to the Theory of Finsler Spaces, Proc. National Seminar on Finsler Spaces, Braşov, Romania, 1980, 131-183.
104. Miron R., Metrical Finsler Structures and Metrical Finsler Connections, J. Math. Kyoto Univ., 23(1983), 219-224.
105. Miron R., Techniques of Finsler Geometry in the theory of vector bundles, Acta Sci. Math., 49 (1985), 119-129.
106. Miron R., Subspaces in Generalized Lagrange Spaces, An. şt. Univ. "Al.I. Cuza" Iaşi, s I a Mat. 33 (1987), 137-149.
107. Miron R., A Lagrangian Theory of Relativity, An. şt. Univ. "Al.I. Cuza" Iaşi, s I a Mat., 32 (1986), 37-62.
108. Miron R., On the Finslerian Theory of Relativity, Tensor, N.S. 44 (1987), 68-81.

109. Miron R., Anastasiei M., *Vector Bundles. Lagrange Spaces. Applications to Relativity* (in Romanian), Editura Academiei Române, 1987.
110. Miron R., Hashiguchi M., Metrical Finsler Connections, Rep. Fac. Sci. Kagoshima Univ. (Math., Pysh. Chem.), 12 (1979), 21-35.
111. Miron R., Izumi H., Invariant Frame in Generalized Metric Space, Tensor, N.S. 42 (1985), 272-282.
112. Miron R., Tatoiu-Radivoiovici M., Extended Lagrangian Theory of Electromagnetism, Rep. Math. Phys. 21 (1988), 193-229.
113. Miron R., Kirkovits M., Anastasiei M., A Geometrical Model for Variational Problem of Multiple Integrals, Proc. Int. Conf. Diff. Geometry Appl., Dubrovnik, 1988, 8-25.
114. Miron R., Watanabe S., Ikeda S., Some Connections on Tangent Bundle and their Applications to the General Relativity, Tensor, N.S., 46 (1987), 8-22.
115. Miron R., Roşca R., Anastasiei M., Buchner K., New Aspects of Lagrangian Relativity, Foundations of Physics Letters, Vol.5, No.2, 1992, 141-171.
116. Miron R., Antonelli P.L., Generalized Lagrange Spaces with the Fundamental Tensor $\exp(2\sigma(x,y))\gamma_{ij}(x,y)$, to appear.
117. Miron R., Kawaguchi T., Relativistic Geometrical Optics, Int. J. Theor. Phys. 30 / 11 (1991), 1521-1543.
118. Miron R., Tavakol R.K., Balan V, Roxburg I., Geometry of Space-time and Generalized Lagrange Gauge Theory, Publ. Math., Debrecen, 42 / 3-4 (1993), 215-224.
119. Motreanu D., Control Vector Fields with Strong Accesibility Property, Tensor, N.S., 49 (1990), 211-217.
120. Okubo T., *Differential Geometry*, Marcel Dekker, New York, 1987.
121. Oshima N., Survey of Rheonomic Aspects in Problems of Dynamics. RAAG Memoirs, Vol. I, B-II (1955), 62-69.
122. Otsuki T., Theory of Affine Connections of the Space of Tangent Directions of a Differentiable Manifold, I, II, Math. J. Okayama Univ., 7 (1957), 1-74, 95-122.
123. Opriş D., Fibrés Vectoriels de Finsler et Connexions Associés, Proc. National Sem. on Finsler and Lagrange Spaces, Braşov, 1980, 185-194.
124. Papuc D., Albu A., *Elements of Global Differential Geometry* (in Romanian), Ed. Didactică şi Pedagogică, Bucureşti, 1973.
125. Pitiş Gh., On the Totally Geodesic Anti-quaternion Submanifolds of Quaternion Kählerian Manifold, Tensor, N.S.,46 (1987), 132-136.
126. Pop I., On Some Homotopy Lifting and Extension Properties for Sequense of Continuous Maps, An. şt. Univ. "Al.I. Cuza" Iaşi, s I a Mat., 19 (1973), 103-111.
127. Popescu I.P., Remarques sur la géométrie de Finsler d'une variété differentiable, An. şt. Univ. "Al.I. Cuza" Iaşi, s I a Mat., 30 (1984), 65-68.
128. Popovici I., Propriétés relativistes d'une métrique de type Sasakien, Rev. Roum. Math. Pures et Appl. 29 (1974), 477-491.
129. Popovici I., Anastasiei M., Sur les bases de la géométrie finslerienne, C.R. Acad. Sc. Paris, 290, s A, 808-810.

130. Răileanu L., *Topological and differentiable manifolds*, Ed. Academiei Române, 1992.
131. Rizza G.B., Strutture di Finsler sulle varieta quasi complesse, Lincei Rend., 33 (1962), 271-275.
132. Roşca R., Espaces-temps possédant la propriété geodésique, C.R. Acad.-Sc. Paris, A 285, 1974, 305.
133. Roxburgh I.W., Post-Newtonian Constraints on the Lagrange Metric $\gamma_{ij}(x) + \alpha y_i y_j$, Tensor N.S., 51 (1992), 59-63.
134. Rund H., *The Differential Geometry of Finsler Spaces*, Springer-Verlag, Berlin, 1959.
135. Rund H., *The Hamilton-Jacobi Theory in the Calculus of Variations*, D. van Nostrand, London, 1966.
136. Rund H., Adjoint Connections on Group Manifolds and Gauge Transformations. The Math. Heritage of C.F. Gauss. World Sci. Publ., Singapore, 1991, 621-644.
137. Sakaguchi T., Subspaces in Lagrange Spaces, Ph. D. Thesis, Iaşi, Romania, 1987.
138. Sasaki S., Hatakeyama Y., On Differentiable Manifolds with Certain Structures which are Closely Related to Almost Contact Structures II, Tohoku Mat. J. 13 (1961), 281-294.
139. Shibata C., On the Curvature Tensor R_{hijk} of Finsler Spaces of Scalar Curvature, Tensor, N.S., 32 (1978), 311-317.
140. Shimada H., On Finsler Spaces with the Metric $L = \sqrt[n]{a_{i_1} \ldots a_{i_n} y^{i_1} \ldots y^{i_n}}$, Tensor, N.S., 33 (1979), 365-372.
141. Szilasi J., Kozma L., Remarks on Finsler-type Connections, Proc. National Sem. on Finsler and Lagrange Spaces, Braşov, Romania, 1984, 181-197.
142. Smaranda D., Immersion d'une modele d'univers avec champ magnetique dans une espace pseudo-euclidian, Bull. Soc. Sci., Liege, 1967, 23-30.
143. Sofonea L., *Representative Geometries and Physical Theories* (in Romanian), Ed. Dacia, Cluj-Napoca, Romania, 1984.
144. Soos G., Uber gruppen von Affinitäten und Bewegungen in Finslerschen Räumen, Acta Mat. Acad. Sci. Hungar, 5 (1954), 73-84.
145. Stavre P., The d-linear connections, Proc. National Sem. on Finsler and Lagrange Spaces, Braşov, Romania, 1988, 375-382.
146. Synge J.L., *Relativity: General Theory*, North-Holland, 1966.
147. Szabo Z.I., Positive Definite Berwald Spaces, Tensor, N.S. 35 (1981), 25-39.
148. Takano Y., Theory of Fields in Finsler Spaces I, Progr. Theor. Phys. 40 (1968), 1159-1180.
149. Takano Y., Gravitational Fields in Finsler Spaces, Nuovo Cimento, 10 (1974), 747-750.
150. Tamassy L., Affine Connections Inducing a Finsler Connection, Coll. Math. Soc. Janos Bolyai, 46, North-Holland, 1988, 1185-1193.

151. Tamassy L., Matsumoto M., Direct Method to Characterise Conformally Minkowski Finsler Spaces, Tensor, N.S., 33 (1979), 380-384.
152. Tamia-Dimopoulou P., A Relationship Between CR-structures and f-structures satisfying $f^3+f=0$, Tensor, N.S., 49(1990), 250-252.
153. Ţarină M., Invariant Finsler Connections on Vector Bundles, An. şt. Univ. "Al.I. Cuza" Iaşi, s I a Mat. 33 (1987) 87-94.
154. Tavakol R.K., Van der Bergh N., Viability criteria for the theories of gravity and Finsler Spaces, GRG, 18(1986), 849-959.
155. Teodorescu N., Géométrie finslerienne et propagation des ondes, Bull. Sci. Acad. Roumanie, 23 (1942), 138-144.
156. Teodorescu P.P., *Mechanical Systems. Classical Models* (in Romanian), Ed. Tehnică, Bucureşti, 1984.
157. Tonnelat M.A., *Les théories unitaires de l'électromagnetisme et de la gravitation*, Gauthier-Villars, Paris, 1965.
158. Udrişte C., Convexity of Finsler Energies, Proc. National Sem. on Finsler and Lagrange Spaces, Braşov, Romania, 1984, 197-210.
159. Vaisman I., *Symplectic Geometry and Secondary Characteristic Classes*, Birkhäuser, Basel, 1987.
160. Vilms J., Curvature of Non-linear Connections, Proc. Amer. Math. Soc., 19 (1968), 1125-1129.
161. Vrănceanu G., *Les espaces non holonomes*, Gauthier-Villars, Paris, 1936.
162. Vrănceanu G., *Lecons de geometrie differentielle*, 3 Vols., Ed. Academiei Române, Bucureşti, 1957.
163. Vrănceanu G., Mihăileanu N., *Introduction to the Theory of Relativity* (in Romanian), Ed. Tehnică, Bucureşti, 1980.
164. Watanabe S., On Some Properties of Generalized Finsler Spaces, Jour. Nat. Acad. Math., 1 (1983), 37-41.
165. Watanabe S., Ikeda S., Ikeda F., On Metrical Finsler Connection of a Metrical Finsler Structure, Tensor, N.S., 42 (1983), 79-85.
166. Watanabe S., Ikeda F., On a Metrical Finsler Connection of a Generalized Finsler Metric $g_{ij}=\exp(2\sigma(x,y))\gamma_{ij}(x)$, Tensor, N.S., 40 (1983), 97-102.
167. Wundheiler A., Rheonome Geometry, Absolute Mechanik Prace Matematyczuofizyczue, Warszawa, 40 (1933), 97-142.
168. Yano K., Ishihara S., *Tangent and Cotangent Bundles*, M. Dekker, 1973.
169. Yasuda H., On TM-connections of a Finsler Space and the Induced TM-connections on its Hypersurfaces, An. şt. Univ. "Al.I. Cuza" Iaşi, s I a Mat, 30 (1984), 99-103.
170. Youssef N.L., Distribution de nulité du tensor de courbure d'une connexion. Thése, Univ. Grenoble, 1978.
171. Yukawa H., Quantum Theory of Non-local Fields, I. Free Fields. Phys. Rev. 77 (1950), 219-226.
172. Zet Gh., Agop M., Manta V., Post-Newtonian Estimation in Relativistic Optics, to appear.

Index

Fundamental Theories of Physics

Series Editor: Alwyn van der Merwe, *University of Denver, USA*

1. M. Sachs: *General Relativity and Matter.* A Spinor Field Theory from Fermis to Light-Years. With a Foreword by C. Kilmister. 1982 ISBN 90-277-1381-2
2. G.H. Duffey: *A Development of Quantum Mechanics.* Based on Symmetry Considerations. 1985 ISBN 90-277-1587-4
3. S. Diner, D. Fargue, G. Lochak and F. Selleri (eds.): *The Wave-Particle Dualism.* A Tribute to Louis de Broglie on his 90th Birthday. 1984 ISBN 90-277-1664-1
4. E. Prugovečki: *Stochastic Quantum Mechanics and Quantum Spacetime.* A Consistent Unification of Relativity and Quantum Theory based on Stochastic Spaces. 1984; 2nd printing 1986 ISBN 90-277-1617-X
5. D. Hestenes and G. Sobczyk: *Clifford Algebra to Geometric Calculus.* A Unified Language for Mathematics and Physics. 1984 ISBN 90-277-1673-0; Pb (1987) 90-277-2561-6
6. P. Exner: *Open Quantum Systems and Feynman Integrals.* 1985 ISBN 90-277-1678-1
7. L. Mayants: *The Enigma of Probability and Physics.* 1984 ISBN 90-277-1674-9
8. E. Tocaci: *Relativistic Mechanics, Time and Inertia.* Translated from Romanian. Edited and with a Foreword by C.W. Kilmister. 1985 ISBN 90-277-1769-9
9. B. Bertotti, F. de Felice and A. Pascolini (eds.): *General Relativity and Gravitation.* Proceedings of the 10th International Conference (Padova, Italy, 1983). 1984 ISBN 90-277-1819-9
10. G. Tarozzi and A. van der Merwe (eds.): *Open Questions in Quantum Physics.* 1985 ISBN 90-277-1853-9
11. J.V. Narlikar and T. Padmanabhan: *Gravity, Gauge Theories and Quantum Cosmology.* 1986 ISBN 90-277-1948-9
12. G.S. Asanov: *Finsler Geometry, Relativity and Gauge Theories.* 1985 ISBN 90-277-1960-8
13. K. Namsrai: *Nonlocal Quantum Field Theory and Stochastic Quantum Mechanics.* 1986 ISBN 90-277-2001-0
14. C. Ray Smith and W.T. Grandy, Jr. (eds.): *Maximum-Entropy and Bayesian Methods in Inverse Problems.* Proceedings of the 1st and 2nd International Workshop (Laramie, Wyoming, USA). 1985 ISBN 90-277-2074-6
15. D. Hestenes: *New Foundations for Classical Mechanics.* 1986 ISBN 90-277-2090-8; Pb (1987) 90-277-2526-8
16. S.J. Prokhovnik: *Light in Einstein's Universe.* The Role of Energy in Cosmology and Relativity. 1985 ISBN 90-277-2093-2
17. Y.S. Kim and M.E. Noz: *Theory and Applications of the Poincaré Group.* 1986 ISBN 90-277-2141-6
18. M. Sachs: *Quantum Mechanics from General Relativity.* An Approximation for a Theory of Inertia. 1986 ISBN 90-277-2247-1
19. W.T. Grandy, Jr.: *Foundations of Statistical Mechanics.* Vol. I: *Equilibrium Theory.* 1987 ISBN 90-277-2489-X
20. H.-H von Borzeszkowski and H.-J. Treder: *The Meaning of Quantum Gravity.* 1988 ISBN 90-277-2518-7
21. C. Ray Smith and G.J. Erickson (eds.): *Maximum-Entropy and Bayesian Spectral Analysis and Estimation Problems.* Proceedings of the 3rd International Workshop (Laramie, Wyoming, USA, 1983). 1987 ISBN 90-277-2579-9

Fundamental Theories of Physics

22. A.O. Barut and A. van der Merwe (eds.): *Selected Scientific Papers of Alfred Landé. [1888-1975]*. 1988 ISBN 90-277-2594-2
23. W.T. Grandy, Jr.: *Foundations of Statistical Mechanics.* Vol. II: *Nonequilibrium Phenomena.* 1988 ISBN 90-277-2649-3
24. E.I. Bitsakis and C.A. Nicolaides (eds.): *The Concept of Probability.* Proceedings of the Delphi Conference (Delphi, Greece, 1987). 1989 ISBN 90-277-2679-5
25. A. van der Merwe, F. Selleri and G. Tarozzi (eds.): *Microphysical Reality and Quantum Formalism, Vol. 1.* Proceedings of the International Conference (Urbino, Italy, 1985). 1988 ISBN 90-277-2683-3
26. A. van der Merwe, F. Selleri and G. Tarozzi (eds.): *Microphysical Reality and Quantum Formalism, Vol. 2.* Proceedings of the International Conference (Urbino, Italy, 1985). 1988 ISBN 90-277-2684-1
27. I.D. Novikov and V.P. Frolov: *Physics of Black Holes.* 1989 ISBN 90-277-2685-X
28. G. Tarozzi and A. van der Merwe (eds.): *The Nature of Quantum Paradoxes.* Italian Studies in the Foundations and Philosophy of Modern Physics. 1988 ISBN 90-277-2703-1
29. B.R. Iyer, N. Mukunda and C.V. Vishveshwara (eds.): *Gravitation, Gauge Theories and the Early Universe.* 1989 ISBN 90-277-2710-4
30. H. Mark and L. Wood (eds.): *Energy in Physics, War and Peace.* A Festschrift celebrating Edward Teller's 80th Birthday. 1988 ISBN 90-277-2775-9
31. G.J. Erickson and C.R. Smith (eds.): *Maximum-Entropy and Bayesian Methods in Science and Engineering.* Vol. I: *Foundations.* 1988 ISBN 90-277-2793-7
32. G.J. Erickson and C.R. Smith (eds.): *Maximum-Entropy and Bayesian Methods in Science and Engineering.* Vol. II: *Applications.* 1988 ISBN 90-277-2794-5
33. M.E. Noz and Y.S. Kim (eds.): *Special Relativity and Quantum Theory.* A Collection of Papers on the Poincaré Group. 1988 ISBN 90-277-2799-6
34. I.Yu. Kobzarev and Yu.I. Manin: *Elementary Particles. Mathematics, Physics and Philosophy.* 1989 ISBN 0-7923-0098-X
35. F. Selleri: *Quantum Paradoxes and Physical Reality.* 1990 ISBN 0-7923-0253-2
36. J. Skilling (ed.): *Maximum-Entropy and Bayesian Methods.* Proceedings of the 8th International Workshop (Cambridge, UK, 1988). 1989 ISBN 0-7923-0224-9
37. M. Kafatos (ed.): *Bell's Theorem, Quantum Theory and Conceptions of the Universe.* 1989 ISBN 0-7923-0496-9
38. Yu.A. Izyumov and V.N. Syromyatnikov: *Phase Transitions and Crystal Symmetry.* 1990 ISBN 0-7923-0542-6
39. P.F. Fougère (ed.): *Maximum-Entropy and Bayesian Methods.* Proceedings of the 9th International Workshop (Dartmouth, Massachusetts, USA, 1989). 1990 ISBN 0-7923-0928-6
40. L. de Broglie: *Heisenberg's Uncertainties and the Probabilistic Interpretation of Wave Mechanics.* With Critical Notes of the Author. 1990 ISBN 0-7923-0929-4
41. W.T. Grandy, Jr.: *Relativistic Quantum Mechanics of Leptons and Fields.* 1991 ISBN 0-7923-1049-7
42. Yu.L. Klimontovich: *Turbulent Motion and the Structure of Chaos.* A New Approach to the Statistical Theory of Open Systems. 1991 ISBN 0-7923-1114-0

Fundamental Theories of Physics

43. W.T. Grandy, Jr. and L.H. Schick (eds.): *Maximum-Entropy and Bayesian Methods.* Proceedings of the 10th International Workshop (Laramie, Wyoming, USA, 1990). 1991 ISBN 0-7923-1140-X
44. P.Pták and S. Pulmannová: *Orthomodular Structures as Quantum Logics.* Intrinsic Properties, State Space and Probabilistic Topics. 1991 ISBN 0-7923-1207-4
45. D. Hestenes and A. Weingartshofer (eds.): *The Electron.* New Theory and Experiment. 1991 ISBN 0-7923-1356-9
46. P.P.J.M. Schram: *Kinetic Theory of Gases and Plasmas.* 1991 ISBN 0-7923-1392-5
47. A. Micali, R. Boudet and J. Helmstetter (eds.): *Clifford Algebras and their Applications in Mathematical Physics.* 1992 ISBN 0-7923-1623-1
48. E. Prugovečki: *Quantum Geometry.* A Framework for Quantum General Relativity. 1992 ISBN 0-7923-1640-1
49. M.H. Mac Gregor: *The Enigmatic Electron.* 1992 ISBN 0-7923-1982-6
50. C.R. Smith, G.J. Erickson and P.O. Neudorfer (eds.): *Maximum Entropy and Bayesian Methods.* Proceedings of the 11th International Workshop (Seattle, 1991). 1993 ISBN 0-7923-2031-X
51. D.J. Hoekzema: *The Quantum Labyrinth.* 1993 ISBN 0-7923-2066-2
52. Z. Oziewicz, B. Jancewicz and A. Borowiec (eds.): *Spinors, Twistors, Clifford Algebras and Quantum Deformations.* Proceedings of the Second Max Born Symposium (Wrocław, Poland, 1992). 1993 ISBN 0-7923-2251-7
53. A. Mohammad-Djafari and G. Demoment (eds.): *Maximum Entropy and Bayesian Methods.* Proceedings of the 12th International Workshop (Paris, France, 1992). 1993 ISBN 0-7923-2280-0
54. M. Riesz: *Clifford Numbers and Spinors* with Riesz' Private Lectures to E. Folke Bolinder and a Historical Review by Pertti Lounesto. E.F. Bolinder and P. Lounesto (eds.). 1993 ISBN 0-7923-2299-1
55. F. Brackx, R. Delanghe and H. Serras (eds.): *Clifford Algebras and their Applications in Mathematical Physics.* Proceedings of the Third Conference (Deinze, 1993) 1993 ISBN 0-7923-2347-5
56. J.R. Fanchi: *Parametrized Relativistic Quantum Theory.* 1993 ISBN 0-7923-2376-9
57. A. Peres: *Quantum Theory: Concepts and Methods.* 1993 ISBN 0-7923-2549-4
58. P.L. Antonelli, R.S. Ingarden and M. Matsumoto: *The Theory of Sprays and Finsler Spaces with Applications in Physics and Biology.* 1993 ISBN 0-7923-2577-X
59. R. Miron and M. Anastasiei: *The Geometry of Lagrange Spaces: Theory and Applications.* 1994 ISBN 0-7923-2591-5

KLUWER ACADEMIC PUBLISHERS – DORDRECHT / BOSTON / LONDON